2002

SCIENTIFIC INTEGRITY

Scientific Integrity

AN INTRODUCTORY TEXT WITH CASES

SECOND EDITION

Francis L. Macrina

Professor and Director, Institute of Oral and Craniofacial Molecular Biology
Virginia Commonwealth University, Richmond, Virginia

ASM PRESS WASHINGTON, D.C.

Publisher's Note

Scientific Integrity: an Introductory Text with Cases (Second Edition) is intended to serve as a text for courses and seminars on responsible conduct in scientific research. The text is not meant in any way to serve as a set of guidelines, rules, or statements officially endorsed by the American Society for Microbiology or any other scientific organization or institution.

The case studies used throughout this text are hypothetical and are not intended to describe any actual organization or actual person, living or dead. The opinions in the text, express or implied, are those of the authors and do not represent official policies of the American Society for Microbiology.

Copyright © 2000   ASM Press
American Society for Microbiology
1752 N Street, N.W.
Washington, DC 20036-2804

Library of Congress Cataloging-in-Publication Data

Macrina, Francis L.
     Scientific integrity : an introductory text with cases / Francis L. Macrina.–
2nd ed.
          p. cm.
     Includes bibliographical references and index.
     ISBN 1-55581-152-3
     1. Research–Moral and ethical aspects.   2. Medical sciences–Research–Moral
and ethical aspects.   3. Integrity.   I. Title.

     Q180.5.M67 M33 2000
     174'.95072–dc21

                                                                99-054234

To Mary

For understanding, patience
And, most of all, love
Beyond our vows of 30 years ago

Contents

Contributors

S. Gaylen Bradley, Ph.D.
Vice President for Academic Affairs
University of Maryland Biotechnology Institute
Baltimore, Maryland
*(Presently: Professor, Department of Humanities,
Pennsylvania State University College of Medicine,
Hershey, Pennsylvania)*

Bruce A. Fuchs, Ph.D.
Director, Office of Science Education
National Institutes of Health
Bethesda, Maryland

Michael W. Kalichman, Ph.D.
Research Ethics Program Coordinator
Associate Professor of Pathology
University of California, San Diego
La Jolla, California

Thomas D. Mays, Ph.D., J.D.
Morrison and Foerster LLP
Washington, D.C.

Cindy L. Munro, Ph.D.
Associate Professor of Adult Health Nursing
Virginia Commonwealth University
Richmond, Virginia

Paul S. Swerdlow, M.D.
Professor of Medicine
Barbara Ann Karmanos Cancer Institute
Wayne State University
Detroit, Michigan

Foreword

MANY STUDENTS ARE DRAWN TO MODERN BIOLOGY and biomedical research by the challenge of solving complex intellectual problems and the opportunity to contribute to human knowledge and the improvement of public health. Training to achieve these goals is lengthy and arduous and requires not only an understanding of important biological and allied scientific principles, but a sense of the social, ethical, and legal context in which modern biology is practiced. Until recently, students received all of their training in these non-scientific disciplines through the words and actions of their mentors and through informal discussions with peers and other scientific colleagues. The enormous power of current biology to alter the world in which we live, and the increasing interest of the public in the activities of research biologists, has necessitated development of more formal training in the ethics, rules, regulations, and laws which govern the conduct of science.

This volume summarizes all of the important areas of scientific integrity and responsible conduct of science, including current ethical issues in modern biomedical research, conflicts of interest, issues related to authorship and collaboration, intellectual property issues and record keeping, and the use of both humans and animals in biomedical research. Following a brief description of the relevant facts, ethical issues, and legal and regulatory requirements, each chapter provides a series of case studies for classroom discussion. Two important principles are carefully followed in this volume: the information is presented in a manner which recognizes the controversial nature of many of the issues and balances opposing points of view, and the case studies are stimulating and likely to result in effective classroom discussions which will interest students in these subjects. Additional teaching tools are found in the appendices, which include sample survey forms for gauging student opinion about a variety of topics associated with responsible conduct of science, some more detailed case studies, the Guidelines for the Conduct of Research at the National Institutes of Health, standard animal and human research protocols, and an example of a U.S. patent.

The target of this text is graduate-level students, but advanced undergraduates and postdoctoral students are also likely to benefit from the information and discussion of the case studies. More senior faculty should be encouraged to become more familiar with more recent information in the book and to participate fully in the case discussions. Although the information in this book is specifically targeted to academic environments, much of it will be useful to trainees and scientists in industry and government. In these cases, institution-specific rules also will apply, and the reader and discussants should be certain they are fully informed about the specific requirements of their organizations.

While it is not surprising that there are some statements in this volume about which well-informed, thoughtful scientists and members of the public will disagree, it is striking how much consensus has developed around some of the important ethical principles which guide the action of scientists in their daily work. Standards related to authorship, collaboration, sharing of research reagents, and mentorship are held commonly by the great majority of practicing scientists. The development of common ethical standards which guide scientific conduct speaks positively for the ability of scientists to acknowledge the important societal role that they play and to effectively oversee their own conduct.

Michael M. Gottesman, M.D.
Bethesda, Maryland
August, 1999

Preface

T HIS TEXTBOOK AND ITS FIRST EDITION PREDECESSOR GREW
from a course in Scientific Integrity I began at Virginia Commonwealth University in the mid-1980s. I was aided again by some of my teaching colleagues in writing the second edition. The book aims to provide a
core of topic areas that can be used to teach trainees about the principles of
scientific integrity. The content of the book is enhanced by the inclusion
of interactive exercises like cases and surveys. We are convinced that the
case study approach to scientific integrity is effective and worthwhile. It
is rewarding for all involved. We instructors often find ourselves learning
new things right along with our students as case discussions unfold and
classroom participants apply personal knowledge and ethical standards to
analyze and solve problems.

So what's new in the second edition? Every chapter of the first edition
has been revised. Some have received important updating; others have un-
dergone significant revision. Some of the end-of-chapter cases have been
refined, and others have been dropped in favor of more timely and topical
ones. New cases have been added: across the whole text, more than 50%
of the short cases are new. Two new chapters have been written for this
edition, one on ethical principles and the scientist and one on collaborative research. The class survey material has been significantly expanded and
revised. Useful appendix material which augments some of the topic areas
has been added. New appendixes include a U.S. patent with narrative, human and animal use protocols, and a verbatim example of an institutional
standards-of-conduct policy. Finally, the advances and availability of the Internet have had an impact. Each chapter now has a resources section noting
the location of on-line information to augment the text's subject matter.

The Internet has also had an impact on how we teach our course at VCU. Along with this book, we now use a Web-based syllabus for our course. The Web site will provide information on course timing and logistics, semester writing assignments, and, in general, how we use our text in teaching scientific integrity. The URL for the course is

http://views.vcu.edu/~macrina/

This book is aimed at graduate and postgraduate trainees in the biomedical and natural sciences. It also will be useful to individuals who train or practice in related scientific fields. Although the subject matter of scientific integrity may be taught at different academic levels, our text is geared for graduate biomedical trainees, preferably those who have had some exposure to the graduate research laboratory. Usually, this means students who are at least beginning the second year of graduate study. There is a culture, language, and mind set that accompanies life in the research laboratory which is not revealed in the classroom setting. We believe that prior laboratory exposure greatly facilitates teaching the principles of scientific integrity. Most important, it gives the students a basis for identifying with the cases. Nonetheless, I have successfully used the first edition of *Scientific Integrity* as a text for teaching advanced undergraduates (with little or no laboratory experience), and I anticipate that the second edition will also be useful for teaching at that level.

In short, this is a book that covers a variety of topics related to the conduct of scientific investigation. But it is *not* a rulebook for the scientist. We discuss guidelines and policies, standards, and codes because we want our readers to be aware that many of these issues are influenced by both written policies and normative standards. We also talk about some of the acceptable ways to do things. Yet, the values of the individual scientist take on major importance in doing scientific research. This will become readily apparent in any case discussion session. Scientists continually make judgments and decisions about their research. Whether the issue is the timely release of experimental materials to a colleague or decisions about authorship on a manuscript, personal and professional standards and values come into play. Thus, definitive, unambiguous advice on dealing with these and other issues cannot be taught in textbooks. There are many acceptable possibilities.

We want to plant the seeds of awareness of existing, changing, and emerging standards in scientific conduct. Applying critical thinking in using the information we present about standards and other normative behavior is equally important. Lifelong learning is as much a part of scientific integrity as it is of any rigorous scientific discipline.

Francis L. Macrina
Richmond, Virginia

Acknowledgments

I THANK THOSE WHO PROVIDED IDEAS FOR WRITING SOME OF
the cases: Gordon Archer, Jan Chlebowski, Scott Diehl, Paul Ferrara,
Patrick Laloi, Roderick Morgan, Grace Spatafora, Glenn Van Tuyle, Alison
Weiss, and Rodney Welch. A few of the short cases were written by students
enrolled in the 1998 VCU scientific integrity course, and their permission
to use the cases here is gratefully acknowledged: David Limbrick, John
Stewart, and Michael Weaver. Very special thanks to Wayne Grody, who
wrote several cases and permitted us to use them in chapter 10. I thank
Todd Kitten for graciously allowing us to reproduce his data book pages
in chapter 11 and Al Chakrabarty for assistance in preparing Appendix V.
John Roberts provided valuable help in the preparation of Appendix IV,
and I thank him and his colleagues Richard Moran and Martha Wellons.
Robert Eiss is gratefully acknowledged for providing information on human
subject experimentation guidelines found in the table in chapter 5. Finally, I
also want to thank Mary McKenney for her expert copyediting and Michael
Gottesman for his helpful comments on the manuscript.

The support and patience of Jeffrey Holtmeier, director of ASM Press,
are deeply appreciated.

My family provided enormous support and inspiration during this book-
writing project, and I thank them and love them for that.

F. L. M.

Notes to Students
and Instructors

M ANY OF THE TOPICS COVERED IN TEACHING scientific integrity
lend themselves to the case study approach. Except for chapters 1 and
2, at the end of each chapter you will find cases for classroom discussion.
Additional examples of case studies in responsible scientific conduct may
be found in the books listed at the end of this note.

How To Use End-of-Chapter Case Studies

This book contains two types of cases. The end-of-chapter short cases
are designed for classroom use. Extended cases that lend themselves to a
written response are contained in Appendix II. The short cases are 200 to
400 words and can be read aloud in a few minutes. Most of the cases in
this book have been used in our course. We assign sets of cases (e.g., from
a single chapter) to small groups of two or three students, asking them to
choose cases to present. Students then individually present their selected
cases in the designated class session. Assigning a case set in advance of the
class provides students with the opportunity to think about their arguments
and to have time to do research or consultation on the topic. For example,
a student might want to consult relevant guideline or policy documents.
Although many cases do not require research, they may not work as well
if the student has not been exposed to a graduate research environment.
In the student evaluation of our course, we regularly ask what factors were
important in the selection of cases for discussion. Student responses indicate
that two of the most important features are: (i) the belief that the case would
promote lively classroom discussion and (ii) the fact that the case had some

personal appeal. That is, students frequently picked cases about which they had some background knowledge or personal experience.

A student leading the discussion of the case begins by reading it aloud in class. He or she then acts as the moderator for the rest of the discussion of the particular case. Discussion of cases is aided by a seating arrangement that allows everyone in the classroom to see one another (e.g., seating around a conference table or arranging chairs into a circle or semicircle). Typical classroom seating arrangements with students facing the front of the room make it difficult for everyone to see who's talking, and this inconvenience can dampen group participation. Case discussions work optimally in small classrooms, with no more than 15 to 20 students. Smaller is even better. A typical case discussion will take 15 to 20 minutes.

Student participation is very important in the process. The instructor should serve only as a facilitator, contributing when clarification is needed, when discussion bogs down, or when closure on a case is needed. The student reads the case and presents his or her impressions, identifying the issues and suggesting a possible solution. The classroom is then open to discussion, and the students air their views on the topic without more than one person talking at once. The instructor or student moderator may have to act as a peacekeeper. Sometimes disputes arise and discussions can become animated, even intense. If the dialogue becomes emotional, insulting or inappropriate comments should not be allowed. Ad hominem comments are unacceptable, and discussants should be cautioned against their use.

Short cases are designed to encourage the discussants to think critically as they analyze and solve the problem at hand. For many cases, this will mean dissecting the facts of the case and separating the relevant issues from the nonrelevant ones. Cases will evoke uncertainties and ambiguities. Sometimes the discussion will begin by students asking questions about the case. If something needs clarification or explanation, it should be provided by the student discussant or by the instructor when needed.

One of the principal features of the cases is that they allow discussants to apply their knowledge and personal standards to problems encountered in doing scientific research. Discussion should lead to one or more acceptable solutions to the same problem. This is important to remember in bringing cases to closure. Much of the time a consensus answer will not emerge. There may be several "right answers," all of which are acceptable. There will always be clear "wrong answers." In proposing solutions, discussants should always be able to arrive at a position that can be defended. Answers may be ranked by merit as part of the case discussion, but usually this is not necessary. A solution is valid as long as it is legal and does not violate what the discussants view as acceptable norms and standards, written or otherwise.

The case reader should evaluate the quality and quantity of the class discussion and bring the case to closure at the appropriate time. Summarizing the discussion helps to do this. Any opposing points of view should be adequately represented in the summary. Occasionally, there may be students who are uncomfortable with the outcomes reached. If this happens, the instructor should encourage continued discussion outside of the classroom with him or her or with the student's mentor.

In summary, case discussion should foster critical thinking as the discussants examine and apply their personal and professional values. The process is one of self-discovery as students formulate answers based on their values and knowledge of professional standards. The application of relevant guidelines, codes, and policies should be brought into play. One final note on case resolution. We have deliberately omitted discussing the possible "answers" to the cases. Cases often have multiple acceptable solutions. Hashing over multiple acceptable answers runs the risk of assigning specific values to the various possibilities, which we feel is not desirable. Further, we believe that "wrong answers" will be obvious. In short, the process of solving the case studies is key to learning from them.

A few cases that are only modestly challenging have been placed at the beginning of each set of chapter cases. Using these cases to get started is recommended. They are designed to acquaint the student and the instructor with the process of case discussion and resolution using examples which lack the complexity of later cases in the set.

Extended Case Studies and Surveys

A different style of case study is longer and usually describes a more complex scenario. The required response is usually guided by specific questions or by a request to complete a written exercise. We call this format the extended case. A number of extended cases appear in Appendix II and explore some of our chapter topics. We have successfully used these extended cases to form the basis of a writing assignment for our course. Students have been required to select cases and write a response of one to two typewritten single-spaced pages per case. In effect, this becomes a term paper upon which part of the course grade is based.

Appendix I contains some surveys which we have found to be useful teaching tools. As with the short cases, we assign these surveys to small groups of students (usually two) on the first day of class. The students collect the completed response sheets from their classmates on an assigned date, collate the data, and present an analysis of the results. Printed response sheets corresponding to each survey are provided by the instructor. Response sheets are submitted anonymously. The assignment includes the date by which the rest of the class must turn in their responses to the students conducting the survey. A date on which the results of the survey will be discussed in class is also set. The student survey-takers then collate the data and prepare a handout or overhead transparency for class presentation of the results. Class time is reserved during a relevant session for discussion of the survey results. This discussion is led by the student survey-takers, and class participation is encouraged. For example, responses to questions that displayed considerable disparity can be explored. A more detailed explanation of the use of surveys as a teaching tool is presented in Appendix I. We have found that these exercises provide some of the same benefits as the short case discussions. Class discussion of survey results can be lively as students come to recognize and appreciate differing points of view on issues related to scientific conduct and training.

Resources

Bailar, J., M. Angell, S. Boots, E. Myers, N. Palmer, M. Shipley, and P. Woolf. 1990. *Ethics and Policy in Scientific Publication.* Editorial Policy Committee, Council of Biology Editors, Bethesda, Md.

Bebeau, M., K. Pimple, K. M. T. Muskavitch, S. L. Borden, and D. H. Smith. 1995. *Moral Reasoning in Scientific Research: Cases for Teaching and Assessment.* Poynter Center for the Study of Ethics and American Institutions, Indiana University, Bloomington.
(This monograph is available on-line at http://www.indiana.edu/~poynter/mr.pdf)

Korneman, S. G., and A. C. Shipp. 1994. *Teaching Responsible Conduct of Research Through a Case Study Approach.* Association of American Medical Colleges, Washington, D.C.

Penslar, R. L. (ed.). 1995. *Research Ethics: Cases and Materials.* Indiana University Press, Bloomington.

chapter 1

Methods, Manners, and Mandates

Francis L. Macrina

Integrity in Science • *Scientific Misconduct* • *Mandates* • *Conclusion*
• *Author's Note* • *References* • *Resources*

Integrity in Science

WHAT DO WE MEAN BY "INTEGRITY IN SCIENCE"? Science is knowledge derived from observation, study, and experimentation. It is systematic and exact: data are collected objectively and tested empirically. The word integrity conjures images of wholeness or soundness, even perfection. For science to provide an understanding of nature and the physical world, the utmost integrity must be woven into both its experimentation and its interpretations.

The term "integrity in science" has made its way into the lexicons of scientists, politicians, news reporters, and others. Integrity is expected, because science is built upon a foundation of trust and honesty. Long before federal agencies published definitions of scientific misconduct, it was obvious that lying, cheating, and stealing in the conduct of research were wrong. We are astonished and incredulous when a scientist admits to falsifying or fabricating research results. Data must be repeatable. Important findings will be checked, and cheating will inevitably be uncovered. Performing experiments, collecting data, and interpreting their meaning constitute a system of auditing often described as the self-correcting nature of science. Fabricated or falsified results cannot escape this process. Bogus results cannot make a contribution to our understanding of a problem.

The landscape of the research environment has changed in the last two decades of the 20th century. Headlines, news shows, books, and magazine articles speak of "stolen viruses," "science under siege," "falsified results," "scapegoats," "whistleblowers," "scientific hoaxes," and "misconduct investigations." What has happened? Are increasing numbers of scientists acting unethically and dishonestly? Can it be profitable to fabricate or falsify results? Has the competitive nature of scientific research placed pressures on scientists that lead to misconduct? Before addressing the issues that stem from such questions, let's talk some about doing research and about researchers.

Perceptions of scientists and science

Understanding, as best we can, how scientists do research is critical if we are to define scientific misconduct. But gaining a feel for how science is done is not easy. Science, after all, is the work of humans, and humans are fallible, impressionable, impulsive, and subjective. They can fall prey to self-deception, rationalizing their actions in ways that mislead themselves and others. The term "sloppy science" is frequently used to describe some behaviors, but the distinction between sloppy science and scientific misconduct can be nebulous. Those seeking clear-cut answers commonly invoke the idea of deliberate deception as the defining element in misconduct. But proving that someone made a conscious decision to falsify or fabricate data or to steal another's ideas can be extremely difficult, if not impossible.

In these times, both scientists and the public have a heightened awareness of the accountability that goes with doing research, especially in the biomedical sciences. Scientists facing the difficulty of acquiring grant funds, justifying their use of animals in biomedical research, and explaining to the public why the "war on cancer" still hasn't been won are sometimes challenged to defend the cause of science and the expenditure of public funds. On the other hand, the public regards science as the definitive vehicle for uncovering truth. They become confused when scientists disagree with one another. They cannot understand how scientific facts can be disputed. Yet, one federal agency's definition of scientific misconduct affirms that scientists will have "honest differences in interpretations of judgments of data" and that "honest error" in science does occur. The public has difficulty understanding that the scientific method can give erroneous results. In advertising, for example, there seems to be no greater virtue than the claim that a product was "scientifically tested" or, better yet, "scientifically proven to achieve results." The public finds the idea of "scientific truth" an attractive one.

When new facts cause scientists to change their previous interpretations and conclusions, the effect on the public is disquieting. Tracking newspaper headlines associated with the effects of oat bran consumption on cardiovascular health illustrates this point. In 1986 typical headlines referred to oat bran as the "next miracle food," and the public was advised to "know your oats." Then in the early 1990s some headlines declared, "oat bran claims weakened," or they spoke of the "rise and fallacies of oat bran." But as that decade progressed, so did our understanding of how oat bran works. Results of peer-reviewed clinical trials began to convince the scientific community that regular consumption of oat bran has positive effects. We learned that the soluble fiber of oat bran absorbs bile salts in the intestinal tract, exerting an effect on cholesterol homeostasis and probably lowering cholesterol blood levels. From this comes the reasonable expectation of decreased atherosclerotic plaque formation in blood vessels — a clear benefit to cardiovascular health. And so the headlines once again changed, reporting that "oat bran study says cholesterol lowered" and "lots of oat bran found to cut cholesterol." One headline reflected the frustration the public must have felt: "Confused about oat bran?" But this seemingly confusing stream of information is just an example of science working the way it's supposed to. The very nature of scientific investigation makes the accumulation of new information and the interpretation of existing data subject to periodic change.

Scientists recognize that this is how science normally works, but in general, people outside of science do not have this same understanding. Disagreements, errors, and new interpretations of results are sometimes reported to the public by the media. It is easy for such reporting to be misinterpreted. The debate about emerging or evolving scientific knowledge can be seen as confusion or interpreted as accusation. This may even cause some to question the integrity of the science. Compounding this problem is the commonly held stereotype that Goodstein (5) calls the "myth of the noble scientist." This myth holds that scientists must be virtuous, upright, impervious to human drives such as personal ambition, and incapable of misbehaving. Goodstein recognizes science as a human activity that has hypocrisies and misrepresentation built into it. As scientists we become accustomed to such behaviors and often don't even recognize misrepresentations. Goodstein argues that this myth of the noble scientist does science a disservice because it blurs the "distinction between harmless minor hypocrisies and real fraud." We will return to this issue in the section "Reporting Science." In summary, the human behavior that is a part of scientific research may influence how that research is done. The effects of behavior patterns may vary, as may the degree to which their perception by scientists or the public is gauged as "good" or "bad." Sorting out these effects is likely to be challenging for scientists and a source of confusion for nonscientists. Failure to appreciate the element of human behavior in the performance of scientific research can lead to misunderstandings that may confuse normal activities with inappropriate behavior.

Scientific method

Textbooks teach us that scientific research proceeds according to "the scientific method." In following such a systematically applied scientific method, a gap in knowledge is identified and questions are posed. Existing information is studied, and a hypothesis — a prediction or "educated guess" — is formed to explain certain facts. Information is gathered, analyzed, and interpreted in the process of testing the hypothesis. Results may support or refute a hypothesis, but a hypothesis cannot be proved. Indeed, a hypothesis can only be disproved. Further testing of specific hypotheses and their derivatives strengthens their support and leads to the genesis of a theory. Theories take into account a strongly supported hypothesis or set of hypotheses and encompass a broadly accepted understanding of a natural concept. It follows that, since they are based on hypotheses, theories can eventually be disproved but they cannot be proved. When hypotheses are not supported, the results obtained to reach this conclusion are often used to refine or construct other hypotheses, and the process begins anew. A hypothesis that has been unequivocally rejected on the basis of the interpretation of experimental evidence can provide the inspiration for a new hypothesis, which may survive the test of repeated attempts to reject it. The value of a hypothesis resides in its ability to stimulate additional thinking and further research, rather than in its initial correctness.

Bauer (1) has written about what he terms the "myth of the scientific method." He contends that scientific research rarely proceeds by the organized, systematic approach that is reflected in textbook presentations. Approaches to solving problems and answering questions involve various blends of empiricism and theorizing. Depending on the scientific discipline and on the intellect and personality of the scientist, research is conducted with considerable variations on the scientific method. Bauer argues that science varies immensely in its characteristics, and he proposes two categories: textbook science and frontier science. Textbook science has withstood the scrutiny of time and is not likely to be subject to frequent change. Frontier science is often termed "cutting-edge" science. It is volatile, sometimes unreliable, and subject to considerable change. Bauer correctly points out that textbook science fails to reveal the true workings of scientific exploration, because it teaches us only about successful science. Hence, it is not an accurate portrayal of the often convoluted pathway that leads to accepted and relatively stable scientific results. Such end products of research are commonly the result of several experimental pursuits that use different lines of intellectual thought and technological approaches. Such efforts can occur over long periods of time, during which corroborative or contradictory evidence must be addressed and, where necessary, reconciled. Textbook science evolves to a point of general acceptance with the caveat that future knowledge may further refine, modify, or even disprove it. To attempt to explain this process as the result of the systematic implementation of a single, prescribed scientific method sheds little light on the way science actually works.

Bauer's concept of frontier science is relevant to scientific integrity. Frontier science invites close examination. Methods, data, interpretations, and conclusions are scrutinized as part of the process. Issues like "honest error" and differences in judgment emerge. Unfortunately, the rigorous analysis of frontier science can lead to erroneous perceptions and misunderstandings that can translate to accusations of scientific misconduct. Scientists' intuition and their judgments and decisions may be subjected to scrutiny in ways that can take on an air of investigation. Who's to say that a scientist's intuition about a problem constitutes bad judgment or sloppy science, as opposed to deliberate deception? Deciding to discount enzyme assay data that were obtained from protein preparations extracted from what a biochemist might call "unhealthy cells" serves as a hypothetical case in point. Can intuition be relied upon to recognize potentially flawed data? Such are the gray areas that scientists, both as practitioners and as critics, must address. Clearly, scientific intuition can be applied to a problem in a way that allows the investigator to make a major conceptual advance.

It is rational to conclude that there is no single scientific method (1, 11, 15). Scientists use many different strategies and methods in their exploration of nature. Rarely, if at all, is the process orderly, even though scientific publications present information in a way that suggests a logical and ordered progression of the research. Bauer submits that we should view the classic description of the scientific method as an ideal rather than a specific formula for performing research. He further suggests that the projection of the concept of a prescribed scientific method provides society with unrealistic expectations of science and scientists.

Last, let us remember that the practice of any form of the scientific method is far from the objective behavior that forms the stereotypical image of research. The objectivity of science that the naive onlooker assumes to be integral to the process begins to evaporate quickly at the stage of formulating the hypothesis. The formation of hypotheses will be affected by the knowledge, opinions, biases, and resources of the investigator. Furthermore, hypotheses are subject to experimental testing by means of technologies and observational methods selected by the scientist. The decision to test a hypothesis means a commitment of time, energy, and money. In the past, these decisions were usually made by an individual, but the increasing complexity and collaborative nature of scientific research, especially biomedical research, frequently means that these decisions are made collectively. In either event, the process is profoundly human in nature, and both "gut feeling" and intellect are used in making decisions.

Thus, defining a universal scientific method with which to measure the integrity of the research process is neither practical nor logical. Howard Schachman's blunt assessment of the prosecution of scientific misconduct carries this message: ". . . it is inappropriate, wasteful, and likely to be destructive to science for government agencies to delve into the styles of scientists and their behavioral patterns" (14). Goodstein (6) further argues that the codification of methods for defining, monitoring, and prosecuting scientific misconduct is dangerous "because it assumes there is a single set of practices commonly accepted by the scientific community and [it] sets up a government agency to root out the deviations from those practices."

Reporting science

In 1963 Sir Peter Medawar wrote a provocative essay entitled "Is the Scientific Paper a Fraud?" (12). Referring to scientific communications published in journals, Medawar's use of the word fraud refers to misrepresentations of the thought processes that led to the work reported. He points out that the results section is written to present facts without any mention of significance or interpretation. These are saved for the discussion section. Medawar snickers that this is where scientists "adopt the ludicrous pretense of asking yourself if the information you have collected actually means anything..." and "...if any general truths are going to emerge from the contemplation of all the evidence you brandished in the section called 'results'." Here Medawar is attacking the idea that scientific discovery proceeds by an inductive process by which unbiased observations are made and facts are collected. From these experimental raw materials, generalizations emerge. He concludes that this inductive format of scientific reporting should be discarded, because it fails to convey the fact that experimental work begins with an expectation of the outcome. This bias extends to which investigational methods are chosen or discarded, why certain experiments are done and others are not, and why some observations are considered to be relevant while others are not. Many years later, David Goodstein's perspective (5) on the scientific paper is captured in his description of the "noble scientist": "...every scientific paper is written as if that particular investigation were a triumphant procession from one truth to another. All scientists who perform research, however, know that every scientific experiment is chaotic — like war. You never know what is going on; you cannot usually understand what the data mean. But in the end, you figure out what it was all about and then, with hindsight, you write it up describing it as one clear and certain step after the other. This is a kind of hypocrisy, but it is deeply embedded in the way we do science."

The research writings and scientific memoirs of François Jacob can be compared to aptly illustrate the contrast between actual research and the reporting of it (8, 9). In his autobiography, Jacob recounts his research with Sydney Brenner and Matthew Meselsen, which was aimed at the identification and characterization of the "X factor" now known as mRNA. Such a factor had been proposed as an intermediary in protein synthesis, despite the absence of a chemical basis for it. Jacob and his collaborators pursued this elusive factor, and he writes in his memoirs that they were "sure of the correctness of their hypothesis." But their initial work was uniformly unproductive as they attempted to demonstrate the X factor attached to ribosomes. So with their "confidence crumbled," Jacob and Brenner retreated to a Pacific Ocean beach where Jacob describes Brenner as suddenly leaping up and shouting: "The magnesium! It's the magnesium!"

Jacob and Brenner returned to the lab and repeated the experiments again, this time with "plenty of magnesium." And, indeed, it was the magnesium that enabled them to demonstrate "factor X" associated with bacterial ribosomes. They had been using too low a concentration of magnesium, resulting in the dissociation of the mRNA from the ribosomes. So Brenner's critical insight on the beach provided the key to demonstrating the existence of this short-lived intermediate that carried the message of the genes in DNA to the ribosomes where protein synthesis occurred. However, the presentation of these results in their 1961 *Nature* paper (**190**:576–581) does not portray events as told in Jacob's autobiography. Instead Brenner's insight is translated into a series of control experiments in which ribosomes, their subunits, and the mRNA were dissociated or associated, depending on the concentration of this divalent cation!

But in the end, Jacob (9) eloquently offers his perspective on such behaviors when he compares writing about research to describing a horse race with a snapshot or penning the history of a war using only official press releases. Jacob says scientific writing transforms and formalizes research. Scientific writing "substitute(s) an orderly train of concepts and experiments for a jumble of disordered efforts.... In short, writing a paper is to substitute order for the disorder and agitation that animate life in the laboratory."

In practice, the scientific method does not often progress systematically and logically, as commonly described in textbooks. But scientific papers read like paragons of logic. They may describe cleverly crafted experimental approaches applied in the most timely and compelling ways. But, in keeping with Goodstein's "myth of the noble scientist," scientific papers often do not represent the true chronology of events or the intricacies of assembling and interpreting facts that have led to the conclusions. Moreover, scientific papers rarely describe or put into perspective the pure luck and mistakes that were also part of the work being reported. Grinnell (7), discussing the writing of Medawar, describes the scientific paper's purpose as providing the necessary information and arguments for accepting a discovery. "Other researchers will expect to be able to verify the data and the conclusions, not the adventures and misadventures that led to them."

Scientific Misconduct

Brief historical perspective

Questionable behavior by scientists is not confined to modern times. Louis Pasteur's pioneering work in the 1880s led to the development of effective vaccines for anthrax and rabies. An examination of Pasteur's data books revealed that the anthrax vaccine used in a famous inoculation trial on sheep was prepared by a chemical inactivation method developed by his competitor, Toussaint. But publicly, Pasteur claimed that in these trials he employed his own method, which used oxygen to inactivate the anthrax bacilli (4). Robert Millikan's selective publication of data on the electric charges of oil drops led to an understanding of the particulate nature of electric charge (1, 5). Millikan intuitively discounted data involving the migration of electrically charged oil drops that did not conform to his expectations, because they had "something wrong" with them. Some have argued that Millikan was simply exercising scientific judgment. Nonetheless, issues of scientific integrity have been raised about Millikan's work in recent times, because facts indicate that Millikan did not publish all of his data. Thus, at issue is not that Millikan discarded certain data gathered from some of the oil drops, but that in his published work on the subject he wrote that he presented all of his available data.

Although allegations of scientific misconduct are not unique to the end of the 20th century, what is unique is their coverage in the news media. Grinnell (7) points out that the public disclosure of scientific misconduct was infrequent during the 1960s and early 1970s, with no more than a handful of cases becoming widely known. But in the late 1970s we began to see a number of cases of alleged misconduct prosecuted publicly, and their coverage by the news media sometimes approached a level of frenzy. The public's eyes were opened to the existence of scandal in science! It was recognized that science could fall victim to the unethical and inappropriate actions of some of its practitioners. The importance of this issue was underscored in the early 1980s with congressional hearings on fraud in biomedical research. During this decade some congressional members aggressively pursued certain cases, further fueling zealous media coverage. The 1990s began with the articulation of definitions and rules about scientific misconduct, and institutions receiving federal research funds had to have policies in place for pursuing allegations of misconduct. Graduate curricula now frequently include courses taught under the rubric of scientific integrity, research ethics, or responsible conduct of research. A federal commission to study research was established in 1993, and after 2 years of work its members proposed a new and considerably more complex definition of misconduct (discussed below), which is currently under study.

Incidence of misconduct

To be sure, scientific misconduct is now commonly discussed and reported in public venues. But is the incidence of misconduct on the rise? What baseline information can we use to make such a measurement? Scientists commonly assert that the incidence of misconduct in research is rare. In fact, an accurate appraisal of the problem is lacking, and more research and reliable data are needed. Published surveys of trainees and scientists frequently reveal a fraction of respondents who claim that they have observed scientific misconduct at some time in their careers. But such studies are subject to the criticism that the responses depend on personal perceptions and interpretations that may differ enormously according to the training and professional experience of the individual. More to the point, the Office of

Research Integrity of the U.S. Public Health Service, Department of Health and Human Services (DHHS), and the Office of the Inspector General of the National Science Foundation investigate scores of misconduct allegations every year. Such investigations have led to the conviction of scientists, trainees, and technicians.

The image of science is tarnished when misconduct is uncovered. Recent history has taught us that even the investigation of alleged scientific misconduct — no matter what the final verdict — can damage the careers of both the accused and the whistleblower and can bring considerable negative publicity to the institutions involved. Preventing misconduct is key in science as in other professions, and it is logical to argue that emphasis needs to be placed on education and appropriate socialization. But even the most rigorous efforts in this regard are not likely to affect someone who is intent on deliberate deception or misconduct.

Perpetrators of misconduct

Arthur Caplan (3) suggests that one who would lie about research data or steal someone else's ideas suffers from failed morals. Training and appropriate socialization in the norms of scientific research are not likely to sway such an individual. And preventing such individuals from entering the research arena or weeding them out once they're in place is challenging.

So who would perpetrate an act of scientific fraud? In this area we are long on speculation and short on well-supported conclusions. Sir Peter Medawar may have summed it up in the fewest possible words. In writing about a case of scientific misconduct, he sought some lesson or truth from the incident but in the final analysis concluded that "it takes all sorts to make a world" (12). Another Nobel laureate, Salvador Luria, suggested that a peculiar pathology exists in the personality of one who would cheat in science (10). He argues that only a distorted sense of reality could account for someone who would falsify or fabricate results. Thinking one could get away with such behavior in science, where external and internal control measures continually demand verification, would be a delusion. Finally, David Goodstein (5) has studied a number of cases of scientific fraud and posits three frequently underlying motives: (i) career pressure, (ii) the belief that one "knows" the answer and can take short cuts to get there, and (iii) the notion that some experiments yield data that are not precisely reproducible.

Toward a definition of misconduct

Scientists do make honest mistakes, and these mistakes ought not to be confused with or interpreted as misconduct. Defining and dealing with behavior that falls between honest error and fraud can be difficult. However, deliberate deception in scientific research — scientific fraud or scientific misconduct — is different. In some circles, including governmental agencies, the term fraud is avoided. Fraud has precise legal meanings, some of which are not relevant or practical when applied to scientific conduct. For example, it can be difficult to prove that damage resulted from a misrepresentation.

Typical definitions of scientific misconduct have two sides. On one hand, the egregious transgressions of fabrication, falsification, and plagiarism are forbidden. But on the other hand, definitions cast a wide net in warning about deviations from accepted scientific practices. Where do we look or whom do we consult to learn about accepted practices? Scientists continually apply standards to the conduct of their research. In certain areas written codes, and in some cases laws, have existed for some time. These include policies for the use of humans and animals in research. In other areas, like conflict of interest, written codes have emerged relatively recently. Codes that may be used to define the basis of authorship on papers are now finding their way into the culture of science. And standards that deal with data sharing and with issues of collaborative research are starting to appear. Guidelines that cover responsible research conduct are becoming commonplace at universities and research institutes. In this book we strive to present a narrative that captures established and emerging thinking about such practices. At the very least we hope to stimulate critical thinking about such matters by providing available points of view. In addition, we aim to challenge the student with cases that require a problem-solving approach.

Whether written or unwritten, standards for doing research provide the foundation for training scientists and for properly conducting science. Bertrand Russell made a cogent point. To paraphrase him, we trace "the evils of the world" to moral defects and lack of intelligence. We know little about eliminating moral defects and unethical behavior, but we can improve intelligence through education. So we seek to improve intelligence rather than morals. Russell's argument is relevant to the teaching of scientific integrity. Both practicing scientists and scientists-in-training must continually examine the subject, play a role in refining existing standards, and contribute to the development of needed standards.

Mandates

DHHS and NSF definitions

During the past three decades, the existence of misconduct in science has been recognized by the scientific community, investigated by governmental agencies, and publicized to society. But the scientific community has not been particularly organized in its analysis of or response to scientific misconduct. In the United States, the two principal funding agencies of the biomedical and natural sciences, the National Institutes of Health (NIH) and the National Science Foundation (NSF), began responding with new or extended initiatives in the 1980s. The NIH, an agency of the DHHS, expanded its Office of Scientific Integrity, which eventually was renamed the Office of Research Integrity. The NSF reaffirmed the role of its Office of the Inspector General in matters of scientific misconduct. Both agencies published definitions of scientific misconduct that have been in force for approximately 10 years.

The DHHS holds that:

"Misconduct" or "Misconduct in Science" means fabrication, falsification, plagiarism, or other practices that seriously deviate from those that are commonly accepted within the scientific community for proposing, conducting, or reporting research. It does not include honest error or honest differences in interpretations or judgments of data.

(*Federal Register* **54**:32446–32451, August 8, 1989)

The NSF definition states that:

Misconduct means fabrication, falsification, plagiarism or other serious deviation from accepted practices in preparing, carrying out, or reporting results from activities funded by NSF, or retaliation of any kind against a person who reported or provided information about suspected or alleged misconduct and who has not acted in bad faith.

(*Federal Register* **56**:22286–22290, May 14, 1991)

Some scholarly societies, universities, and research institutes have prepared their own definitions of scientific misconduct. Often, the DHHS or NSF definitions have been directly adopted or modified for such purposes.

Both the DHHS and the NSF definitions clearly forbid "fabrication, falsification, and plagiarism." This is commonly referred to as the FFP core. Fabrication means making up results; falsification means tampering with results; and plagiarism means passing off another's ideas as your own. In other words, in science as in life, it is wrong to lie, cheat, or steal. Both DHHS and NSF definitions warn against serious deviations from accepted research practices.

These definitions of misconduct have stirred debate and controversy. For example, the phrases "accepted practices" and "serious deviations" are open to broad interpretation. Some practices are written down, and deviation from them is readily identified. It can be argued that there is an unwritten but widely practiced code of conduct for scientific investigation and that scientists know serious deviations when they see them. But one person's accepted practice may be unacceptable to another. Consistent with this, a National Academy of Sciences panel expressed concern that an interpretation of such phraseology might result in someone's being accused of scientific misconduct based on "their use of novel or unorthodox research methods" (13).

The discussion and implementation of the DHHS and NSF definitions have raised questions about another aspect of misconduct, namely, those transgressions for which there are existing avenues of prosecution. If someone breaks the law while doing scientific research, is that misconduct? Consider the scientist who, while doing research, embezzles grant funds, vandalizes laboratory equipment, or sexually harasses a colleague. All of these transgressions are covered by various civil or criminal laws. The DHHS definition has held in both word and deed that such offenses are not unique and thus are not appropriately treated as scientific misconduct, but as violations of state or federal laws. The uniqueness of scientific misconduct lies in the fact that the expert judgment of scientists is needed to prosecute and resolve such problems.

However, the investigation of misconduct under the aegis of the NSF definition has covered a broader scope. NSF's position has been that if the transgression does harm to science, then it should be pursued as scientific misconduct, and this philosophy has been put into practice. This strategy is articulated by Buzzelli (2), who argues that in such cases federal funding agencies have an enforcement role that must protect the integrity of the research process and its funding mission. His thesis is that some actions must be classified as scientific misconduct regardless of whether they are covered by general laws or regulations or whether they are unique to science and scientific research.

The DHHS definition allows that scientists may disagree about the interpretation of results. It even says that scientists may make errors in their work. However, it goes on to clarify that neither of these things is scientific misconduct. The NSF definition considers the whistleblower, the person who reports an alleged incident of misconduct. Indeed, it makes retaliation against that person ("who has not acted in bad faith") an act of scientific misconduct.

A universal definition?

In 1993, the NIH Revitalization Act established the Commission on Research Integrity (CRI) in response to the belief in Congress that there were still deficiencies in the scientific community's ability to deal with research misconduct. The 12 appointed members of the CRI represented a diversity of disciplines, including science, administration, law, sociology, and ethics. Their report, which has been under study since late 1995, recommends new wording for the definition of scientific misconduct.

1. Research Misconduct

Research misconduct is significant misbehavior that improperly appropriates the intellectual property or contributions of others, that intentionally impedes the progress of research, or that risks corrupting the scientific record[a] or compromising the integrity of scientific practices. Such behaviors are unethical

---

[a]The record encompasses any documentation or presentation of research, oral or written, published or unpublished.

and unacceptable in proposing, conducting or reporting research or in reviewing the proposals or research reports of others.

Examples of research misconduct include but are not limited to the following:

Misappropriation: An investigator or reviewer shall not intentionally or recklessly

(a) plagiarize, which shall be understood to mean the presentation of the documented words or ideas of another as his or her own, without attribution appropriate for the medium of presentation; or

(b) make use of any information in breach of any duty of confidentiality associated with the review of any manuscript or grant application.

Interference: An investigator or reviewer shall not intentionally and without authorization take or sequester or materially damage any research-related property of another, including without limitation the apparatus, reagents, biological materials, writings, data, hardware, software, or any other substance or device used or produced in the conduct of research.

Misrepresentation: An investigator or reviewer shall not with intent to deceive, or in reckless disregard for the truth,

(a) state or present a material or significant falsehood; or

(b) omit a fact so that what is stated or presented as a whole states or presents a material or significant falsehood.

2. Other Forms of Professional Misconduct

a. Obstruction of Investigations of Research Misconduct
The Federal Government has an important interest in protecting the integrity of investigations into reported incidents of research misconduct. Accordingly, obstruction of investigations of research misconduct related to Federal funding constitutes a form of professional misconduct in that it undermines the interests of the public, the scientific community, and the Federal Government.

Obstruction of investigations of research misconduct consists of intentionally withholding or destroying evidence in violation of a duty to disclose or preserve; falsifying evidence; encouraging, soliciting or giving false testimony; and attempting to intimidate or retaliate against witnesses, potential witnesses, or potential leads to witnesses or evidence before, during, or after the commencement of any formal or informal proceeding.

b. Noncompliance With Research Regulations
Responsible conduct in research includes compliance with applicable Federal research regulations. Such regulations include (but are not limited to) those governing the use of biohazardous materials and human and animal subjects in research.

Serious noncompliance with such regulations after notice of their existence undermines the interests of the public, the scientific community, and the Federal Government and constitutes another form of professional misconduct.

This proposed definition would cover all misconduct involving the use of federal funds, thus obviating the need for separate NSF and DHHS definitions. The Commission's report also stressed that proposing hypotheses that ultimately turn out to be false, offering conflicting interpretations, and making erroneous observations and analyses are all part of free scientific inquiry. These practices are held as being essential to the advancement of science and are never to be considered misconduct. While stressing that the CRI seeks a uniform federal definition of misconduct, the report also leaves open the door to academic and research institutions' adopting more demanding standards. The report further affirms that, once implemented, the definition will develop its full meaning only when tested in actual circumstances. There is an explicit reliance on societies, institutions, individuals, and case law to "develop the interpretive context."

The CRI's proposed definition of scientific misconduct has met with debate and controversy. Criticisms have ranged from failure of the definition to take into consideration the ambiguity inherent in the practice of science to the use of terms that have specific legal meaning in place of those that have an obvious research context. The so-called FFP core is now replaced with an MIM core (misappropriation, interference, and misrepresentation). Some critics have argued that the broad language and expansive scope of the proposed definition will create new problems and open further debates during the prosecution of cases. Other critics charge that the >400-word definition is unnecessarily expansive and likely to force investigations of virtually every accusation of dishonesty, unfairness, and inappropriate professional actions.

In any event, definitions of misconduct have become part of the scientific culture. Any institution receiving research support from the NIH is required to have a policy in place for dealing with scientific misconduct, which is why a definition of misconduct is needed. Nonetheless, definitions have been controversial and vigorously debated, and this debate is likely to continue with the CRI's definition now on the table. Current definitions need to be part of the working vocabulary of the scientist, and scientists need to be part of the debate that will establish a background of knowledge, allowing the definitions to work. Specifically, the methods of science and the styles and behavior of scientists create a mix that makes it difficult to conceptualize accepted practices. But that difficulty shouldn't stifle dialogue, the search for knowledge, or, most important, education about the principles of scientific integrity.

Conclusion

The practice of science has always encompassed values that include honesty, objectivity, and collegiality. The progress of science in the latter half of this century is but one fitting tribute to the success of the research enterprise. There is nothing fundamentally wrong with the conduct of science. However, emphasis on the workings of science and the conduct of scientists has shifted considerably in recent years. Governmental oversight and definitions of scientific misconduct sometimes lead one to believe that scientific integrity is a new concept. It is not. The codes that govern conduct in science are likely to grow more explicit in years to come, as witnessed in part by the recent focus on articulation of misconduct definitions. These are practical matters that cannot be ignored by scientists. This book attempts to crystallize a number of issues related to scientific conduct and to encourage both students and practitioners of science to think about them in both theoretical and practical ways.

Author's Note

As this book was in the final stages of production there were new developments related to a possible U.S. government-wide definition of scientific misconduct. This definition, developed by a committee of the White House's Office of Science and Technology Policy (OSTP), contains the "fabrication, falsification, and plagiarism" core. This language is in contrast to the proposed CRI definition detailed in this chapter. The OSTP definition is part of a policy document which has other sections titled: "Findings of Research Misconduct," "Responsibilities of Federal Agencies and Research Institutions," "Guidelines for Fair and Timely Procedures," and "Actions." The definition was published in the October 14, 1999, *Federal Register* (vol. 64, no. 198, p. 55722–55725) along with OSTP's request for public comments. The proposed definition and policy can be accessed on the Office of Research Integrity (ORI) web site at: http://ori.dhhs.gov. The ORI web site should be consulted periodically to monitor developments in the consideration and possible adoption of this definition.

References

1. **Bauer, H.** 1992. *Scientific Literacy and the Myth of the Scientific Method.* University of Illinois Press, Chicago.

2. **Buzzelli, D. E.** 1993. A definition of misconduct in science: a view from NSF. *Science* **259:**584–585, 647–648.

3. **Caplan, A.** 1998. *Due Consideration: Controversy in the Age of Medical Miracles.* John Wiley & Sons, Inc., New York, N.Y.

4. **Geison, G. L.** 1995. *The Private Science of Louis Pasteur.* Princeton University Press, Princeton, N.J.

5. **Goodstein, D.** 1991. Scientific fraud. *Am. Scholar* **60:**505–515.

6. **Goodstein, D.** 1992. What do we mean when we use the term scientific fraud? *Scientist*, March 2, p.11–13.

7. **Grinnell, F.** 1992. *The Scientific Attitude.* The Guilford Press, New York, N.Y.

8. **Grinnell, F.** 1997. Truth, fairness, and the definition of scientific misconduct. *J. Lab. Clin. Med.* **129:**189–192.

9. **Jacob, F.** 1988. *The Statue Within.* Basic Books, Inc., New York, N.Y.

10. **Luria, S.** 1975. What makes a scientist cheat. *Prism*, May, p. 15–18, 44. (Reprinted in J. Beckwith and T. Silhavy. 1992. *The Power of Bacterial Genetics*. Cold Spring Harbor Laboratory Press, Cold Spring Harbor, N.Y.)

11. **Medawar, P. B.** 1984. *The Limits of Science*. Oxford University Press, Oxford, U.K.

12. **Medawar, P. B.** 1991. *The Threat and the Glory: Reflections on Science and Scientists*. Oxford University Press, New York, N.Y.

13. **National Academy of Sciences.** 1992. *Responsible Science*, vol. I: *Ensuring the Integrity of the Research Process*. National Academy Press, Washington, D.C.

14. **Schachman, H. K.** 1993. What is misconduct in science? *Science* **261:**148–149, 183.

15. **Wolpert, L.** 1993. *The Unnatural Nature of Science*. Harvard University Press, Cambridge, Mass.

Resources

The report of the Commission on Research Integrity cited may be found on-line at

http://gopher.faseb.org/opar/cri.html/

The Web site for the DHHS Office of Research Integrity may be found on-line at

http://ori.dhhs.gov/

The Professional Ethics Report published by the American Association for the Advancement of Science, which runs articles dealing with professional ethics in science, may be accessed on-line at

http://www.aaas.org/spp/dspp/sfrl/per/per.htm

The monograph *On Being a Scientist: Responsible Conduct in Research* can be accessed on-line by clicking on the Science and Ethics link at the Web site of the National Academy Press Reading Room:

http://www.nap.edu/readingroom/

chapter 2

Ethics and the Scientist

Bruce A. Fuchs and Francis L. Macrina

*Overview • Underlying Philosophical Issues • Utilitarianism
• Deontology • Critical Thinking and the Case Study Approach
• Moral Reasoning in the Conduct of Science • Conclusion
• References • Resources*

Overview

M ANY OF THE DECISIONS THAT SCIENTISTS make in their day-to-
day activities are pragmatic ones. That is, they are made after consideration of facts that can be interpreted on the basis of well-established
knowledge and accepted principles. For example, when planning a surgical
procedure involving a rabbit, one must decide on the type and dose of anesthetic to be used. This decision is determined by professional judgment,
published recommendations, and consultation with the appropriate animal
experts. It is also strongly influenced by the formal rules and policies that
govern the use of animals in research. On the other hand, the decision to
use a rabbit in the first place has both pragmatic and moral components.
Most scientists conduct a particular medical experiment on animals because
the risk to humans is unacceptably high. Although some members of our
society question whether this decision is an ethical one, the majority accept
the necessity of animal research but insist that it be conducted in a humane
manner. Here, we have entered the realm of moral reasoning. These decisions are based on our judgment of what we ought to do — and we want to
do the right thing. But determining what is *morally* (as opposed to *legally*)
right and wrong in such cases is not always assisted by guidelines or a policy
manual. There are a number of past research studies that, while conducted
in accordance with acceptable practices at the time, are widely viewed today
as having been unethical. To avoid repeating such errors, we must all strive
to carefully examine the moral dimensions of our current research practices.

Today we commonly encounter codes and policies that guide scientists
in decision making. Institutional standards of conduct, codes of ethical

behavior adopted by scientific societies, and instructions to authors published in scholarly journals are but a few examples of the kinds of written guidance increasingly available to scientists. On the other hand, there are many examples of decision making in science that are not underpinned by clear-cut accepted standards. For example, which of our data do we publish? In this connection a National Academy of Sciences panel report (4) asserts that "the selective use of research data is another area where the boundary between fabrication and creative insight may not be obvious." With whom and under what circumstances do we share our research data? When is it acceptable not to share research data? Another area where no clear-cut standards exist involves the responsibilities of mentoring trainees. If a mentor provides little guidance to a floundering trainee, claiming that the trainee must "sink or swim," is the mentor neglecting his or her responsibility?

In contrast to the pragmatic decisions about the choice of anesthetic in an experiment, these are ethical decisions. Ethics is a branch of philosophy that is concerned with morality and seeks to provide guidance as to how we ought to live. It seeks to conceptualize right and wrong, drawing on personal and societal values. Morality enables one to distinguish and decide between conduct that is right and conduct that is wrong. In common use, morality often implies conformity with a behavioral code that is generally accepted in some defined setting or culture. Ethical behavior in the workplace implies the adherence to a collection of moral principles that underlie some specific context or profession and is commonly referred to as applied ethics.

The case studies included throughout this book will give rise to discussions that will help students reason through problems that require ethical decision making. In this chapter we shall briefly discuss some aspects of ethical decision making, focusing first on two general ethical theories. We shall also discuss elements of moral reasoning and critical thinking that are likely to facilitate the analysis and resolution of the cases.

Underlying Philosophical Issues

It is unfortunate that many of those working in the biomedical sciences have had little formal introduction to the field of ethics, because they may as a consequence have little appreciation for its power as a discipline. Occasionally, scientists are suspicious that "soft" disciplines such as moral philosophy lack the same type of academic rigor displayed by their own fields. It is not uncommon for scientists to criticize animal rights activists for being excessively emotional and insufficiently rational. Yet, scholars like the animal rights activists Peter Singer and Tom Regan are respected for their rational, not emotional, arguments in favor of granting animals far more moral weight than society currently allows them.

Some people believe that ethical opinions are mere preferences akin to expressing a taste for a flavor of ice cream or a type of music. For these people there is little basis (or reason) for differentiating between ethical positions. However, few philosophers would seriously argue for such a strongly subjective view of ethics. We make rational decisions about our ethical positions in a way that we do not make decisions about ice cream. If a friend expresses a preference for strawberry, none of us would feel compelled to argue the merits of chocolate. This would not be the case for a friend expressing intent to commit murder. However, ethics are also not strongly objective in the manner of many scientific principles. Scientists anywhere around the world, or at anytime throughout history, who seek to measure the density of pure gold will find, within the error of their instruments, the same result. Yet, there is no comparable experiment that we could perform to assess the morality of a practice, such as polygamy, that is acceptable in some cultures and taboo in others. Ethics fall in between these extreme positions. Ethical issues are neither matters of taste nor immutable physical constants that can be objectively determined irrespective of time and culture.

Ethics is usually subdivided into two areas known as normative ethics and metaethics. Normative ethics seeks to establish which behavior is morally right or wrong; that is, it seeks to establish norms for our behavior. Normative ethics is persuasive in that it attempts to set out a moral theory that can be used to determine which views are acceptable and ought to be adopted. This differs from metaethics, which concerns itself with an analysis of fundamental moral concepts, for example, concepts of right and wrong, or of duty. We will not discuss metaethics but will focus instead on some of the normative ethical theories that attempt to persuade us to redefine our behavior.

While not all philosophers advance identical ethical theories, this fact should not be attributed to any inherent weakness in the discipline. It is not at all uncommon for two biomedical scientists to disagree on the implications of a certain data set. It is quite possible that the two scientists are approaching the problem with different hypotheses in mind. Likewise, given an ethical dilemma, one will often find ethicists who reach differing conclusions as to the best course of action. The difference of opinion may be attributable to the fact that each ethicist has tried to solve the dilemma using a different normative ethical theory. Alternatively, each may have used a similar ethical theory and yet differed greatly in the amount of weight each ascribed to the various components of the problem. In addition, there can be disagreements over the empirical facts of a case (for example, whether an animal feels pain during a given procedure). The point is that moral problem solving, like biological problem solving, is an extremely complex process, and we should not be surprised to find that different people do not always arrive at the same conclusion.

However, it is equally important to realize that while many ethical dilemmas may not have a single "right" answer, there are answers that are clearly wrong. Who would seriously argue that a coin toss should decide ethical questions, or that abortions should be considered moral on Mondays and immoral on Tuesdays? Ethical positions can be evaluated and compared using techniques that are not entirely foreign to those in the sciences. Ethical theories can be evaluated on their rationality, their consistency, and even on their usefulness.

While the evaluation of competing ethical theories is a difficult task, there are areas of general agreement where we might begin (3). Ethical theories, like any other, are expected to be internally consistent. No theory should be allowed to contradict itself. Similarly, theories that are unclear or incomplete are clearly less valuable than theories that do not suffer from these flaws. Simplicity could also be considered an advantage. If all else were equal, it would be preferable to employ a simple theory over one that is complex or difficult to apply. We should also expect an ethical theory to provide us with assistance in those dilemmas where intuition fails to give us a clear answer. Most real-life ethical dilemmas are considered as such precisely because compelling moral arguments can be made in support of each side of the issue. These types of situations are those in which we most require the guidance of a moral theory. Additionally, ethical theories should generally agree with our sense of moral intuition. Who would wish to adopt an ethic that, although consistent, complete, and simple, advocated murder for profit? However, it is more difficult to decide about a theory that runs counter to our moral intuition in an area less clear-cut than murder, or in a number of minor areas. This is where the evaluation process becomes extremely difficult (2). How are we to decide whether it is the theory or our intuition that is out of line? We may decide that if a theory is rational, is well designed, and gives answers that correspond to our moral intuition on a large range of issues, then in a particular instance it is our intuition that is in error.

We in the natural sciences have something of an advantage over moral philosophers. Usually, we can design an experiment to discern which of two competing hypotheses is more correct. Philosophers do not have the luxury of performing an experiment and letting the data decide between the competing theories. However, ethicists do continually subject their own philosophies, and those of their colleagues, to "thought experiments" involving real or hypothetical ethical dilemmas. This process involves using a particular ethical theory to perform the moral calculus needed to answer a problem. It is sometimes found that the rigorous application of an ethical theory will lead to an outcome that is unacceptable, either to the philosopher or to the larger society. The philosopher may then decide to modify the theory in hopes of increasing its acceptability, or may choose to stick with the theory and instead suggest that society itself ought to be modified.

Utilitarianism

Ethical theories are generally divided into two major categories. The first of these is called either teleological or consequentialist, and the second is referred to as deontological. Teleological theories focus exclusively on the consequences of an action in order to determine the morality of that action. Thus, to determine if a particular act is moral or immoral, one determines whether the consequences of that act are considered good or bad. Those theories that do not exclusively evaluate the consequences of an act to determine its morality are called deontological. Deontological theories, considered in the next section, are commonly referred to as "duty-based," in contrast to the "outcome-based" nature of teleological-consequentialist theories.

The best-known example of a teleological theory is utilitarianism. Jeremy Bentham (1748–1832) was the first person to articulate the theory under that name, and John Stuart Mill (1806–1873) was also influential in its development. Utilitarianism acknowledges the fact that many acts do not produce purely good consequences or purely bad consequences, but some combination of the two. To decide whether a particular act is moral, a person must sum up all of the consequences, both good and bad, and assess the net outcome. Moral actions are those that cause the best balance of good versus bad consequences.

In addition, utilitarianism requires a person to consider the interests of everyone. It is not permissible to merely consider what is best for you personally. Suppose that you are considering lying about the results of an experiment that you have performed. You reason that lying about the experiment will greatly increase the chance of your paper's being accepted into a prestigious journal. This will, in turn, enhance your career, your salary, and your family's security. However, utilitarianism requires that you also consider the impact of your decision on other people. You must consider the fact that the scientists who read your paper and are misled by its fabricated results may be harmed by your decision. Some of them may decide to initiate a new series of experiments or to cease a line of investigation based on your fabricated data. If your research has direct clinical relevance, it is possible that patients may be directly injured or killed by your deceit. If you are caught in your lie, still more harm will accrue both to you directly and to the public's confidence in science. If you consider the cumulative negative impact of your lying, and not just the positive benefits that you are seeking, it will become apparent that the net outcome is a bad one. According to utilitarian theory, this act of deceit is immoral and you ought not to carry it out.

Now let's imagine a very different situation. A relative of one of your colleagues has escaped from a mental institution and shows up at the lab where you both work. Waving a scalpel and screaming that he wants to kill your friend for "ruining his life," he asks you to tell him where she is working. Although you know exactly where she is, what should you do? After performing the same type of utilitarian calculus as above, it is clear that you should lie to the escaped patient. The good and bad consequences that will flow from this particular act of deceit provide a net outcome that is markedly different from the scenario described above. Thus, in utilitarianism we find ethical decisions that change as circumstances change. An act that is deemed immoral under one set of circumstances can become morally obligatory under another. But exactly what are we to consider when we try to evaluate good and bad consequences? According to Mill, the only good is happiness and the only bad is unhappiness. Bentham thought that pleasure was the only good and that pain was the only bad. These terms are defined somewhat more broadly than you might imagine. Pleasure includes satisfaction of desires, attainment of goals, and enjoyment, while pain includes, in addition to physical discomfort, things such as the frustration of one's goals or desires.

Utilitarianism, like all other ethical theories, has its critics. One criticism is that it is excessively burdensome to employ. Utilitarianism requires that we all evaluate how each of our actions will impact everyone. How is it possible to actually do this? How is it possible to predict the consequences of even a fairly simple action on everyone? If we are required to do this for each of our actions, how will we be able to get anything accomplished? The advice to use our common sense does not seem to be very helpful. Another criticism of utilitarianism is that it would appear to condone, or even mandate, some actions that most of us would find horrendous. Suppose we find a patient who has a lymphoma that is producing a substance of tremendous use in the treatment of AIDS. However, the patient is totally uncooperative, refusing either to accept treatment for his illness or to allow samples of his cells to be taken for research purposes. Utilitarianism might allow us to kill this person and divide his cells among the interested research labs. While one person would die, many AIDS patients would live. Utilitarianism is potentially at odds with our concept of individual human rights.

Deontology

The second of the two major categories of ethical thought, deontology, does not depend exclusively on the consequences of an action to determine its morality. This does not necessarily mean that consequences play no role whatsoever in deontological theories. Those theories that admit to the relevance of consequences, in addition to other considerations, have been referred to as "moderate" theories, while theories that maintain that consequences must not be considered at all are called "extreme." The best-known deontological theory is that developed by the German philosopher Immanuel Kant (1724–1804). His theory is an example of an extreme deontological position in that the consequences of an action are not considered in establishing its morality. Kant believed that using the utility of an act to determine whether it is right or wrong is a terrible mistake. He realized, as we have already seen, that such a standard compels the moral person to perform a particular act in one situation, while forbidding it in another. This changing standard of morality was unacceptable to Kant, and so he developed a theory based on a principle that, unlike utility, would not change from one situation to another.

The principle that Kant developed to accomplish this purpose is called the categorical imperative. Kant formulated this principle in a number of different ways that he maintained were all equivalent (3). One of these formulations advises us to "act only on that maxim through which you can at the same time will that it should become a universal law." How does this principle guide and constrain our actions? To determine if a particular act is moral, we must first ask ourselves if we would wish that the rule governing our action be made a universal law — that is, if we would wish for everybody to follow the same course of action. If we cannot truthfully desire that anyone else be permitted to perform the action that we are considering, that act is immoral.

Let's, once again, suppose that you are considering lying about the results of certain experiments that you have performed. Before doing this, the categorical imperative requires that you first ask yourself whether or not you can honestly wish that your deed be universalized into a rule. This rule would permit all scientists to submit fraudulent data as genuine. Clearly, such a rule would destroy the credibility of all scientists and preclude the ability of the scientific community to make organized advances (as well as having much broader implications for the general concept of truthfulness). No one could legitimately wish that such a rule be universalized — therefore the act is immoral. Note that there is no consideration of the consequences of your contemplated act of deception. Whether or not you might benefit from your deed never enters into the moral calculus.

A second formulation of Kant's categorical imperative is more frequently encountered in discussions of medical ethics (3). This formulation advises us to "act in such a way that you always treat humanity, whether in your own person or in the person of any other, never simply as a means, but always at the same time as an end." This statement makes it more clear that Kant's principle also requires a certain respect for persons. Note that Kant does not demand that we *never* use a person as a means to an end, just that we do not use a person *solely* as a means. When a physician treats paying patients, she is clearly using them as a means through which she can achieve an end for herself (earning a living). Yet if this is the physician's sole consideration in treating patients, she will be acting immorally toward them. Patients, and all other persons, are to be treated as ends as well as means. Patients have interests independent of those of the physician from whom they have sought treatment. In other words, patients are their own ends. A physician who prescribes "snake oil" is acting immorally because she fails to treat the patient as an end. The physician who provides her patients with the best care available treats them as both a means and an end.

While it is interesting and useful to understand how moral philosophers approach ethical problems, it is not essential to understand the intricacies of utilitarian or deontological theory to make good moral decisions. Most of the rest of this book will be devoted to considering the types of real-life ethical dilemmas encountered by working scientists. By discussing the issues involved and solving the problems posed in the case studies, students will be better prepared to make positive contributions in their chosen profession.

Critical Thinking and the Case Study Approach

Scientists should strive to make certain that each of their professional decisions, whether pragmatic or ethical, is sound. Ideally, ethical decisions will, like strong hypotheses, endure the test of time. But we must also acknowledge that ethical standards are sometimes revised over time as a result of continuing scrutiny and reinterpretation in the face of emerging knowledge. To analyze and deal with the problems that challenge us in our daily activities, we need to be well grounded in the rules and standards of conduct expected of us as scientific professionals. A good start is to find and become familiar with the written codes that govern scientific behavior. Documents on human and animal experimentation, authorship, conflict of interest, and general codes of conduct are critical resources. But knowledge of such resources is only the first step in fostering responsible research practices. An understanding of how to apply the existing codes, as well as an ability to reason beyond their explicit language, is needed for problem solving in the real world. The instructional format embraced in this book affords opportunities to improve these skills by providing short case studies. The discussion of these cases will allow students to practice solving realistic problems by interpreting and correctly applying ethical standards.

These short case studies are designed to get the discussants to think critically as they analyze and problem-solve. "Critical thinking" has become a mantra in some academic circles as the problem-based learning approach has permeated the curricula of undergraduate, graduate, and especially professional programs. But what do we mean by critical thinking? Why is it important that we be critical thinkers?

Critical thinking is a cognitive process that clearly identifies issues and evidence related to a problem, thereby allowing defensible conclusions to be made. When discussing case studies like those found in this book, students should first separate the relevant issues from the nonrelevant ones. Relevant issues must then be analyzed, and the factual matters, backed up by evidence, must be distinguished from nonfactual ones. Students must also decide how to weigh the nonfactual matters, such as statements of opinion or expression of personal values.

Critically thinking about cases means that one must apply both factual knowledge and an understanding of appropriate scientific behavior to the problems encountered. It is important to remember when discussing cases that a consensus answer may not emerge. Nevertheless, several acceptable solutions to the problem may be found. Acceptable solutions must always be in compliance with standards related to global considerations (e.g., issues related to plagiarism or human rights). Solutions to cases always need to be examined to be sure they cannot be misinterpreted. In other words, they should not contain any loopholes. Examples of unacceptable solutions include violations of specific standards, guidelines, or rules and regulations. Solutions that are inconsistent with the written or unwritten ethical standards for scientific conduct generally accepted by the profession are also unacceptable. (See "Note to Students and Instructors" at the front of this book for a detailed discussion of how to approach case studies.)

Moral Reasoning in the Conduct of Science

The cases in this book will challenge you to analyze situations and make decisions based on information and evidence. Many of them will also require you to employ moral reasoning to reach your decision. In their monograph *Moral Reasoning in Scientific Research* (1), Bebeau and her colleagues suggest four psychological processes that are consistent with behaving morally. These were initially proposed by Rest, Bebeau, and Volker (5) and have been referred to as Rest's Four-Component Model of Morality. These components are:

> **Moral sensitivity:** The individual faced with a situation makes interpretations concerning what actions are possible, who would be affected by these actions, and how these actions would be regarded by the affected parties.
>
> **Moral reasoning:** The individual makes a judgment about what course of action is morally right (or fair, or just, or good), thus prescribing a potential course of action regarding what ought to be done.
>
> **Moral commitment:** The individual makes the decision to do what is morally right, giving priority to moral values above other personal values.

Moral perseverance (or moral implementation): The individual implements the moral course of action decided upon, facing up to and overcoming all obstacles.

Bebeau et al. (1) point out that, although these four processes can interact and even influence each other, in practice they also can be independent of one another. For example, a person may be quite adept at interpreting the ethical issues of a situation but unable to develop good arguments for the proposed moral judgment. When discussing cases, we can usually recognize and appreciate the skills involving moral sensitivity, moral reasoning, and moral commitment. In fact, the case discussions can enhance these skills. Because the cases reflect realistic situations, practice will improve the ability to recognize and reason through actual moral dilemmas in scientific research. For example, one can be expected to discover and use written codes of conduct and to better appreciate and apply normative standards. On the other hand, evaluating moral perseverance (implementation) is usually not possible when discussing case studies. Obviously, the true measure of this crucial component lies in what an individual actually does — something that is very difficult to play out in a case study. Nevertheless, it is sometimes possible to guess what an individual would do in a situation. We have encountered case discussions and write- ups in which a student, acting as the protagonist in the scenario, displays appropriate moral sensitivity, reasoning, and commitment. But then, in bringing the case to closure, the discussant describes some personal action that, in effect, portrays him or her as "walking away" from the situation. In other words, the discussant discloses an action that clearly indicates an unwillingness to implement the plan (and suffer its consequences).

Conclusion

Moral sensitivity, reasoning, commitment, and perseverance will all be needed in addressing the dilemmas raised in the cases found in subsequent chapters. We affirm the guidance provided by the criteria of Bebeau et al. (1) for making well-reasoned moral responses to dilemmas in scientific research. First, your response to the case should address all issues and points of ethical conflict. Move beyond just labeling issues and clearly articulate the conflicts emanating from the various elements of the case. Second, be sure your response considers the legitimate expectations of all interested parties. Keep in mind that parties may be affected who are not specifically invoked in the case narrative. Third, recognize that your proposed actions will have consequences. Clearly describe the probable consequences, their effects, and how they were incorporated into your decision. Fourth, identify and discuss the obligations or duties of the protagonist of the case.

What professional duty is at issue, and why does the scientist have that duty?

References

1. **Bebeau, M., K. Pimple, K. M. T. Muskavitch, S. L. Borden, and D. H. Smith.** 1995. *Moral Reasoning in Scientific Research: Cases for Teaching and Assessment.* Poynter Center for the Study of Ethics and American Institutions, Indiana University, Bloomington. (This monograph is available on-line at http://www.indiana.edu/~poynter/mr.pdf)

2. **Elliot, C.** 1992. Where ethics come from and what to do about it. *Hastings Center Rep.* **22:**28.

3. **Mappes, T. A., and J. S. Zembaty.** 1991. Biomedical ethics and ethical theory. *In* T. A. Mappes and J. S. Zembaty (ed.), *Biomedical Ethics*, 3rd ed. McGraw-Hill Book Co., Inc., New York, N.Y.

4. **National Academy of Sciences.** 1992. *Responsible Science*, vol. I: *Ensuring the Integrity of the Research Process.* National Academy Press, Washington, D.C.

5. **Rest, J.** 1986. *Moral Development: Advances in Research and Development.* Praeger Publishers, New York, N.Y.

Resources

Center for the Study of Ethics in the Professions at the Illinois Institute of Technology has a Web site that contains many codes of ethics of professional societies, corporations, government, and academic institutions. The Codes of Ethics Online Project may be found at

http://216.47.152.67/codes/

The Ethics Center for Engineering and Science Web site provides access to codes of conduct, cases, and a variety of other topics related to scientific integrity. It may be found on-line at

http://www.cwru.edu/affil/wwwethics/

The Web site of the Survival Skills and Ethics Program based at the University of Pittsburgh may be found at

http://www.pitt.edu/~survival/

A monograph entitled *Developing a Code of Ethics in Research: A Guide for Scientific Societies* (1997) may be obtained from the Association of American Medical Colleges in Washington, D.C.

http://www.aamc.org/

chapter 3

Mentoring

Francis L. Macrina

*Overview • Characteristics of the Mentor-Trainee Relationship
• Selection of a Mentor • Mentoring Guidelines • Conclusion
• Case Studies • References • Resources*

Overview

THE WORD MENTOR HAS ITS ORIGINS in the poetic epic *The Odyssey*, written by Homer more than 2,500 years ago. In Homer's story, Odysseus, king of Ithaca, sails off with his army to do battle in the Trojan War. Before leaving, Odysseus entrusts the care and education of his son Telemachus to his faithful friend Mentor. Mentor's responsibilities become enormous in scope and duration. The war lasts 10 years, and Odysseus's return trip takes another decade as he encounters one astounding adventure after another. Meantime, Penelope, Odysseus's wife, is being courted by noblemen in her husband's absence. Thinking that Odysseus will never return, these suitors waste his possessions by staging numerous feasts and parties. Throughout all of this, Mentor faithfully performs his oversight duties. His efforts are manifested in the young man Telemachus, who ultimately demonstrates he is worthy to be the son of Odysseus. Interestingly, sometimes Athene (daughter of Zeus and goddess of wisdom) appears disguised as Mentor to deliver critical advice to Telemachus. This image further underscores the value of Mentor's guidance. And so the word mentor has come to mean a loyal and trusted friend, enlightened advisor, and teacher.

In the research training setting, a mentor is defined as someone who is responsible for the guidance and academic, technical, and ethical development of a trainee. Mentoring is more than simple advising. It emerges from an extended relationship that allows the mentor to provide advice built on a foundation of both professional and personal knowledge. Mentoring extends beyond the training phase of one's professional development. Mentor-protégé relationships may continue throughout the better part of a career, in science and in other professions. This chapter primarily discusses predoctoral mentoring, referring to the participants as mentor and trainee. Some of what will be said applies to mentoring of the postdoctoral trainee as well. Career mentoring at later professional stages depends on many of the basic principles and strategies that are employed in mentoring trainees at the predoctoral level.

The canons of scientific integrity derive their life from effective mentoring in graduate training programs. Mentors inform, instruct, and provide an example for their trainees. The actions and activities of mentors affect the intellect and attitude of their trainees. The educational transfer process may be obvious or subtle, but the effects are rarely in dispute: trainees emerge from their programs with an intellectual and ethical framework strongly shaped by their mentors. Indeed, trainees often assume the traits and values of their mentors. Thus, mentors are the stewards of scientific integrity. Yet, the young faculty member who has just accepted his or her first trainee into the lab is not likely to have had much formal education in the principles of mentoring and is very likely to have no experience at all. The direct experience of dealing with trainees improves mentoring skills. To be sure, the skills and responsibilities of mentoring sometimes elude precise articulation and definition, because, as a human activity, there is great variation in the practice of mentoring trainees. There are many effective styles and, although common traits may be shared, there is no one prescribed method.

Characteristics of the Mentor-Trainee Relationship

A growing body of material discusses and analyzes mentoring in a variety of professional settings, including the research laboratory and the university (see references 5 to 7 and the resources section). In the sciences and related professions there are at least four general categories of activities that describe what mentors do. These apply universally — to a greater or lesser extent — to mentor-trainee or mentor-protégé relationships at any level.

Mentors demonstrate and teach style and methodology in doing scientific research

Especially during formal training, mentors share their talents for defining problems, asking questions, and selecting the means for solving problems and getting answers. This may be done in a very calculated way wherein a novice is guided through a problem with considerable assistance from the mentor. Alternatively, mentors may convey their style and methods for problem solving by example, allowing trainees to observe the process. What is learned may range from how the mentor formulates a hypothesis to how he or she keeps up with the literature and developments in the field. It is rare for the mentor not to make an impression in this setting, and the trainee usually assimilates some of the ways in which the mentor deals with the theoretical and practical aspects of doing research. These mentoring issues can remain in play throughout one's career, applying both to the young scientist doing postdoctoral training and to the seasoned faculty member doing a research sabbatical.

Mentors evaluate and critique scientific research

There are many opportunities for mentors to convey to trainees "how things are going." Whether reviewing results in a data book; listening to a presentation, lecture, or seminar; or critiquing a manuscript or dissertation, the mentor can and should provide constructive criticism. Such activities give the mentor a chance to identify problems and propose remedies and to challenge the trainees to refine their research skills. In practical terms, these opportunities often allow the mentor to help improve the trainee's communication skills. These activities also continue throughout a career. For example, scientists may develop mentor-protégé relationships with colleagues who read and critique their proposals, manuscripts, or other writings.

Mentors foster the socialization of trainees

Mentors provide information to trainees about the workings of science. This may involve familiarizing trainees with policies, guidelines, and regulations about the conduct of research. Normative standards pertaining to authorship, peer review, data sharing, and collaboration are things that trainees may hear about first from their mentors. Mentors make trainees aware of the ethical responsibilities of scientists and provide by example and instruction the tenets of responsible conduct in research. In short, the trainees' entry into the profession involves learning appropriate behaviors, and mentors take an active role in this process.

Mentors promote career development

Mentors are advocates. They look out for the professional health and well-being of their trainees. Mentors can help with insight, information, and advice about career planning. They can help trainees understand and practice networking by encouraging them to communicate with other scientists and by introducing them to other scientists. Mentors help trainees develop and refine appropriate interpersonal skills like negotiation, mediation, persuasion, and poise. Later in a career, mentors may promote protégés by suggesting their names as speakers or conference organizers or by recommending them for service assignments that are part of good professional citizenship. Nominating trainees or protégés for awards can also be done to foster and enhance their career.

Mentors perform different duties at different times

The primary duties of a mentor change over time, and at any moment they may involve different aspects of the relationship. Switching from the role of mentor-advisor to that of mentor-confidant or mentor-critic might occur over the span of a day. Being responsive to the changing demands of the mentor role requires critical attention and oversight. Mentoring is a one-on-one activity. It is typically depicted as an intense relationship that demands continued personal and intellectual involvement on the parts of both mentor and trainee. Mentoring relationships work best in an atmosphere of mutual respect, trust, and compassion. Mentoring is dynamic and complex. Attempting to simplify the scope of mentoring duties and responsibilities is misleading and counterproductive. Guston (2) correctly points out "what the mentor role is *not*." A mentor is not just a patron (resource advisor), not just a supervisor (one who oversees a dissertation), not just an institutional linkage between student and the academic administration, and, finally, not just a role model. Mentoring roles overlap and receive different emphasis depending on specific circumstances and the changing needs of the trainee.

Trainees depend on mentors

One of the unique aspects of predoctoral mentoring is the degree to which the trainee is dependent on the mentor. In many cases, this dependence is grounded in finances, because the mentor's grant provides stipend support and often tuition and fee payments. Almost always in the biomedical sciences, the mentor's grant provides the resources that are critically needed for trainees to perform and complete their dissertation research. Moreover, the mentor is usually directly or indirectly involved in providing or securing the resources for trainees to attend meetings or workshops that are important to their graduate training experience. Finally, trainees are critically dependent on their mentors for a position when they finish their programs. Such positions might entail postdoctoral training or employment in universities, industry, or government. Such dependence on the mentor's evaluation continues well into the trainee's career; for example, applying for a position beyond a postdoctoral training experience usually means that the predoctoral mentor provides a letter of recommendation.

Thus, a graduate trainee is profoundly dependent on his or her mentor. (A similar dependence is encountered in the postdoctoral trainee-mentor relationship.) This dependence means that the trainee is vulnerable to abuses of power. Although such abuse would seem antithetical to the basic premise of mentoring, trainees do fall victim to such circumstances. Abuses of power can take the form of acts of commission as well as acts of neglect. The trainee usually finds himself or herself in a difficult position when such situations arise. The very person who should be available to solve the problem at hand turns out to be at the heart of the problem. Nonetheless, the mentor should be directly approached if such problems are perceived by the trainee. Communication between mentor and trainee can be an effective way to resolve the situation. In addition, graduate advisory committees, other faculty, and departmental chairs can usually help. Dependence on their mentors at a time when they feel abused by the same person presents trainees with a dilemma that is not easily resolved. However, avoiding the problem is a virtual guarantee that the problem will get worse.

The mentor-trainee relationship is an exclusive one

Although graduate programs usually mandate that each predoctoral trainee be guided by an advisory committee, the mentor usually chairs the committee and is the trainee's principal advocate in this forum. The exclusive and intense nature of the mentor-trainee association in science is underscored by the usual longevity of such relationships. Predoctoral mentoring in graduate biomedical research usually marks the beginning of a relationship that significantly outlives the time spent in formal training. Trainees often continue to rely on their graduate mentors for advice and counsel as they progress through the beginning stages of their professional careers. Mentors are often viewed in the light of the performance of their former trainees, and this association can extend the mentor's responsibilities indefinitely. Djerassi (1) points out that evidence of the mentoring relationship can sometimes be found even after the death of the mentor.

Staying aware of the academic status, intellectual development, and practical research progress of a trainee requires regular oversight, information exchange, and frequent and regular interpersonal communication. One critical issue is the size of the research group. As the number of people in a research group increases, there is less time to conduct a proper and effective mentor-trainee relationship. Mentors need to face up to this reality as they weigh commitment, take on additional responsibilities, and develop their research training programs. There is a point of diminishing return in the number of trainees who can be effectively mentored. When that threshold is crossed, the ability to responsibly guide trainees is compromised, and the viability of the training experience is put in jeopardy. Poorly mentored trainees can unknowingly cut corners, make mistakes, or not recognize errors. Over time, such behavior can come back to haunt the mentors by jeopardizing the credibility of their research programs. Thus, neglect of mentoring responsibilities and duties can harm both mentors and trainees.

At times, members of the graduate advisory committee or even other faculty may assume transient mentoring roles. For example, a trainee in biochemistry may need to produce antibodies against a protein she has isolated. To achieve this goal, she may be scientifically mentored by an immunologist who is a member of her advisory committee. Mentoring activities in this case might involve instruction and advice regarding compliance with regulations concerning the use of animals in research, the handling of animals in the called-for experiments, and relevant immunological methods needed to do the work.

The mentor-trainee relationship requires trust

Certain fundamental characteristics must be evident in the actions of both the mentor and the trainee. Personal respect is absolutely necessary on both sides of the mentoring relationship. Mutual trust is another essential ingredient of a successful mentoring relationship. Throughout the relationship, trainees must trust their mentor's advice and actions that bear on their training programs. Most students at the early stages of their programs depend strongly, if not exclusively, on their mentor's knowledge and expertise in helping them select a viable dissertation research project. A mentor who has developed a reputation for recommending changes in a trainee's dissertation project at the least sign of failure may have difficulty attracting and keeping students in the lab. Such actions tend to lessen confidence and undermine trust in the mentor's scientific decision-making style. Mentors, for their part, must cultivate a trust in the caliber of work performed by the trainee over the course of the dissertation research project. In an active mentoring relationship, the mentor is able to gauge a trainee's performance by three principal means: (i) direct laboratory observation, (ii) viewing the trainee's raw and analyzed research data, and (iii) listening to trainees present their ideas and data in both informal and formal settings. Over time, the mentor develops a degree of confidence in the trainee's operating style based on these observations. Direct laboratory observation is usually a significant component in the early stages of training but may wane or even disappear as the trainee progresses and matures. Data observation and related discussion take place throughout the course of graduate training. This activity should be characterized by regular face-to-face meetings, with data books and other relevant materials at hand. The mentor should observe trainees as they give seminars, write research reports, or lead journal clubs. This activity, which should persist throughout the training experience, serves two functions: (i) it allows for continuing assessment of student progress in scientific thinking and analysis, and (ii) it provides an excellent forum for the mentor to critique the scientific communication skills of the trainee.

Free and open communication flows from an atmosphere of mutual respect and trust in a successful mentor-trainee relationship. Good mentors are critical and demanding of their trainees, and these characteristics should be explicit in all forms of communication with the trainee. When combined with compassionate personal support and enthusiasm for the work, trainees are likely to recognize helpful criticism and guidance and not confuse these messages with displeasure, hostility, or intimidation. Such an interchange, in turn, cultivates a collegial relationship between the participants as together they share and analyze information, critique each other's ideas, and solve problems with each other's help. Attribution of credit and recognition of accomplishments should be clearly articulated. Taken together, these activities are the important first steps in the broad-based socialization of a young scientist. Indeed, such mentoring activities set the stage for a career in science and continue throughout the scientist's career in different contexts and to different degrees.

Selection of a Mentor

There are both subjective and objective criteria that may assist graduate trainees in selecting a mentor (5, 6), as in the following list.

- Active publication record in high-quality, peer-reviewed journals (consider using *Science Citation Index* to determine the frequency of citation of selected papers)
- Extramural financial support base: competitiveness and continuity of support
- National recognition: meeting and seminar invitations, invited presentations, consultantships
- Rank, tenure status, and proximity to retirement age
- Prior training record: time it takes trainees to complete a degree, number of trainees, and enthusiasm for previous trainees' accomplishments
- Current positions of recent graduates
- Recognition for student accomplishments (e.g., coauthorship practices)
- Organizational structure of the laboratory and direct observation of the laboratory in operation

In the past several years, electronic communication has greatly facilitated the collection and evaluation of information about potential mentors. The Internet has dramatically lessened the need to seek information in curricula vitae and departmental reports. Academic departments and individual faculty members frequently maintain Web sites displaying their research interests, publications, accomplishments, and other useful information. Bibliographic databases accessed through PubMed (http://www.ncbi.nlm.nih.gov/PubMed/) or Grateful Med (http://igm.nlm.nih.gov/) provide immediate access to publication records and abstracts. Similarly, the Computer Retrieval of Information on Scientific Projects (CRISP) database maintained by the National Institutes of Health (NIH) affords a preliminary evaluation of a faculty member's sponsored research program (access at http://commons.cit.nih.gov/crisp/owa/CRISP.Generate_Ticket).

Selection of a mentor is usually based on three principal activities. The first is education; trainees can read research descriptions in advertisements as well as published works of the prospective mentor to determine if his or her interests coincide with their graduate training goals. The second important activity in mentor selection is interpersonal interaction, both with the potential mentor and with members, including trainees, of his or her research group. It is in the best interest of the potential mentor and the trainee to meet on several occasions and to thoroughly discuss the practical issues of dissertation research possibilities and the logistics of selection of a project. It is also appropriate to discuss issues such as mentoring style (supervision, general expectations, and goal setting) and other personal and academic issues related to graduate training. Candid discussion at this point not only provides the basis for an intelligent decision on the part of both the prospective mentor and the trainee, but it also sets the stage for the free and open communication that must support the trainee-mentor relationship during formal academic training and dissertation research. Talking with lab members about their view of the training environment provides a valuable perspective for the trainee seeking a mentor. The training climate, enthusiasm of other trainees, and corroborative information on the mentoring style of the laboratory head can make the trainee more comfortable with the prospect of selecting this person as a mentor or can raise more questions that will need to be answered by the prospective mentor. A third useful activity in the mentor selection process stems from the existence of so-called lab rotation programs now found in many graduate biomedical science departments. Such programs typically have the first-year graduate student doing a specific research project (or learning specialized methodology) over the course of a few weeks. These programs provide a firsthand view of the operation of the lab and its personnel dynamics, including mentor-trainee relationships. For many entering graduate students, this encounter is often their first exposure to the day-to-day workings of a research environment. This exposure allows prospective trainees to directly assess the climate they will encounter in their training experience. Does the mentor provide much direct supervision, or are technological skills and data analysis and interpretation relegated to another senior lab member? Has the training environment changed much over time in the experience of the current trainees? Have the methods of training used by the mentor been successful over time? The rotation system also allows the mentor to view the prospective trainee at the research bench and thus to acquire useful, albeit brief and casual, impressions of the trainee's potential.

In summary, selection of a mentor requires both formal and informal activities coupled with thoughtful analyses on the part of both mentor and advisee. However, even the most thoughtful decisions, based on the careful collection of facts and data, can result in mentor-trainee relationships that do not work. Conflicting personal styles that emerge over time, disenchantment with a general area of research, and evolving changes in aspects of mentoring responsibilities or discharge of duties all can cause a mentor-trainee relationship to degenerate. When this happens, resolution at an early stage is the best course of action for all involved. Candid mentor-trainee discussion of problems may need to be augmented by third-party mediators (e.g., departmental chairs or graduate program directors). Intractable problems should be recognized and accepted; switching mentors in a predoctoral training program can and should be implemented to solve such problems. Prolonging problems by failing to face up to them often creates tension in the training environment and unnecessarily lengthens the duration of the training program for the trainee.

Mentoring Guidelines

Mentoring guidance found in responsible research conduct guidelines

Guidance on mentoring has emerged in various forms. Frequently, institutional policies on the responsible conduct of research include discussions of mentoring. Some of these documents begin with a preamble that affirms the importance of mentoring in scientific training. Examples of such policy statements that are available on the Internet may be found in the resources section at the end of this chapter. A distillation of the common points covered by representative policies follows.

- Specific assignment of trainees to faculty mentors must be made, with responsibility for the trainee residing unambiguously with the faculty member

Mentoring relationships in predoctoral training begin with the assignment of a temporary advisor and continue throughout the training program with the selection of a dissertation advisor. The duties of the temporary advisor and the permanent advisor should be clearly articulated.

- The ratio of mentor to trainees in a given laboratory should be small enough to foster scientific interchange and to afford supervision of the research activities throughout the training program

Few will argue with the assertion that, at some point, the size of a laboratory research group curtails and may even preclude responsible and effective mentoring. However, defining that point is difficult, because it depends on such factors as the type of trainee (entry level or advanced, predoctoral or postdoctoral), the nature of the work being performed, the overall time commitments of the mentor, and the mentor's management skills. Most would argue that active mentoring of more than 10 to 12 trainees is not possible. Larger groups must have a secondary mentoring network in place, wherein senior members of the lab also serve as mentors. Such an infrastructure may enable the laboratory head to delegate mentoring duties, but this practice can be argued against on the grounds that such systems are not in keeping with mentoring guidelines as put forward by some institutions. Specifically, mentoring is predicated on mentor-trainee interchange and, as such, does not afford the latitude for delegation of such responsibility.

- The mentor should have a direct role in supervising the designing of experiments and all activities related to data collection, analysis and interpretation, and storage

The emphasis here is on close supervision of the trainee's progress, highlighted by personal interaction. Some of the standards-of-conduct documents underscore the importance of direct, active supervision by providing a contrasting statement: mentors who limit their roles to the editing of manuscripts do not provide adequate supervision.

- Collegial discussion among mentors and trainees should pervade the relationship, and this should be highlighted by regular group meetings that contribute to the scientific efforts of the group and, at the same time, expose trainees to informal peer review.

In fact, regular group meetings should be augmented by mentor-trainee meetings that are held regularly and privately. Individual attention provides the mentor and the trainee with a unique opportunity for uninhibited communication, critical analysis, and problem solving on matters that may be unique to the trainee or the specific project.

- The mentor is responsible for providing trainees with all relevant rules, regulations, and guidelines that may apply to the conduct of research (e.g., human and animal use documents, radioactive and hazardous substance use documents)

The mentor has a responsibility for oversight and enforcement in this area, too. Trainees must comply with rules and regulations as observed directly or monitored indirectly by the mentor. The breach of any established policy will come to rest with the mentor as the individual with overall responsible for the laboratory group.

Some responsible-conduct guidelines have included behavioral assessment-type issues that fall in the category of mentoring responsibilities. These include the formulation of realistic expectations for the trainees' performance, which should be made explicitly known to the trainee. The mentor then has an obligation to provide a realistic appraisal of a trainee's performance. Mentors have also been advised to be alert to behavioral changes in trainees that may be due to stress or to substance abuse. Another document emphasizes that the research laboratory experience should be at all times a learning experience, engendering appreciation of proper methods and conduct and appropriate ethical consideration of all those touched by the research. Yet another document talks about the duty of a mentor to provide a meaningful training experience for the trainee; this is followed by the admonition that projects in which the mentor has a monetary stake or other compelling interest are not acceptable training experiences.

Institutional documents on mentoring

Some institutions have prepared position papers or guidelines that specifically address training and mentoring (see the resources section). These documents list and explain responsibilities applying to both sides of the mentoring relationship. Some specific instructions for the mentor include the following: providing a clear map of expectations leading to the academic goal; providing enough time for the trainee; providing training and instruction in professional communication skills; engaging the trainee beyond the classroom and laboratory; placing value on diversity and dealing with mentoring issues attendant to diversity; providing ethical guidance; and assisting the trainee with career planning. Creating and fostering an environment of collegiality is mandated: both mentor and trainee need to properly recognize and acknowledge their respective contributions. Similarly, open communication in both directions is extolled. Mentors are cautioned against conflicts of interest that may interfere with their duties (e.g., familial or personal relationships). Trainees are encouraged to devote appropriate time and energies to achieving academic excellence (and their graduate degrees); be aware of the demands imposed on faculty members; and recognize the responsibilities of the mentor in monitoring the integrity of the trainee's research.

Expectations for predoctoral trainees

Those new to mentoring often rely on the departmental or graduate program guidelines to help them formulate the expectations of the trainee. Seasoned mentors can benefit from using such documents as well. The language in these documents can be translated into easily understandable goals. When wedded to a time frame, the guidelines create a perspective that affords clear milestones for evaluation by the mentor, while at the same time providing motivation for the trainee. Although this aspect of mentoring is general and rather pragmatic, it provides the foundation for communication and correctly transmits, from an early point, the active involvement of the mentor in the training process. The International Union of Immunological Societies (IUIS) has recognized a need for uniform standards in graduate training in immunobiology. A proposal describing such guidelines (4) serves as a useful model for discussion. This and similar documents (3) can help guide the articulation of expectations and standards, which, in turn, can be made clear to trainees. Ultimately, this process greatly assists the mentor, especially during the early evaluative steps of a training program.

Although the timing of achievement of specific outcomes may vary according to experience and individual preference, the general standards of a given program might parallel those recommended by the IUIS report. These standards are presented and briefly annotated here to provide useful examples.

- The candidate should demonstrate a general knowledge of basic immunology

  This means understanding experimental methods in a way that fosters appreciation of basic concepts rather than simple acceptance of conclusion reached by others. It also means reading and comprehending the primary and secondary scientific literature in the discipline. In the IUIS report, specific examples of appropriate journals are listed. Mentors and graduate advisory committees should likewise offer guidance in journal reading to the new trainee. The candidate's level of understanding can be evaluated by direct observation of the mentor and by performance in courses and comprehensive examinations.

- The candidate should be familiar with the immunology literature and be able to acquire a working background knowledge of any area related to immunology

  The ability to read critically and to use such information to ask further questions and propose relevant research problems is essential. Active journal clubs (where students present and discuss the primary literature), seminars, and the writing of research proposals and research papers all provide good indicators of students' appreciation and comprehension of the literature.

- The candidate should possess technical skill

  Through course work or independent training and study, the student must display the ability to master techniques needed to conduct the assigned dissertation research project. Evaluation in courses and direct observation of laboratory techniques and resulting data are needed to assess performance.

- The candidate should ask meaningful questions

  Subjective evaluation by the mentor as well as reviewing the student's critical evaluation of the literature as components of seminar presentations and writings provide an indication of performance. The IUIS report (4) points out that "acquisition of the ability to formulate meaningful questions is a major step in the candidate's transition from a passive to an active role in research."

- The candidate should demonstrate oral and written communication skills

   Seminar and journal club presentations are expected; regular participation may be specifically defined. Verbal communication skills are best honed through regular practice. Informal or, if appropriate, formal evaluation of performance should be provided by the mentor, graduate advisory committee, and other faculty. Comments and guidance should be constructive, candid, and provided at every opportunity. Writing skills are also improved through practice, and informal peer review involving mentors, faculty, and scientific colleagues is essential.
- The candidate should demonstrate skill in designing experimental protocols and in conducting productive independent research

   These skills are evaluated by the mentor on a frequent basis and by the student's graduate advisory committee on a periodic basis. With time, evidence of progress in this area becomes apparent to the mentor; less description and detail are needed to launch the student into specific aspects of the project.
- The candidates and supervisor should adhere to the ethical rules accepted by the scientific community

   This point embraces the expectation that the mentor-trainee relationship is the principal vehicle for the scientific and professional socialization of the trainee. It further expects that students will have available to them appropriate relevant training opportunities (e.g., good laboratory practice, appropriate use of animals and human subjects).

In summary, careful articulation of expectations is essential to graduate training. Clear expectations provide a perspective for the trainee and can motivate her or him. Equally important, they help mentors carry out their duties by holding students to explicit, recognized standards that can be easily communicated, readily monitored, and reasonably enforced. Clear expectations are a special help to inexperienced mentors by providing them with specific parameters for guidance and evaluation. They can and should be worded in general terms, taking into account the variances of differing programs and disciplines while at the same time not inhibiting the creative processes that underlie graduate dissertation research.

Conclusion

Mentor-trainee relationships in science are critical to both the technical training and professional socialization of young scientists. The mentor-trainee selection process should involve an informed decision on the part of both participants. The mentor-trainee relationship must be built on mutual trust and respect. It is a dynamic interpersonal relationship, with both parties having distinct responsibilities. Recognizing the responsibilities of mentors has become more formal at some institutions, with written documents providing guidance. Guidance on the responsibilities of trainees can often be found in departmental documents such as graduate program policies and in the writings of scholarly societies.

Case Studies

3.1 Graduate student Stan Mayfair has been meeting with potential dissertation advisors. Stan becomes interested in Professor McGuire's research program, and they have several lengthy discussions about potential projects available for Stan's dissertation research. Late in these discussions, Stan asks Dr. McGuire about the financial support of various projects. For each project, Stan insists on knowing the source of support, the duration of the support, and McGuire's plans for continuing sustained support. McGuire is put off by these questions and indicates that these are inappropriate concerns for the student. He refuses to provide the information. Are these appropriate questions for the student to ask? Is the faculty member's response justified?

3.2 Professor Eastwood frequently receives letters that announce junior-level positions for scientists. Sometimes Eastwood posts the letters on the departmental or laboratory bulletin board. Other times he distributes them to faculty or postdoctoral fellows using a routing list. On occasion he directly gives a letter to a postdoctoral trainee. Drs. Smith and Jones are currently postdoctoral trainees in Eastwood's lab. Dr. Jones discovers that Dr. Smith has applied for a job at a prestigious university. In a subsequent conversation, Dr. Smith tells Dr. Jones that Professor Eastwood provided him with a preadvertisement letter inviting applicants to apply for the position. Dr. Jones confronts Eastwood and indicates that she is upset that she was not notified about this position. Eastwood asserts that his policy is to deal with such letters selectively. He states that he could not support Dr. Jones for the position in question so he did not provide her with the letter in advance of the published advertisement. Comment on this situation, specifically discussing the mentor's policies. What are the mentor's responsibilities in such matters? If you were a postdoctoral trainee in the laboratory, what would be your expectations regarding such matters? As a mentor, what would be your policy on such matters? Why?

3.3 John Brandt and Professor Woodworth have met several times to discuss possible projects that John might take on as a doctoral dissertation project. During the last discussion Woodworth recites a series of rules that he applies uniformly to his advisees. He indicates that he wants John to know the rules of his laboratory fully before making a decision to join the lab. Most of the issues covered are straightforward, reasonable, and come as no surprise to John. However, one rule surprises and concerns him. Woodworth says that he does not permit his laboratory advisees to enter into romantic relationships with one another. Should such a relationship develop, he insists that one of the members of the relationship find a new advisor and a new laboratory. John argues that this is direct interference with personal matters and that such relationships are of no concern to the advisor. Woodworth counters with the fact that twice in the past 5 years his laboratory has been significantly disrupted by romantic relationships between his student advisees. These situations have resulted in ill will, diminished productivity, and a negative effect on the overall morale of his laboratory group. The faculty member indicates that he has carefully considered the implications of such relationships and has decided that the only reasonable thing to do is to prevent the problems they create by asking those involved to decide which of the two of them will leave the laboratory. Discuss the issues of mentorship responsibilities, ethics, and conflicts of interest that you feel are important to this scenario.

3.4   Milton France, a senior-level graduate student, is seen less and less during the day by his mentor and other members of the laboratory. It becomes apparent to the mentor, Dr. Wise, that Milton is working very long hours during evenings and nights when most of the other laboratory workers are not there. This persists for several weeks, and Dr. Wise does not think the pattern is a good one. Dr. Wise approaches Milton and requests that he spend more time during "standard working hours" in the lab. Dr. Wise argues that interaction with him and with other members of the laboratory is important and that it is best for all to talk about science regularly. Milton argues that he can work much more efficiently when fewer people are around. He cites the fact that a piece of equipment he was using in his research was continually busy throughout the daytime hours and this was not conducive to his performing needed experiments in a timely fashion. Milton discloses that this was the "straw that broke the camel's back," forcing him into working unconventional hours. Both the faculty advisor and the student hold tight to their arguments, and over the next several days the situation between them grows tense. Comment on this situation and consider what avenues might be pursued to bring about resolution of this conflict.

3.5   Jennifer Roberts, a predoctoral student, has cloned a lymphocyte gene that encodes a membrane protein. The discovery of this gene was unexpected, and Jennifer and her mentor, Dr. Beth Hillary, agree that further characterization is warranted. Dr. Hillary indicates that the nucleotide sequence of the gene should be determined, and Jennifer agrees. Dr. Hillary wants the nucleotide sequence to be determined by a commercial laboratory that does such analyses on a fee-for-service basis. Jennifer argues that she has had no experience in determining DNA sequences and would like to learn the technology. Dr. Hillary comments that this would be time-consuming and would unnecessarily slow down the student's progress in other areas. Dr. Hillary indicates that if Jennifer is interested in learning DNA sequencing, she should enroll in a techniques course at a later time. The student is not receptive to this suggestion. Can the mentor's decision be justified in your view? Can you suggest resolutions to this problem?

3.6 Maureen has prepared a research proposal in the form of an NIH grant application as part of the requirements for her Ph.D. degree. Maureen came up with the idea for the proposal after reading her mentor's funded grant application. Maureen has developed the idea thoroughly, and her mentor has provided only minimal assistance in the development of her grant proposal. Several weeks later, Maureen learns that verbatim sections of her grant proposal have been included in a new grant application being submitted to the NIH by her mentor. Comment on the appropriateness of the mentor's action in using Maureen's written material. Maureen wishes to confront her mentor over this issue and has come to you for advice.

3.7 Robin Carvell has been a postdoctoral fellow in a large research group for 3 years. He has accepted a job at a university and is in the last month of his formal training. Dr. Eleanor Hunt, his mentor, requests to meet with him privately shortly before his departure. Dr. Hunt produces a typewritten document that summarizes Robin's contributions during his training. Moreover, the document lists biological materials that Robin will not be allowed to remove from the laboratory when he leaves. Finally, it spells out several areas not yet under investigation in Dr. Hunt's laboratory that Robin is forbidden to work on in his new position. There is a signature line for Robin at the end of the document to indicate his agreement with its language. Dr. Hunt asks Robin to take the document home, read it carefully, and return the signed copy to her in the morning. Robin leaves the office and is quite upset with this situation. He believes his mentor is acting selfishly and unethically. He comes to you seeking advice.

3.8 Dr. Mitchell Conrad has received a grant from an industrial source to do basic research that has long-term implications for commercialization. A new graduate student, Michelle Lawless, has just joined his lab after completing one semester of graduate coursework. Dr. Conrad outlines several projects that can be pursued by Michelle in the industrially sponsored research program. Dr. Conrad indicates that there is a proviso listed in the industrial grant agreement which says that all material to be submitted for publication first be reviewed by the company. This review must always be completed within 120 days. Dr. Conrad points out that this presents only a minimal disruption to the normal publication process as compared with the unrestricted publication of material gathered under federal research grants. He also mentions that the positive aspects of working on this proposal include the fact that there is money in the grant for Michelle to travel to at least two meetings per year. Also, the grant application provides money for a personal computer that will be placed at Michelle's lab station while she is working on the project. Dr. Conrad emphasizes that working on the project will likely give Michelle an "inside track" with the company should she want to pursue job possibilities there following graduation. Michelle agrees to work on the project. Comment on the ethical and conflict-of-interest implications of this scenario.

3.9 Ron Archer is the graduate advisor for several predoctoral students. One of his students, Gordon Polk, shows Ron data that describe a novel property of an enzyme under study. Both Ron and Gordon believe this work has major implications for expanding the knowledge of this enzyme. At Ron's request, Gordon repeats the experiments successfully. Then, because of the important implications of this work, Ron approaches another predoctoral student in the lab and asks her to perform the same experiments in order to double-check the results. Ron instructs the student not to discuss the experiments with anyone else in the lab in order to obtain independent data to confirm Gordon's potentially important findings. Are the advisor's actions justified in this case?

3.10 Jim Allen has been a postdoctoral fellow in your lab for 3 years. He is in final negotiations for a tenure-track assistant professorship at another university. He is excited about taking this job, and you are pleased that the position will allow him an excellent opportunity to grow into an independent scientist. At the request of Dr. Wiley, his prospective departmental chair, Jim has been preparing an equipment list needed to set up his laboratory. Jim has come to you for advice several times while preparing this list. This morning he shows up in your office and you immediately sense he is upset. Last night Dr. Wiley called and asked him to be sure to include several additional equipment items on his list. Dr. Wiley told him, "Setting up faculty is our best opportunity to get equipment money for the department from the dean and vice president's office. The department desperately needs a new FPLC chromatography unit, a phase-contrast microscope, a scintillation counter, and an ultra-low-temperature freezer. So please add these to your set-up list. I promise that asking for these items won't compromise our ability to secure the money for the equipment you actually need for your lab." In Jim's present or planned research, he has no need for an FPLC or a phase-contrast scope. Jim feels he is being asked to falsely represent his needs to the university administration. He is worried that if he objects to or refuses Dr. Wiley's request, he may not be offered the job. He asks you for advice on how he should proceed.

3.11 Hal Sloan, a junior faculty member, has developed a mentor-protégé relationship with Chet Alexander, a professor in his department. Over coffee one morning, Hal tells Chet that he is "seeing" a graduate student in the department. Hal refers to her by a fictitious name, Diane. Hal tells Chet that he is currently delivering a series of five lectures in a cell biology course in which Diane is enrolled. Chet cautions Hal that this may be a conflict of interest. Hal says he already thought about this and proposes to solve the problem as follows. He intends to meet with the course director and give him an answer key to his questions on the upcoming cell biology midterm test. Hal will ask the course director to use this key to grade Diane's answers to Hal's questions on the midterm. Hal will, of course, volunteer to grade the answers written by the rest of the students in the class. Finally, Hal tells Chet that he intends to alert the departmental chair about his relationship with Diane and ask the chair to avoid making any assignments that put Hal and Diane in any type of working or academic relationship (e.g., committee work, other courses). As Chet, what comments, advice, or suggestions do you have for your protégé?

3.12 Mike Morton is a third-year graduate student at Big West University, where he is immersed in his dissertation research in cell biology. One fall Saturday afternoon you are working in the lab when Mike arrives to do some work, having just attended a Big West home football game. He seems in a jovial mood as he shuts down a high-voltage electrophoresis apparatus and prepares his gel for processing. He then prepares some samples and starts an ultracentrifuge run that will take 3 hours. As he works near your bench you can smell alcohol, and you conclude that although Mike may not be drunk, he has clearly been drinking. You have some passing concern that Mike could be endangering himself and others by operating potentially dangerous lab equipment following alcohol consumption. The next day you visit the lab to change some cell culture media, and you discover Mike's centrifuge has completed its run and is sitting idle with Mike's samples still in it. You phone his apartment but get no answer, so you send him an e-mail alerting him to the problem. The next morning the centrifuge is still not in operation, but Mike's tubes are no longer in the rotor. Sensitized to these events, you take a keen interest in Mike's behavior. You notice that you can sometimes smell alcohol on his breath in the mornings when he comes to the lab. Are you obliged to act on these observations? What actions, if any, do you take?

References

1. **Djerassi, C.** 1991. Mentoring: a cure for science bashing? *Chem. Eng. News*, November 25, p. 30–33.

2. **Guston, D. H.** 1993. Mentorship and the research training experience, p. 50–65. *In Responsible Science — Ensuring the Integrity of the Research Process*, vol. II. National Academy Press, Washington, D.C.

3. **International Union of Biochemistry, Committee on Education.** 1989. Standards for the Ph.D. degree in biochemistry and molecular biology. *Trends Biochem. Sci.* **14:**205–209.

4. **Revillard, J.-P., and F. Celada.** 1992. Guidelines for the Ph.D. degree in immunology. *Immunol. Today* **13:**367–373.

5. **Smith, R. V.** 1998. *Graduate Research — A Guide for Students in the Sciences*, 3rd ed. University of Washington Press, Seattle.

6. **Stock, M.** 1985. *A Practical Guide to Graduate Research.* McGraw-Hill Book Co., New York, N.Y.

7. **Yentsch, C., and C. J. Sindermann.** 1992. *The Woman Scientist — Meeting the Challenge for a Successful Career*, p. 145–159. Plenum Press, New York, N.Y.

Resources

Examples of responsible research conduct guidelines that deal with mentoring issues may be found on-line at

http://www.hms.harvard.edu/integrity/scientif.html — Harvard University

http://www.nwu.edu/research/policies.html — Northwestern University

http://www.pitt.edu/~provost/ethresearch.html#Res — University of Pittsburgh

http://www.library.ucsf.edu/ih/VI.html#VIa — University of California, San Francisco

http://www.nih.gov/news/irnews/guidelines.htm/ — National Institutes of Health

Resources relevant to mentoring in science

The NIH Mentoring Guide can be accessed from the NIH Intramural Resource Sourcebook Page at

http://www1.od.nih.gov/oir/sourcebook/

Adviser, Teacher, Role Model, Friend: On Being a Mentor to Students in Science and Engineering. 1997. National Academy Press, Washington, D.C. The full text of this book is also available on-line at

http://www.nap.edu/readingroom/books/mentor/

The University of Oregon's Guidelines for Good Practice in Graduate Education may be found on-line at

http://darkwing.uoregon.edu/~gradsch/guidelines.html

The University of Arizona Position Paper: Mentoring: *The Faculty-Graduate Student Relationship*. Contact Dr. Maria Teresa Vélez, Associate Dean, Graduate College, University of Arizona, Tucson, AZ 85721. (520) 621-7814.

Mvelez@u.arizona.edu

Other resources

The Web site of the National Electronic Industrial Mentoring Network for Women in Engineering and Science may be found at

http://www.mentornet.net/

The Faculty Mentoring Guide of Virginia Commonwealth University's School of Medicine may be found at

http://www.vcu.edu/teaching/bestpractices/medicinementoring/

chapter 4

Authorship and Peer Review

Francis L. Macrina

Scientific Publication and Authorship • The Pressure To Publish
• The Need for Authorship Criteria • Instructions to Authors
• Guidelines for Authorship • Peer Review • How Peer Review
Usually Works • Being a Peer Reviewer • Conclusion • Case Studies
• Resources

Scientific Publication and Authorship

PUBLICATION OF EXPERIMENTAL WORK accomplishes several things. In addition to reporting new scientific findings, it allows evaluation of results and places them in perspective against a larger body of knowledge. Published work also credits other scientists, whose contributions and ideas have been built upon in the research. It also enables others to extend or repeat work by providing a description of experiments performed. Finally, the author byline attributes credit for the work and, equally important, establishes who accepts responsibility for it.

Scientists, their professional societies, and the publishers and editors of scholarly journals all agree that the determination of authorship is an important matter. In general, authors must contribute to the reported work in some way; what is recognized as an appropriate contribution varies, however. Defining the responsibilities of authorship looms as an even larger problem. Suppose a published paper contains an honest mistake that has a major effect on the paper's scientific message. It is determined that the mistake is attributable to one of the four coauthors on the paper. Are the other three authors responsible for the mistake as well? Or, does their responsibility simply stop with an adequate explanation of why they could not have detected the mistake? If the "mistake" is the result of fraudulent behavior on the part of just one coauthor, are your answers to these questions still the same?

Historically, the scientific community has relied on rather informal, often unwritten, and sometimes vague or ill-defined criteria for determining

49

authorship on scientific papers. That approach has not served science well. It can breed misunderstanding, hard feelings, and confusion. But the climate is changing as institutions, societies, editorial boards, and publishers seek to clarify and define the criteria used to assign authorship and its responsibilities.

The Pressure To Publish

Publications are the stock in trade of the scientific researcher. In academic settings, publishing helps scientists win grants, promotions, tenure, higher salaries, and professional prestige. For these reasons, there is pressure to publish. Unfortunately, scientists may sometimes react to these pressures in ways that lead to irresponsible actions. The need for that "one more paper" to add to the progress report of a grant application (to get a grant award) or an employer's activity report (to get a raise) or the curriculum vitae (to get a job) creates pressure to publish. Scientific research is competitive, so there's a need to be "first." And establishing the priority of one's scientific contributions is accomplished through publication. Keeping research funding means being productive, and prestige in science is often associated with being on the cutting edge of the field. Papers also publicize research activities, allowing principal investigators to recruit new trainees and junior investigators to their groups.

The large number of scientific journals provides many options for submitting papers. Journal quality and reviewing standards vary, so there is always likely to be a place where research findings can be published. The pressures to publish have given rise to euphemisms that describe what sometimes happens in scientific publishing. "Salami science" refers to the publication of related results in "slices": data sets are split and published separately instead of being presented in a unified way. This practice increases the number of published papers from the same body of data, giving the impression of increased productivity. Another phase used to describe a related practice is "the least publishable unit," the smallest amount of data that can be written as a manuscript and published. Some publications and editors may be contributing to these practices. Publication categories termed variously "Notes," "Short Communications," or "Preliminary Reports" accept brief reports of important findings that can stand on their own. When editors and reviewers do not heed their journal's policies, such brief publication formats open the door to the "salami slicers" and the "reductionists." The ethics of publishing data in a way that maximizes the number of papers is open to debate. Most would argue that it is not inherently wrong and that scientists must have the freedom to publish how and what they see fit. However, the fragmentary nature of such publications sometimes makes them difficult to evaluate. They can mislead the reader and create confusion in the field by giving inappropriate emphasis to one piece of work. Finally, unjustified multiple publications put undue strain on the peer review process.

The Need for Authorship Criteria

Today in the biomedical sciences, single-authored research publications have become a rarity. Even at the most fundamental level — the training of students and postdoctoral fellows — the multiauthored paper is appropriate and expected. Interdisciplinary approaches mandate collaboration. This makes multiauthorship the norm, and there is no expectation that the number of coauthors has to be limited. Authors on the byline of a paper all have a stake in their published work. Defining that stake can be elusive, however, without rational guidelines.

Scientists agree that it would be wrong to include as an author on a paper someone who had made no experimental, technical, or intellectual contributions to the work. Similarly, if someone thought of and performed a key experiment and provided an interpretation of the results, authorship for that person would be obligatory. These extremes have never really been in question. But decisions on authoring scientific papers frequently fall in between these examples. And the responsibilities of individuals whose names appear on multiauthored papers are not always clear, although increasingly debated. "If you are willing to take the credit, you have to take the responsibility" is a much-used statement that is not so simple to deal with in every case of coauthored scientific publication. Many now believe that guidelines, if not policies, that deal with assigning authorship and defining responsibilities are needed.

The dimensions of authorship are being increasingly explored. For example, the number of publications on the subject of authorship criteria has jumped from a handful in the 1970s to hundreds in the 1990s. The titles of these papers speak of "spelling out authors' roles," "accountability," "new requirements for authors," and "irresponsible authorship." Some journals have featured multiple papers on the various aspects of scientific publication. Institutions and professional societies have implemented guidelines dealing with authorship and publication. One professional organization has deployed a task force to study authorship.

Clearly, the tradition of written communication in science is being subjected to increasing study. The kind of change we can expect from this attention remains to be seen, but it is likely to be for the better as criteria and practices are defined and clarified. Against this backdrop of introspection, this chapter reviews some of the commonly accepted standards of publication and authorship, including the process and responsibilities associated with the peer review of scientific publication.

Instructions to Authors

A good place for the novice author to begin learning about the standards for authorship is in the "Instructions to Authors" section of a scientific journal. These instructions are published regularly, sometimes in every issue but at least once in every volume or every year. Instructions to authors also are commonly found on-line at the journal's home page. These instructions provide the details of manuscript preparation required by the journal, its general policies, and often its philosophy of publication. These latter points, although different from journal to journal, are indeed standards for publication. Sometimes these issues are reaffirmed after the paper is submitted; for example, they may be stated in the letter acknowledging receipt of the manuscript, in the acceptance letter, or in a note sent with the page proofs. Instructions to authors should always be read before beginning manuscript preparation. In fact, consulting these instructions can assist in the decision on which journal to publish in. Journal publishers often use this space to state the kinds of research considered appropriate for publication. This information, along with perusal of the published material that appears in the journal, helps with the decision on where to submit a paper. In addition, it is helpful to seek the advice of experienced colleagues on where to publish.

Details of manuscript preparation

Instructions to authors contain essential information needed to prepare and submit the manuscript. Details on format, space constraints or word limitations, preparation of figures, use of abbreviations and symbols, and proper chemical, biological, and genetic nomenclature are found there. For information on symbols and nomenclature, many journals use various authoritative reference books or guides as their accepted standards. Instructions to authors often contain housekeeping details such as how many copies to submit, where to mail the manuscript, and the cost of page charges. Finally, some journals provide guidance on the preparation of the various sections of the scientific paper: the abstract, introduction, materials and methods, results, and discussion. Such guidance is useful to the novice writer. Reading the journal's guidelines on how to prepare these sections is preferable to trying to deduce them from reading papers published in the journal (although that can help, too).

Authorship

Increasingly, journals provide some guidance on the definition of authorship and its responsibilities. The words ultimately come down to the same two issues in the majority of examples. First, an author has to make a significant contribution to the work. Most statements like this leave plenty of room for interpretation and thus are flexible (a trait favored by some and opposed by others). Second, statements defining authorship often mention that all authors on a manuscript take responsibility for its content. Some statements indicate all of the authors consent to its submission, or that they all have read the manuscript. Some journals require letters of submission signed by every coauthor. Others limit the number of names that may appear in the paper's byline.

Copyright

A usual condition of manuscript acceptance for almost all scientific journals is that the authors assign the copyright to the publisher (see chapter 9). Usually the senior author (sometimes called the corresponding author) is empowered to do this for all the coauthors. Also, many journals require the authors to obtain permission to use any copyrighted material that is included in their manuscript, e.g., a diagram from a previously published paper. This is usually a formality that involves writing to the publisher who holds the copyright for the work to be included and describing its intended use. Many publishers have forms that can be completed in lieu of a letter.

Manuscript review

Matters relating to the peer review of the manuscript often are found in the "Instructions to Authors." Some journals allow authors to suggest the names of impartial reviewers, either ad hoc referees or members of the editorial board. This helps the editors do their job, and it is wise to take advantage of the opportunity. Who qualifies as an impartial reviewer? Opinions vary, and criteria are subjective. Often excluded as impartial reviewers are (i) people at the author's institution, (ii) people who have been associated with the author's laboratory, and (iii) the author's collaborators or coauthors at other institutions. Individuals in the latter two categories are considered in view of the time that has elapsed since the author's last interactions with them.

Often a description of the peer review process is found in the "Instructions to Authors." The process also may be described in the letter or card sent acknowledging receipt of the manuscript. Authors need to read about this process and know how it works. It can vary significantly for different journals. Understanding the process helps authors in dealing with the manuscript during peer review. The typical path of a manuscript through the review process is discussed later in this chapter.

Prior publication

In 1968, the Council of Biology Editors (see Robert Day's *How to Write and Publish a Scientific Paper* in the resources section) defined a "primary scientific publication" as the following:

An acceptable primary scientific publication must be the first disclosure containing sufficient information to enable peers (1) to assess observations, (2) to repeat experiments, and (3) to evaluate intellectual processes; moreover, it must be susceptible to sensory perception, essentially permanent, available to the scientific community without restriction, and available for regular screening by one or more of the major recognized secondary services (e.g., currently Biological Abstracts, Chemical Abstracts, Index Medicus, Excerpta Medica, Bibliography of Agriculture, etc., in the United States and similar services in other countries).

Despite this and similar definitions, agreeing on what qualifies as prior publication is arguable to some. There is ambiguity when considering, for example, papers published in monographs (invited short papers or meeting proceedings). It is not easy to determine how "readily available" a source may be. How many copies of a monograph have to be sold or distributed to qualify as available? If all copies of the monograph have been distributed in the United States, is it acceptable to submit essentially the same work to a journal published in Europe? Some argue that original work published in conference reports, symposium or meeting proceedings, or equivalent monographs is by definition preliminary owing to considerations of format and space. Often methods cannot be fully described, and such work is usually not subjected to peer review. Self-deception may be at work in these arguments, however. Scientists generally agree that it is wrong to publish the same material as a primary publication in two different places. Using that philosophy as a guide is highly recommended.

Unpublished information cited in manuscripts

Some journals require proof of permission to cite the unpublished work of others. Information provided by a colleague as a "personal communication" may require a letter granting permission. The same is usually true for preprints or submitted manuscripts provided by your colleagues. This practice is increasing among publishers of scientific journals. Although a colleague may have provided a manuscript that has been submitted for publication, she may not feel comfortable allowing that work to be cited in another paper before she knows that hers is accepted. By formally asking her permission, any prospect of misunderstanding is eliminated.

In the case of the author's unpublished work — "in press" or "submitted" manuscripts — an increasing number of journals require that copies of such manuscripts accompany the new submission so that they can be used if needed during peer review.

Sharing research materials

In natural science and biomedical journals it has become standard for publishers to include statements about sharing research materials. This includes cell lines, microorganisms, mutants, plasmids, antibodies, and other biologicals and reagents. There are usually conditions stated for the release of such materials. For example, materials must be available at cost (e.g., preparation and shipping), they must be requested in reasonable quantities, and they must not be used for commercial purposes. Scientists also usually request materials from the authors of the publication in which the material was initially described. For example, it is not acceptable to request a bacterial strain from a third party, even though it may be convenient to do so. A mutant needed for work in Chicago may have been constructed by a scientist in Japan, but a colleague in a nearby city already has it. It is not appropriate to ask the stateside colleague to provide the mutant. Ask the Japanese investigator who made it and published the results. At the very most, you could suggest that he allow you to get a culture from your conveniently located neighbor.

Also included in many author's instructions is the request that authors properly deposit specialized data, e.g., nucleic acid sequences and X-ray crystallographic data, in appropriate databases. Sharing research materials and proper deposition of results into databases are usually listed as conditions of acceptance for the paper.

Conflict of interest

Sometimes scientific journals ask authors to disclose their financial associations with companies whose activities might be affected by the results of the paper. If the paper is accepted, how this disclosure is presented is usually handled on a case-by-case basis. The potential conflict-of-interest disclosure is likely to become a more prominent issue as research sponsored by the biotechnology industry continues to increase in academic institutions and in noncommercial research institutes. Thus far, such language in author's instructions has been found primarily in biomedical journals that publish research with clinical implications. Other related issues undoubtedly will have to be explicitly addressed in the future. For example, what about parallel conflicts involving members of editorial boards who serve as reviewers for such papers? What about editors themselves who may have financial associations with companies whose activities may overlap with the content of journal articles? How can such issues be monitored, and how are conflicts handled when they are identified? There are no ready answers, but the need for dialogue, careful consideration, and the development and implementation of policies is apparent.

Imperfections

Some journals also include policies on the handling of disputes once papers are published. Occasionally, journals are explicit about the option of having their editors examine original data in the process of dispute resolution. In addition, many journals describe policies for publishing corrections, errata, or retractions of papers.

Guidelines for Authorship

In seeking the definition of authorship, what is the test for telling the difference between "earned authorship" and "honorary authorship"? Scientists generally hold the former to be right and the latter wrong, but there's a continuum between the two. Making decisions along that continuum is difficult to do when arguments still persist over the definitions of the ends — the unqualified acceptable and unacceptable ways. There seems to be considerable agreement on what is unacceptable. However, there can be great differences of opinion on what earns someone the right to have his or her name on the author byline. Although some journals have provided guidance on defining authorship and its responsibilities, this usually has been done with sweeping statements that afford broad interpretation. Usually such broad-based guidance is of limited value to the novice writer or trainee.

The answers to questions about authorship and its responsibilities must come from the scientific community. In recent years, these answers, or at least attempts at making them, have come from the institutions in which scientists practice. Some institutions now publish guidelines on the meaning and definition of authorship. In addition, there is an increasing body of literature that addresses the subject. A sampling of these writings, some of which include publication and authorship guidelines, may be found in the resources section.

Following is a review of the general points often covered in the writings on authorship. Authors need to understand and abide by relevant guidelines. Consider the topics covered here to be a general substrate for thinking about authorship.

The senior author

An often-used term is that of senior author (sometimes primary author). Guidelines often define this person as the principal investigator, leader of the group, or laboratory director. If the byline of a paper lists a faculty mentor along with two of her predoctoral trainees and one postdoctoral trainee, then the mentor is the senior author.

The senior author may be the first author listed on the byline. Most agree that the first author is defined as having played a major role in generating the data, interpreting the results, and writing the first draft of the manuscript. In many cases, however, the first author and the senior author are different. When this is so, it is customary for the senior author's name to be last in the byline. Sometimes more than one of the authors has senior status. Most of the time it is still possible to define one of them as the senior author of the paper, based on the respective contributions of the rest of the coauthors (e.g., in which senior author's lab the work was done). If not, it is possible for senior authorship to be shared; this designation and the position of the names of the senior authors on the byline should be decided by their mutual consent.

Responsibilities of the senior author

Guidelines often vest senior authors with overarching responsibilities. What follows is an amalgamation of the typical responsibilities listed in several documents from universities and research institutions.

The senior author along with the first author decides who else will be listed as coauthors. General criteria for making these decisions are discussed below. The senior author is responsible for notifying all coauthors of this decision and for facilitating discussion and decision making about the order of appearance of the coauthor's names on the byline.

The senior author, usually with the help of the first author and sometimes other coauthors, decides on the people to be listed in the acknowledgments section of the paper. The senior author should notify the individuals to be acknowledged. The senior author also is responsible for listing in the acknowledgments all sources of financial support for the work. In short, the senior author is responsible for appropriately acknowledging all contributions to the work reported in the paper.

Senior authors review all data contained in the paper and, in doing so, assume responsibility for the validity of the entire body of work. This assertion, which is commonly being written into institutional guidelines, presents problems in regard to specialized work that may be outside the senior author's area of expertise. In such cases, a suggested way of handling this is for the senior author to gain a reasonable understanding and verification of the data from the appropriate coauthor. Still, this problem persists as interdisciplinary research abounds and researchers from highly technical and specialized fields collaborate and copublish their results. Nonetheless, some of the guidelines in effect today are very specific on this point: the senior author must "understand the general principles of all work included in the paper."

The senior author has a responsibility to facilitate communication among coauthors during the preparation of the manuscript. This means reviewing raw data and discussing new ideas for additional work. It certainly means reaching agreement on the part of all coauthors as to interpretation of results and conclusions.

The senior author should be able to describe the role and contributions of all coauthors in the work. At some institutions, doing this in writing is urged or even required, and the documentation is retained in departmental files. Also in this vein, some institutions require a signed document from coauthors indicating their approval of the manuscript and their permission to submit it. As mentioned above, some journals now require that the letter covering the submitted manuscript indicate essentially the same thing, and the letter must be signed by every coauthor.

The senior author makes sure that the logistics of manuscript submission are properly followed. Such things as manuscript format and related material and local editorial review (if required) are included here. The senior author is also responsible for all dealings with the publisher. This would include things like correspondence, execution of copyright assignments and authorship agreement forms, and, where appropriate, financial matters such as page charges and reprint costs. The senior author should establish a policy regarding distribution of reprints of the paper, including the provision of a reasonable number of reprints to all coauthors.

The senior author coordinates and oversees the responses to the peer reviewers' comments if the manuscript has to be revised. He or she is responsible for involving the coauthors in this process as appropriate and for seeking the approval of all coauthors to submit the revised manuscript. The senior author is responsible for acting on and honoring requests to share materials from the research once the paper is published. The senior author is responsible for coordinating and making responses to general inquiries or challenges about the work. The senior author assumes responsibility in dealing with the publication of corrections, errata, or retractions. This includes coordinating preparation of such items by seeking the comments and agreement of all coauthors. Finally, the senior author is responsible for the appropriate retention and storage of all data used to prepare the manuscript.

The first author

The first author is the author whose name appears first in the byline of the paper. This person may also be called the principal author (a confusing term in some documents because the senior author is usually a principal investigator). As mentioned above, the first author is the person who participated significantly in the work by (i) doing experiments and collecting the data, (ii) interpreting the results, and (iii) writing the first draft of the manuscript.

The submitting author

The submitting author is usually the author who sees the manuscript through the submission process, e.g., letter writing, coordinating responses to the editor, responding to peer review comments. Sometimes this person is called the corresponding author. This is usually the senior author, but it can be the first author. For example, a mentor (senior author) may want his postdoctoral fellow (first author) to gain experience in dealing with the peer review process. It should be remembered that certain responsibilities will fall on this author (see above). Many publishers indicate the submitting author on the first page of the published article. The responsibilities of the senior author with respect to correspondence after publication will then fall on the submitting author. When the submitting author and the senior author are not the same person, there should be a clear understanding of how follow-up correspondence related to the manuscript will be handled (e.g., requests for biologic materials).

Other coauthors

Coauthors whose names appear between the first and last author in the byline of a paper are usually determined by the senior author and the first author. The order of these coauthors can be based on the importance of their contributions to the work in descending order from the first author. Decisions on authorship need to be made before the paper is written. It may be appropriate to change the order of the authors as the manuscript preparation progresses. The senior author and the first author should take the lead in any decision to revise author order, but such decisions should involve all the coauthors.

What counts toward authorship

Authorship encompasses two fundamental principles: contribution and responsibility. An author must make a significant intellectual or practical contribution to the work reported in the paper. With such authorship goes the responsibility for the content of the paper. By keeping such concepts simple, the qualitative and quantitative aspects of these contributions and the precise nature of the responsibilities are left open to interpretation. This is preferred by many in the scientific community. In any event, the articulation of authorship contributions and responsibilities by institutions provides clarity that aids both the seasoned and the novice author.

The meaning of authorship has been presented in a variety of ways. Significant contributions are frequently described as those that have an effect on the "direction, scope, or depth" of the research. They have also been stated in terms of "conceptualization, design, execution, and/or interpretation" of the research. The development of necessary methodologies and data analysis essential to the conclusions of the project are also sometimes listed as contributions that justify authorship. Sometimes the language is specific, and contributions to the project are linked to having a "clear understanding of its goals." This leads to the issue of responsibility. Some have addressed this issue in defining authorship by invoking the need "to take responsibility for the defense of the study should the need arise" or "to present and defend the work in context at a scientific meeting." The challenge of coauthor responsibility where disparate contributions have been made was addressed in one case by saying that exceptions to this rule will need to be made when "one author has carried out a unique, sophisticated study or analysis." In other words, in certain collaborative studies, it may not be possible for every author to be able to rigorously present and defend all aspects of the work.

What does not count toward authorship

In many guidelines, naming the contributions that do not merit authorship have been as helpful as naming those that do. Merely providing funding for the work, or having the status of group or unit leader does not alone justify authorship. Neither does providing lab space or the use of instrumentation. Finally, doing routine technical work on the project, providing services or materials for a fee, or manuscript editing in themselves are not sufficient justification for authorship.

Peer Review

Many scientists are called on to review manuscripts. This happens in two ways. First, scientists may be appointed as members of editorial boards of scientific journals, in which case their duties as reviewers are formalized. Editorial board members regularly receive papers to review, and their names appear in each issue of the journal designating them as reviewing editors or editorial board members, or an equivalent term. Second, scientists may be asked to be ad hoc reviewers. In this case they receive papers from editors to review and are asked to evaluate them as a courtesy. Usually, ad hoc reviewers are acknowledged in the last journal issue of the year. Many scientific journals rely heavily on the services of ad hoc reviewers.

Editorial board members and ad hoc reviewers provide a critical service. They prepare written evaluations that help the editor decide on the acceptability of the submitted manuscripts. Equally important, their comments usually allow the authors to improve their manuscript if it is not acceptable for publication in its current form. Reviewers may suggest improvements in writing style, presentation of data, or even further experiments to be done.

A scientist named to an editorial board is likely to receive guidelines from the editor or publisher on how to prepare a review. Over time, the editor may provide additional advice on how to write reviews so they are more helpful to the editor and the authors.

Ad hoc reviewers are often asked to serve before a manuscript is mailed to them. At that time it is a good idea for potential reviewers to check with the editor's office to see if guidelines for ad hoc reviewers are available. They are usually brief and can be very helpful. If they exist, reviewers should secure a copy before beginning a review. If not, reviewers writing their first ad hoc review are likely to have a single frame of reference: reviews they have received on their own manuscripts.

The peer review of scientific papers has come under scrutiny in recent years, with some arguing for its radical change or complete abolishment. This is not the place to take up that debate. Instead, in the belief that peer review is an important element of responsible scientific conduct, the process will be presented here in two parts. First, we'll examine the flow of a manuscript through a typical cycle of peer review. Then, we'll discuss the duties and responsibilities of the peer reviewer.

How Peer Review Usually Works

Typical peer review begins with submission of a manuscript by mail to an editor or to a central office of the publisher of the journal. In the latter case, the office then assigns the manuscript to an appropriate editor. Usually, scientific journals have multiple editors who represent the various subspecialties of the subject matter. The editor then reads the paper (or enough of it) to decide on whom to ask to review it. Editors may select editorial board members or ad hoc reviewers for this job. Typically, a single paper is mailed to two and sometimes three peer reviewers (also termed referees). Some journals have special forms on which to prepare manuscript reviews, but these frequently consist of lots of blank space for the reviewers to write comments. There may also be a separate form for comments that are intended only for the eyes of the editor. The editor asks the reviewers to complete their evaluations in a specific period of time, usually less than a month. When the completed reviews are returned to the editor, he or she reads them. The editor then makes one of three decisions: (i) accept the paper, (ii) reject the paper, or (iii) return the paper to the authors for revision. In all cases the editor sends the authors a letter indicating the basis of his or her decision. Obviously, in the case of outright acceptance, the letter is brief. However, editors are usually specific in their decision letters when explaining rejection or the need for revision. Such letters reflect the editor's own opinions of the paper, along with the reviewers' comments and recommendations. Along with this letter to the authors go the verbatim copies of the reviewers' comments. Parts of the review forms that indicate the reviewers' recommendation ("accept," "reject," or "revise") as well as any comments made to the editor are not sent to the authors. Editors may use comments sent to them separately by reviewers to help in composing their decision letter.

For most scientific journals in the biomedical and natural sciences, the comments of the reviewers are anonymous. However, some journals do reveal the identity of reviewers to the authors. This can be done as a matter of policy or by encouraging reviewers to sign their written reviews.

Authors consider the reviewers' and editor's comments in revising their papers. They may make changes based on comments they agree with.

Alternatively, authors have the right to rebut any and all criticisms of the reviewers. The basis for handling each of the reviewers' comments must be explained to the editor in a letter that accompanies the revised manuscript. It is then the editor's job to reach a final decision on the paper and to notify the authors.

Being a Peer Reviewer

What to do when the manuscript arrives

There are a number of housekeeping chores that reviewers must do when a manuscript is received by mail. It is important and courteous to attend to these quickly. First, the reviewer must scan the paper and decide whether he or she is qualified to review it. The review deadline must be evaluated: can the reviewer complete the review in the time allotted by the editor? If the reviewer is uncomfortable with either of these criteria, the manuscript should be sent back to be reassigned. Also, reviewers should check that they have a complete version of the manuscript. Are all the pages, figures, and tables there? If anything is missing or illegible (e.g., photocopies of micrographs instead of originals), the editor or editorial office should be contacted to get the needed material.

Reviewers must be comfortable with the job of impartially reviewing the work. Their review of the paper must not constitute a conflict of interest, real or perceived. Some journals have guidelines for this. Typically cited conflicts include papers from investigators at the reviewer's institution, trainees who have recently been in the reviewer's lab, or collaborators of the reviewer at other institutions. Commercial interests also create conflicts. For example, is the paper authored by scientists at a company that pays the reviewer as a consultant or has made a grant or gift to the reviewer's research program? Conflict-of-interest decisions of this type usually rest with the reviewer. Most of the time, the information that points to the conflict is known only to the reviewer, and the editor may never become aware of it. The reviewer has to decide whether there is conflict or whether others might perceive specific actions as conflict. A simple rule is: "When in doubt, don't review the paper." The reviewer may contact the editor to seek advice on matters of potential conflict. In general, any extensive rationalization for overcoming what might be a perceived conflict is usually a signal to both the reviewer and the editor that a real conflict may exist or may be perceived by others. In such cases reassignment of the manuscript to another reviewer is necessary.

If a reviewer returns a manuscript for reassignment, it is a courtesy to tell the editor the reason for doing so. It is also customary to suggest the names of potential substitute reviewers. Such help is valuable, and editors appreciate it.

Some of the guidance frequently found in peer reviewer guidelines follows.

Philosophy of review

The peer reviewer's job has two aims: (i) to help the editor make a good decision on the acceptability of the paper, and (ii) to help the authors communicate their work accurately and effectively. The peer reviewer does not have to be an adversary to do either of these jobs. Especially in the latter case, the reviewer should be an advocate for the authors. Indeed, guidelines sometimes tell reviewers to take a positive attitude toward the manuscript, and this is good advice. Reviews that are confrontational are distressing to authors and often make things difficult for all involved. Meaning sometimes gets lost in impolite and ill-considered language, and this can make the editor's job of evaluating the reviewer's comments confusing. It can distract and mislead authors as they prepare their rebuttals. Authors may "miss the point" and in doing so fail to improve their manuscript. Additionally, time is often wasted when authors feel the need to respond in kind to offensive language in their rebuttal letters to editors.

Confidentiality

A manuscript sent to a reviewer for review is a privileged communication. It is confidential information and should not be copied by any means or shared with colleagues. Under no circumstances should the reviewer get assistance from colleagues in performing the review without explicit permission of the editor.

A customary policy is that a peer reviewer should never contact an author directly about the manuscript under review. This sounds like unnecessary advice because most journals use anonymous review. However, even if journals allow disclosure of the reviewers' identity to the authors, direct contact between the two during the review process is usually forbidden. The reviewer's opinion about the merit and acceptability of a manuscript is considered by the editor, who makes the final decision. By talking to authors, reviewers may communicate misleading messages that can make the editor's job more difficult. Thus, reviewers who need clarification or additional information should contact the editor and let him or her obtain it from the author.

Common criteria for evaluating merit

The manuscript should contain a clear statement of the problem being studied, and it should be put in perspective. Reviewers should evaluate this perspective in the context of appropriate literature citations. In other words, are the authors giving appropriate credit to prior work in the field, especially those contributions upon which the present report is built? The originality of the work should be carefully weighed. The reviewer should consider whether the manuscript reports a new discovery or if it extends or confirms previous work.

Experimental techniques and research design should be appropriate to the study. Did the authors use the right tools and techniques to test their hypotheses? Description of methods is very important. This is the part of scientific communication that permits verification of the work. The description of the materials and methods should provide sufficient detail that other investigators in the field can repeat the work. It is acceptable for some methods to be mentioned briefly and then cited in the references. However, such citations should be the correct ones. Papers should not be used as methods citations if they contain incomplete descriptions or if they refer to yet another paper for the details of the method.

The reviewer should examine the presentation of data for clarity and effectiveness, keeping in mind several questions. Is data presentation cluttered or confusing? Are figures and photographs unclear? What about the organization of the data seen in tables and figures? Are there too many tables or figures? Can some be deleted? Would data given in tabular form be better presented in figures? Should data in tables be combined or single-panel figures be redone as multipanel ones?

Interpretations of the data need to be sound and clearly worded. The discussion of the work should be appropriate: arguments should be logically presented, and any speculation should be built on data in the paper or the existing literature.

The writing in the manuscript should be clear, easy to follow, and grammatically correct. Many guidelines affirm that the peer reviewer's job is not to rewrite the manuscript. However, citing examples of writing deficiencies will help the authors in making global revisions. The reviewer should also note whether the authors are adhering to correct scientific nomenclature and abbreviations as specified by the journal.

The reviewer should evaluate the title and abstract after reading the paper. Are they adequate and appropriate? As electronic communications increase, the availability of abstracts is becoming widespread. The abstract has become the first line of scientific communication in this medium. So the abstract needs to clearly describe the essence of the problem, how it was approached, and the outcome of the research.

Writing the review

The format for preparing a manuscript review varies from journal to journal. However, it is typical for a review to begin with a paragraph or two that summarizes the major findings and highlights of the paper. If there are overriding considerations, either positive or negative, they should be presented here. Shortcomings or flaws that have influenced the reviewer's assessment of the paper should be stated in general terms; specific comments can be included later in the review.

Following this narrative, it is customary for the reviewer to list specific, numbered comments. Numbering makes it easier for the authors to respond to the critique and for the editor to make a final decision. Specific comments should offer guidance to the authors on how to improve their work. Problems should be identified and solutions suggested where possible.

Finally, it is customary for the reviewer not to indicate in the narrative or in the specific comments the ultimate recommendation for the paper. Instead, this should be clearly transmitted to the editor. As mentioned earlier, it is commonly done with a specific form or in a brief note. There is a reason for this. Rarely do editors send a paper to just one reviewer; using two or three experts is the norm. Reviewers can and do disagree about the merits of the same paper. When this occurs, it is the editor's job to sort out the reviews and then write his or her final disposition in a decision letter to the author. It is frustrating to the authors to read two reviews of the same work when one recommends acceptance and the other rejection.

Conclusion

Written communication is an essential part of scientific research. Science benefits society only insofar as its findings are made public and applied. Indeed, biomedical scientists have a moral obligation to share new knowledge in order to advance and improve the health and well-being of humankind. Scientific knowledge is accepted only when the published research results that support it hold up under scrutiny and independent corroboration.

The duties and responsibilities of authorship are not to be taken lightly by scientists. In the past, many of the decisions about authorship on scientific papers were based on unwritten norms and standards. In recent years, written guidelines for authorship have been promulgated by institutions, societies, and journal publication boards. These provide guidance to authors and can be especially informative to the novice writer.

Providing peer review of scientific publications is an obligation that is shared by scientists. While peer review must be scholarly and rigorous, it must also be timely, respectful, and courteous. Above all, peer review must be constructive. Peer review plays a vital role in the publication of research findings, although the process is being increasingly challenged. Its workings and effectiveness are likely to be the subjects of continuing debate among scientists for years to come. Nonetheless, the process of peer review is performed under both written and unwritten guidelines. Explicit descriptions of duties and responsibilities of peer reviewers are now frequently published by scientific journals. In part, they aim to foster consistency and integrity in the process.

Case Studies

4.1 Jim Morris, a faculty member in the Department of Chemistry at Research University, has written a manuscript for submission to a prestigious journal in cell biology. Bill Burdock, a colleague of his in the Department of Biology at the same institution, is a member of the journal's editorial board. Jim submits the manuscript to Bill for consideration and review. They briefly discuss the conflict-of-interest implications of this action but decide that, because the manuscript will be reviewed by at least two outside referees, it can be appropriately handled with Bill acting as editor without a real or perceived conflict of interest. Do you agree with this decision? Why or why not?

4.2 Walt Miller, a member of an NIH study section, is reviewing his assigned grant proposals before the study section's meeting. While writing his reviews, he receives by mail a manuscript from a journal editor and is asked to review it as an ad hoc referee. The author of the manuscript turns out to be the principal investigator of one of the grant proposals Walt is assigned to review. Indeed, the manuscript in question has been submitted as an appendix to the proposal. Walt decides to review the manuscript sent to him by the journal editor. Comment on his decision.

4.3 Helen Louis has published the description of three new bacterial mutants in a peer-reviewed journal. Mutants 1 and 2 were exhaustively characterized and described in the report, but mutant 3 was mentioned only briefly. Larry Savage writes her to request mutant 3. Larry clearly describes his intended use for the mutant in studies that are currently under way in his laboratory. Helen refuses to release the strain. Helen affirms that mutant 3 was described only in a preliminary way in the paper and states that another major manuscript is in preparation in which mutant 3 will be the central focus. She says she will be happy to release the mutant after the second manuscript has been accepted for publication. Larry refuses to accept this rationale and presses his request for the mutant strain. Comment on this situation.

4.4 Sara Nichols had a very productive postdoctoral training experience. With her mentor she coauthored four important papers on oncogene expression. She was the first author on all of these papers. Jacob Smith, her mentor, conceived of the ideas for the work, but Sara did all the experiments, interpreted the results, and wrote the papers. Sara is now an assistant professor struggling to get her first grant in order to continue her oncogene research. She reads a new review article on oncogene expression in which the author repeatedly cites her four papers as being very important. However, the author of the review continually refers to these papers as the contributions of "Smith and coworkers." Sara is offended and upset by this. There were no other "coworkers" who contributed to this work, and she believes that the papers should be referred to as the work of "Nichols and Smith." She is worried that the inappropriate reference to her work will undermine her contributions and deprive her of credit that can promote her career advancement. She writes to you, the editor of the journal that published the review. How do you respond to her? What, if anything, will you do about this situation?

4.5 Dr. George Adams receives a manuscript for ad hoc review from the editor of a scientific journal. George gives the manuscript to Al Nance, his senior postdoctoral fellow. He asks Al to read the manuscript and prepare some written comments critiquing it. One week later, Al provides Dr. Adams with one page of written comments and also gives him an extensive verbal critique of the paper. Dr. Adams then prepares a written review and submits it to the editor of the journal. A few weeks later, Dr. Adams learns that Al made photocopies of the entire literature section of the manuscript because it contained "some useful references." Dr. Adams verbally reprimands Al, telling him that no part of a manuscript received for review should be copied. Comment on the behavior of both the faculty member and the postdoctoral fellow in this scenario.

4.6 Melvin Evans, a graduate student in cell biology, has purified two recombinant proteins as part of his dissertation research. These proteins differ only at a few key amino acid positions. Based on other biochemical data, Melvin believes the proteins are virtually identical. After a discussion with Jeff Lee, a graduate student in biochemistry, Melvin concludes that it would be reasonable to compare these two purified proteins by circular dichroism. Jeff offers to collaborate on the project by analyzing the two proteins by this technique. Dr. Dawson, Jeff's advisor, approves of this and alerts Melvin's advisor that this will be a fruitful collaboration that should result in a coauthored publication. His rationale for this is based on (i) Jeff's intellectual contribution in presenting the data and operating highly technical instrumentation and (ii) on his own intellectual support of Jeff's work and financial support of the circular dichroism instrument facility. Melvin's advisor is opposed to a coauthored paper, arguing that Jeff's contribution is largely technical and does not merit coauthorship. He suggests that Jeff and his advisor be acknowledged in the paper along with the grants used to support the circular dichroism facility. Discuss the relevant issues of authorship in this case.

4.7 Professor Don Mills develops a DNA probe as a side project under NIH grant funding. Although not immediately applicable, this DNA probe has potential in the diagnosis of a latent viral disease of humans. He publishes his results in a peer-reviewed scientific journal. After publication of this work, Dr. Mills is called by John Banner, the director of research at a large U.S. pharmaceutical firm. Banner requests a plasmid carrying the probe sequence for use in his company's research. Banner assures Mill that the company has no intention of commercializing the DNA probe. Mills refuses to comply with the request, claiming that the potential for commercialization is always present in the research environment of a for-profit company. Banner counters that Mills has published his results and must release the material under the standards of publication set by the peer-reviewed journal. Banner contacts you, the journal editor, and asks you to resolve this problem. What, if anything, do you do?

4.8 You have edited a book on a widely used class of antihypercholesteremic drugs. Chapters have been contributed by several leading authorities in the field. Dr. Brad Murray wrote an excellent chapter that includes some very dramatic data on the comparative pharmacology of one of these drugs. His chapter contains two photographs of electrophoretic activity gels that demonstrate enzyme levels in normal and transgenic mice. About 6 months later you are reading Murray's latest paper in the *Biochemical Research Proceedings*, a prestigious peer-reviewed journal. The paper contains the same two activity gels included in his chapter for your book. These data are key elements of the paper, which has serious implications for the design of more effective antihypercholesteremic agents. In looking at the receipt and publication dates of the paper, you find that it was accepted for publication one month before your book went to press and appeared 3 months before the book was published. Neither publication is cited in the reference section of the other. Were Murray's actions legal? Were they ethical? As editor, what, if anything, are you obliged to do about this?

4.9 Dave Clubman completes his Ph.D. program and leaves the laboratory immediately to attend to personal matters. An important manuscript based upon his dissertation exists only in a preliminary draft. During the next year, Henry Franks, his former advisor, attempts to contact Dave to complete the manuscript. After some months, Dr. Franks edits the manuscript, prepares the figures, and sends the updated version to Dave. Dave acknowledges receipt of the manuscript but provides no comments and does not sign a memorandum acknowledging consent to submit the manuscript. During this period, some results similar to Dave's are published by another laboratory. Dr. Franks and a postdoctoral fellow extend the work and prepare a new manuscript with Dave as first author and the postdoctoral fellow as an additional coauthor. The manuscript is sent to Dave by certified mail, but he does not provide any comments or return a signed memorandum agreeing to submission for publication. A third party hears that Dave blames Dr. Franks for the delay and is trying to "give him a hard time." Dave was supported by federal funds, and his results were included in annual progress reports to the granting agency. Can Dr. Franks submit the manuscript and publish it if it is accepted by the journal? What should be the authorship of the paper? Should any comments be included in the acknowledgment section?

4.10 Marvin Brian, a faculty member at a major research university, is funded by a contract with a relatively well-defined work scope. Brian is the advisor of an advanced doctoral student, Henry Ruth, and a beginning graduate student, Mark Butterworth. Henry serves as the lead investigator and prepares and presents reports to the funding agency. Mark works on the same project, sharing his data with Dr. Brian and Henry. After working on the project for about 2 years, Mark submits his master's thesis, which is reviewed and approved by Dr. Brian but not seen by Henry. A year or so later, when Henry is finishing the text of his doctoral dissertation, he discovers that Mark's thesis contains at least one complete table representing his work in exactly the same format that Henry has used to express his results. The master's thesis contains a general acknowledgment of Henry, among others, but no specific attribution is given for the verbatim table. All parties are aware that this research is supported by a contract with a defined work scope. Does this sponsorship justify duplicate publication in a master's thesis and a doctoral dissertation without explanation? If not, how should the matter be handled? Once duplicate publication occurs, what should be done and who is responsible for initiating remedial action?

4.11 Suzanne Booth is recruited as a postdoctoral fellow in a laboratory where research centers on the cell biology of a specific mammalian cell type. Suzanne's training has been in eukaryotic gene cloning and molecular genetics; no such technology is available in this laboratory. Suzanne completely trains a senior-level graduate student working in the group. Under Suzanne's supervision, the student proceeds to build a cDNA library and isolates by molecular cloning a gene for a membrane protein. Several months later, a manuscript describing this work is prepared for submission. The principal investigator of the laboratory, Professor Jack Taylor, and the student are listed as coauthors. Suzanne is listed in the acknowledgments section of the paper. She is upset with this disposition and confronts Dr. Taylor. Taylor says that he has strict rules about authorship and that Suzanne's contribution was a technical one that does not merit authorship. Taylor quotes from several different standards-of-conduct documents indicating that authorship must be strictly based on intellectual and conceptual contributions to the work being prepared for publication. Technical assistance, no matter how complex or broad in scope, is not grounds for authorship. Does Suzanne have a case for authorship?

4.12 Dr. Colleen May is a participating neurologist in a clinical trial to assess the efficacy and toxicity of a new anticonvulsant medication. For the duration of the 2-year study, each neurologist is to meet with each of his or her patients for an average of 30 minutes each month. In Dr. May's case, this amounts to an average of 20 hours per month. During each visit, the physicians administer a variety of specialized tests, requiring judgments dependent on their experience and training in neurology. At the completion of the study, the results are to be unblinded and analyzed by the project leaders. It is anticipated that at least two publications will be prepared for the *New England Journal of Medicine*. Dr. May has just learned that she will be listed in the acknowledgments but not as a coauthor of the manuscript. Dr. May argues that she has provided nearly 500 hours of her expert time, far more than needed to complete a publishable study in her experimental laboratory. Does Dr. May have a case for authorship?

Resources

Suggested readings

Bailar, J., M. Angell, S. Boots, E. Myers, N. Palmer, M. Shipley, and P. Woolf. 1990. *Ethics and Policy in Scientific Publication*. Editorial Policy Committee, Council of Biology Editors, Bethesda, Md.

Booth, V. 1993. *Communicating in Science*, 2nd ed. Cambridge University Press, New York, N.Y.

Day, R. A. 1998. *How to Write and Publish a Scientific Paper*, 5th ed. Oryx Press, Phoenix, Ariz.

Lancet. 1998. **352:**894–900. This issue contains a series of short articles on publication and professional development.

National Academy of Sciences. 1993. *Responsible Science*, vol. II: *Ensuring the Integrity of the Research Process*. National Academy Press, Washington, D.C.

This volume contains guidelines on responsible conduct, including criteria for authorship, used at several institutions.

Tipton, C. M. 1991. Publishing in peer-reviewed journals — fundamentals for the new investigator. *Physiologist* **34:**275–279.

Waser, N. M., M. V. Price, and R. K. Grosberg. 1992. Writing an effective manuscript review. *BioScience* **42:**621–623.

Other resources

The American Chemical Society has guidelines covering the ethical obligations of authors, editors, and manuscript reviewers, and of scientists publishing outside the scientific literature. The guidelines are available on-line at

http://pubs.acs.org:80/instruct/ethic.html

The Society for Neuroscience has published guidelines entitled *Responsible Conduct Regarding Scientific Communication* (1998). Section headings in this document are as follows:

1. Authors of Research Manuscripts
2. Reviewers of Manuscripts
3. Editors of Scientific Journals
4. Abstracts for Presentations at Scientific Meetings
5. Communications Outside the Scientific Literature
6. Dealing with Possible Scientific Misconduct

These guidelines may be found on-line at

http://www.sfn.org/guidelines/

Examples of institutional guidelines on authorship may be found at

http://www. library.ucsf.edu/ih/: the investigators' handbook of the University of California, San Francisco

http://www.pitt.edu/provost/ethresearch.html: University of Pittsburgh Guidelines for Ethical Practices in Research

Also consult the *Guidelines for the Conduct of Research at the National Institutes of Health* found in Appendix III of this book.

Proceedings of the Third International Congress on Peer Review (1998) (*JAMA* **280:**213–302) featuring numerous articles on biomedical peer review may be found on-line at

http://www.ama-assn.org/sci-pubs/journals/archive/ jama/vol-280/no-3/toc.htm#top

The Web site of the Council of Biology Editors Authorship Task Force contains a useful collection of references and links along with a collection of provocative writings and a task force white paper. The site has an interactive forum for soliciting comments on authorship and the task force's activities.

http://www.sdsc.edu/CBE/

The 5th edition of the International Committee of Medical Journal Editors' *Uniform Requirements for Manuscripts Submitted to Biomedical Journals* may be found on-line at

http://www.acponline.org/journals/resource/unifreqr.htm

Use of Humans in Biomedical Experimentation

Paul S. Swerdlow

Overview

T HERE ARE MANY IMPORTANT ETHICAL ISSUES in scientific
endeavors, but none has been better codified than experimentation
involving human beings as subjects. Much of early medicine undoubtedly
involved experimentation, most of which was not regulated. In fact, the
rules for experimentation with people were initially summarized in the
Nuremberg Principles that came out of the Nuremberg war criminal trials at the end of World War II. These trials held accountable those involved in human experimentation performed without the consent of the
subjects. Although largely of historical significance today, the Nuremberg
Principles (also called the Nuremberg Code) provided the foundation for
future guideline documents, most notably the Declaration of Helsinki (discussed below). The 10 Nuremberg Principles included statements about
protection of human subjects, experimental design based on previous animal studies, careful risk-to-benefit analysis in the context of the importance
of the problem being studied, performance of experiments only by scientifically qualified persons, subject-initiated withdrawal from the research at
any stage, and investigator-initiated cessation of the experiment in the face
of possible injury, disability, or death.

Unfortunately, a significant number of ethically questionable studies
have been performed (2), before and after promulgation of the Nuremberg

Principles. A particularly egregious example is the Tuskegee study conducted at the Tuskegee Institute with funding from the U.S. Public Health Service (6). The aim of the 1932 study was to determine the course of untreated syphilis in African Americans, a disease that was widely believed to be a distinct entity from that in whites. The arsenic- and mercury-based therapy then in use was quite toxic but generally believed to be beneficial. No patient consent was obtained in this study wherein spinal taps were disguised as "free treatment." Even the scientific basis of the study was flawed, since most of the 412 infected men had received some initial treatment as an inducement to participate in the study. It was later decided that, since their treatment had been inadequate, follow-up as an untreated cohort was warranted. The study clearly documented a 20% decrease in life span for the infected men as compared with the control group of 204 uninfected men.

In the 1940s, when penicillin was found to be effective therapy, the study was nonetheless continued. It was reasoned that this was the last chance to study untreated syphilis because of soon-to-be-widespread antibiotic use. Patients were not informed about the potential new therapy, although their infections could have been cured by penicillin. As late as 1969, a review panel allowed the study to continue. The Macon County Medical Society, which included African American physicians, promised to assist in the study and to refer all patients before using antibiotics for any reason. In 1972 the study was finally reported in the public press. In 1973, more than 20 years after penicillin was in widespread use, the government finally took steps to ensure treatment of the few surviving infected patients. Even 25 years after the closure of this study, and with numerous safeguards in place, many people remain reluctant to trust clinical research studies.

In 1964 the World Medical Association sponsored a conference in Helsinki, Finland, to formalize guiding principles for the ethical use of humans in biomedical experimentation. The Declaration of Helsinki, prepared at this conference, has prevailed as the international standard for biomedical research involving human subjects. At subsequent conferences in 1975, 1983, 1989, and 1996, the Declaration of Helsinki has been amended and affirmed as a guiding force in experimentation with human subjects. The text of the most recently amended Declaration of Helsinki is reprinted at the end of this chapter. Table 5.1 presents a history of events and documents that are related to human subject research.

Are You Conducting Human Subject Research?

Before discussing the major areas of human subject research procedures and regulations, it is useful to address those activities that define this type of investigation. The legal requirements that govern human subject

Table 5.1 A chronology of international guidelines for human subject experimentation[a]

Year	Document	Authority
1947	Nuremberg Code	
1948	Universal Declaration of Human Rights	United Nations General Assembly
1964	Declaration of Helsinki (1)	World Medical Association
1966	International Covenant on Civil and Political Rights	United Nations General Assembly
1975	Declaration of Helsinki (2) — Tokyo	World Medical Association
1982	Proposed International Guidelines for Biomedical Research Involving Human Subjects	Council of International Organizations of Medical Sciences/World Health Association
1983	Declaration of Helsinki (3) — Venice	World Medical Association
1989	Declaration of Helsinki (4) — Hong Kong	World Medical Association
1991	International Guidelines for Ethical Review of Epidemiological Studies	Council of International Organizations of Medical Sciences/World Health Organization
1993	International Ethical Guidelines for Biomedical Research Involving Human Subjects	Council of International Organizations of Medical Sciences/World Health Organization
1996	Declaration of Helsinki (5) — South Africa	World Medical Association

[a] The information in this table was compiled and kindly provided by Robert Eiss of the Fogarty International Center. Table contents are limited to major international guidelines. A number of policy documents dealing with human subject experimentation emanating from various U.S. agencies or initiatives may be found in reference 12. Such documents as the Belmont Report, The Common Rule, and policy documents from the U.S. Food and Drug Administration, the Department of Health and Human Services, and the Centers for Disease Control and Prevention can be found in this reference.

experimentation are broad and may cover research based on materials being used or activities that create an interface between a human subject and the researcher. The Web sites of both the Office for Protection from Research Risks (OPRR) and the Department of Energy (DOE) provide helpful information in this regard (see Resources at the end of this chapter). In particular the OPRR site contains decision charts that graphically clarify whether a research activity is subject to federal regulations governing human subject experimentation. The DOE site contains a series of questions that allow an investigator to define research as human subject experimentation. The DOE site also contains an on-line course that provides guidance on human subject experimentation. The key points from the OPRR and DOE sources may be summarized as follows.

Human subject research includes all studies where there is an intervention or interaction with a living person that would not be happening outside of the conduct of the experimentation. Even if this is not the case, the activities may still be subject to regulations if identifiable data or information gathered during the research — or collected outside of the study in question — may be linked to the human subjects. Data collected through intervention include direct methods, such as drawing venous blood, and indirect methods, such as manipulating the environment of the human subject. Federal regulations also apply to human subjects that are used to test devices, materials, or products that have been developed through research.

The use of existing human subject data or specimens may be subject to federal regulations regardless of whether they were generated as part of the study in question. In general, the use of any materials of human origin creates the need for prior evaluation of the research and the associated regulatory obligations. Tissues, blood, organs, excreta, secretions, hair, nail clippings, and materials derived from these sources (e.g., DNA) generally define the activity as human subject experimentation subject to regulatory compliance. A related generalization is that such research activities may be exempt from regulations if the information derived from data or specimens is recorded and maintained in a fashion that precludes linking it to its human subject origin. When in doubt, investigators should consult with the office of their institutional review board for advice and guidance on whether the definition of human subject research has been met.

The Issue of Informed Consent

Key among the principles of experimentation on human subjects is the concept of informed consent. Several elements are required for informed consent. The person must first be "competent to consent" — to understand consequences and to make decisions. The decisions do not have to meet any particular criteria for "good" decisions — he or she may enter a study for the "wrong" reason or may make a decision someone else thinks is "bad." In other words, one must simply need to be able to understand the consequences of various decisions and have the capacity to make such a decision. In practice, many people who are clearly competent routinely make bad decisions regarding relationships, employment, medical care, and many other matters. The standard of competence for medical research is no different.

Consent must also be voluntary, that is, free from coercion. Coercion to participate in studies, however, can be very subtle and at the same time powerful. Coercion can come from many sources, including the patient's family, the researcher, the physician, the institution, and even the health care system itself. While most researchers and institutions avoid coercing study participants, subtle coercion may not be apparent to those conducting the research, let alone the potential subjects for the research. Some of these elements are difficult to control. Family coercion to participate in some form of therapy is often strong, even when no clear benefit exists. This is often seen in cancer chemotherapy where, even though prolongation of survival may be minimal and treatment fraught with side effects, familial pressure to take treatment can nevertheless be intense. This is usually related to standard therapy, but the same factors may pertain in research situations. Studies of genetic pedigrees for inherited conditions are much more likely to be revealing if more family members participate. Family pressure can be extreme in these situations and even extend to those who do not wish to know if they carry a certain gene (such as that for Huntington disease).

Different aspects of coercion can become part of the health care system, as illustrated in the following two examples. First, people without insurance may join studies to receive basic care that would otherwise be unavailable. While this has often been a problem in underdeveloped countries, it is now an increasing problem in the United States, as nearly one-sixth of the population is currently uninsured. Under some health insurance plans, in an effort to decrease costs, physicians have not been allowed to present certain standard medical alternatives to their patients. Thus, patients in such situations may face subtle coercion to join a study because all medical options presented seem inadequate. The second example derives from situations where only marginally effective standard therapies exist and therapeutic research is felt by many to be a patient's best option. Such research compares the most promising new therapy with the best current (but usually far from ideal) therapy. In aggressive attempts to control costs, health insurance plans are limiting a patient's freedom to embark on therapeutic clinical trials by calling such trials "experimental." Nearly all health care policies specifically exclude experimental expenses. Such denials occur even when the costs of the study are no greater than those for the standard therapy. An ethical dilemma arises when all potential therapies for the disease in question are experimental. The result may be that even those willing to enroll in large peer-reviewed clinical trials may not be allowed to participate.

Coercion by the basic researcher (one not licensed to treat patients), physician, or institution must also be controlled. Researchers are often reimbursed in clinical studies on a per-patient basis. The per-patient fee covers the experimental costs and often a portion of the researcher's salary and even the departmental budget as well. There is thus great incentive to enroll as many patients as possible. While the basic researcher usually has little to use to coerce people into participating (other than reimbursement for the activity), a physician-researcher has much more power. To a large and ever-increasing extent, the physician controls the patient's access to the U.S. health care system and is often totally entrusted to make medical decisions for the patient. Many patients refuse to even question their physicians about these decisions, in part because they trust them since they possess requisite specialized knowledge and in part because of paternalistic (or maternalistic) attitudes held by many physicians. Under such circumstances, it is easy for patients to feel that if they decline to participate in a study, they may lose a precious doctor-patient relationship and even access to the health care system. Such issues must be addressed through consent forms and patient education, or coercion may occur. This is especially likely if the physician is a participant in or will benefit from the research (e.g., the department employing the physician conducts the research). Cases of outright fraud and deception have been reported, although this happens more often outside of academic medical centers where oversight is greater (7, 8). It is also important to regard the circumstances of the study and how the study will be employed in special populations where coercion is more likely (see below).

Consent must also be informed. The participant must have adequate information to make a valid decision. The participant has the right to hear about all known risks of the study, including risks that are even beyond what would normally be discussed for medical informed consent. When routinely informing a patient about potential risks of a procedure or course of treatment, the physician makes an effort to reveal all realistic risks that are likely to affect the decision making of the patient. However, known risks of extremely small magnitude are often not mentioned. They are confusing and may adversely affect decision making to the detriment of the patient. For example, risks significantly less than dying in a car accident on the way to the doctor's office are often not disclosed. With a study, however, particularly one that is not of therapeutic intent, all known risks should be disclosed for truly informed consent.

Merely presenting the information is not sufficient. Informed consent requires comprehension of the risks by the participant. The investigator should verify that the person really understands the various options and risks and potential benefits of the study. For this reason, many institutions encourage the participant to have a relative or friend witness the signature on the consent form. This provides the person with an ally who hears the same information, can ask additional questions, and can ensure that the concepts were presented in an understandable way.

Who must ensure that the above obligations are fulfilled? It is the obligation of all who participate in the research to ensure that informed consent is obtained. This duty is not restricted to those who obtain the informed consent or to those involved solely with the clinical parts of the study. It is an obligation of all involved. It can be delegated to parts of the group but should not be delegated lightly; that is, all involved are responsible to see that it is done correctly. It is essential for all involved to read the consent form and then to ensure that the study, its risks, and its benefits are fairly and understandably presented.

Institutional Review Boards

Institutions receiving federal support in the United States are required to have an institutional review board (IRB) to approve and oversee research on human subjects. The Office for Protection from Research Risks is the arm of the Department of Health and Human Services charged with the oversight of federally funded institutions in the United States. Committees similar to the IRB are found in other countries, but their rules and composition vary (11). Rules pertaining to the formation of U.S. IRB committees are relatively simple. Most academic institutions have larger committees than required. The committee must include at least five members, and the membership list must be filed with the U.S. Secretary of Health and Human Services. All five members cannot have the same profession, and there must be at least one member with primary concerns in nonscientific areas (often a lawyer, ethicist, or member of the clergy). There must also be at least one member not affiliated with the institution or with family so affiliated. The nonaffiliated member may also be the nonscientific member.

Approval of projects requires a simple majority vote. At least one nonscientific member of the committee must vote but does not have to vote for approval. No member is allowed to participate in the review of a project in which he or she has a personal interest. The committee may invite experts to appear, but such experts may not vote. Proposals must be rereviewed yearly, and there must be written procedures that prescribe the operations of the committee. Serious or continuing noncompliance with the process must be reported to the Secretary of Health and Human Services.

The committee is charged with specific criteria with which to review proposals. First the risks to subjects must be minimized consistent with the aims of the research. Ideally, proposed procedures would be those already being performed for diagnostic or therapeutic purposes. For research to be ethically valid, it must first be technically valid. Even a study with minimal risk requires that valid scientific results are to be obtained, or it cannot be justified. This is most often a problem with small clinical studies in which statistically valid data may be difficult to obtain. Common reasons for such small studies include:

- *Study of a rare disease or disorder.* These are often called "orphan diseases" and are commonly ignored by pharmaceutical companies. The small potential market often cannot justify the drug development costs. The Food and Drug Administration, however, periodically sponsors studies of drugs for orphan diseases.
- *Pilot studies of new therapies.* It is often difficult to get funding for large and therefore expensive clinical studies. Pilot studies test the feasibility of new treatments but are generally not sufficient to establish efficacy. They provide the information needed to properly design and obtain funding for the larger study.

These types of studies must have clearly defined endpoints so the IRB can determine their risk-to-benefit ratio. Valid endpoints can include determination of treatment toxicity, patient compliance, or drug pharmacokinetics. Attempting to determine efficacy of treatments with too few patients, however, will likely create problems at the IRB. Statisticians, in particular, will instantly realize that the chances of determining efficacy with a small population are nil unless dramatic changes are found in easily measured outcomes. A good statistical analysis is often essential for proper study design and can save time and unnecessary effort with the IRB, with granting agencies, and subsequently with data analysis.

Most important, the risks to subjects must be reasonable in relation to anticipated benefits. Study benefits include benefits to the research subject as well as the importance of the knowledge that may reasonably be expected to result. In assessing the risk-to-benefit ratio of the project, only the risks and benefits of the research should be considered. Risks of procedures that would still be performed if not included in the study should not be considered. Similarly, a beneficial procedure performed as part of a study cannot be considered a benefit if the same procedure would be performed without the study.

For clinical studies in which two different treatments are being compared, there must be a valid null hypothesis that the two arms are equivalent. This is the concept of equipoise, that neither of the two treatments is known to be better. The researcher should be able to honestly say that there is no convincing data that one arm is better. If one arm is known to be better, the point of the study is moot and the research is no longer ethical. This includes placebo studies in which the test treatment is compared with no treatment at all. Such studies may be reasonable if the efficacy of the treatment being tested is not known and there is no known efficacious therapy.

The committee is prohibited from considering long-range effects of research on public policy that may result from the research. For example, in reviewing a study of an expensive therapy for dissolution of gallstones, the committee should not take into account the potential bankrupting of the health care system if the procedure were eventually used on all gallstone patients.

Selection of subjects must be equitable. For example, it is not appropriate to restrict a study to people with health insurance in the hopes that such patients will eventually financially support the hospital should they return to have other medical problems treated. There is also a national effort to ensure that minority populations and women are not excluded from studies, as has been done in the past. One reason often used to exclude women from studies was the issue of pregnancy. A new drug has likely not been tested in human pregnancy and will pose an unknown risk to such pregnancies. It was often felt simpler not to include women so as not to have to worry about pregnancy. Currently, most studies will allow women using medically approved birth control to participate. Furthermore, if the research will be of potential medical benefit to the woman, pregnancy does not necessarily exclude her from the research (see below).

The Institutional Review Board and the Informed Consent Issue

IRBs must ensure that informed consent is sought from each prospective subject or his or her legally authorized representative. Those unable to consent but who have an appropriate legal representative or guardian may participate if such representative gives informed consent. All consents must be documented and signed by a witness. To avoid questions of conflict of interest, it is important that the witness not be part of the investigating team. The best witness is a friend or relative of the participant. Such a person often has a background similar to the participant's and can help ensure that the study is explained in terms both can understand. In addition, he or she will be able to ask questions and sometimes even help explain the study to the participant.

The research must make adequate provision for monitoring data to ensure the safety of subjects. The Food and Drug Administration also requires such information for all new agents. Adequate provisions must be made as well to protect privacy. Records containing identifying information should be maintained in locked locations and restricted to those who have a need to use the information and who are trained in medical confidentiality or privacy. It is especially important not to discuss such information in public places such as hallways, elevators, or lunch rooms where comments might be overheard. It is often a good idea to create a second database lacking identifying information for ease of use and convenience.

Special provisions must be made for studies in which some or all of the subjects are likely to be vulnerable to coercion or undue influence. This includes people with acute or severe physical or mental illness and those who are economically or educationally disadvantaged. One such safeguard could be to have a patient representative to ensure that, when studies are complicated and involve acute medical situations or include people with limited education, subjects completely understand all implications. Consent forms must be read to those who cannot read (or read well) and should be written so they are easy to understand. It often helps to have the consent reviewed by those used to dealing with the educationally disadvantaged.

There is an increasing trend for consent forms to be approved by central authorities for large projects involving substantial numbers of people. While this may seem intrusive, such efforts have so far yielded high-quality consent forms by employing people with expertise in the creation of such forms and who are skilled in presenting complex topics in lay language. With large double-blind studies, a separate data and safety monitoring committee is often used.

Certain types of research are exempt from consent requirements by the federal government and most IRBs. Consent is not needed for research conducted in educational settings involving normal educational practices. This includes research on education instructional strategies, the efficacy or the comparison of instructional techniques or curricula, or classroom management methods. Research involving the use of educational tests (cognitive, diagnostic, aptitude, achievement) is also exempt if the information taken from these sources is recorded in such a manner that subjects cannot be identified directly or indirectly.

Research involving surveys, interviews, or observations of public behavior does not need consent unless responses are recorded in such a manner that the subjects can be identified directly or indirectly, or the responses or behaviors could place the subject at risk for criminal or civil liability, or the research deals with sensitive aspects of behavior (such as illegal conduct, drug use, or sexual behavior). Research involving surveys or interviews is also exempt from consent when respondents are public officials or candidates for public office.

What should be included in an informed consent? Consent forms fulfill several roles in human research. They are designed to describe the study in detail, including risks and benefits. They can, however, also be a contract and include compensation for participation in the study. Consent forms must describe the compensation for participation in the study. Consent forms must explain the participants' rights, including the right to withdraw from the study at any time. They must also reassure participants that they will not forfeit any other rights because of refusal to participate or withdrawal from the study. The form should specify what happens if a participant becomes pregnant and whether birth control is required to participate. The consent form also provides the participant with the phone number of the investigator and that of the IRB, should a participant with concerns not wish to speak with the investigator. Each institution has its own format, but uniformity of protocol and consent formats aids in the review process.

One particularly sticky area for informed consent is that of stored DNA samples. Such samples contain the full genome of the donor, including information on predisposition to genetic diseases and other potential health or employment problems. Such information could be tremendously damaging for a participant. He or she could be denied health, life, or disability insurance or even employment based on the information. The protection of this information must be considered by the IRB and explained in the consent form. If such samples are to be stored for future use, the types of use must be specified. If a new use is found in the future, a new consent might be required from the donors for this use. Such consents are difficult to obtain, especially given our mobile society. One alternative is to make the samples anonymous by stripping off any identifiable information so the samples cannot be tracked back to the donor. The difficulty here is that no further information about the samples is then available.

The Institutional Review Board and Expedited Review

Many committees have procedures for expedited review for specific types of research involving no more than minimal risk. These include procedures listed below, adapted from the Code of Federal Regulations (45 C.F.R. 46).

Prospective collection of:

- Biological specimens for research purposes by noninvasive means such as: (a) hair and nail clippings in a nondisfiguring manner; (b) deciduous teeth at time of exfoliation; (c) permanent or deciduous teeth if routine patient care indicates a need for extraction; (d) excreta and external secretions (including sweat); (e) uncannulated saliva collected either in an unstimulated fashion or stimulated by chewing gumbase or wax or by applying a dilute citric acid solution to the tongue; (f) placenta removed at delivery; (g) amniotic fluid obtained at the time of rupture of the membrane prior to or during labor; (h) supra- and subgingival dental plaque and calculus, provided the collection procedure is not more invasive than routine prophylactic scaling of the teeth and the process is accomplished in accordance with accepted prophylactic techniques; (i) mucosal and skin cells collected by buccal scraping or swab, skin swab, or mouth washings; (j) sputum collected after saline mist nebulization.
- Blood samples by finger stick, heel stick, ear stick, or venipuncture collected no more than twice weekly (a) from healthy, nonpregnant adults who weigh at least 110 pounds in amounts not to exceed 550 ml in an 8-week period or (b) from other adults and children, considering the age, weight, and health of the subjects, the collection procedure, the amount of blood to be collected, and the frequency with which it will be collected, but the amount drawn may not exceed the lesser of 50 ml or 3 ml per kg in an 8-week period.
- Research involving materials (data, documents, records, or specimens) that have been collected or will be collected solely for nonresearch purposes (such as medical treatment or diagnosis).
- Data obtained through noninvasive procedures (not involving general anesthesia or sedation) routinely employed in clinical practice, excluding procedures involving X rays or microwaves. Any medical devices must already be approved for marketing and not currently being tested for safety and effectiveness. Examples: (a) physical sensors that are applied either to the surface of the body or at a distance and do not involve input of significant amounts of energy into the subject or an invasion of the subject's privacy; (b) weighing or testing sensory acuity; (c) magnetic resonance imaging; (d) electrocardiography, electroencephalography, thermography, detection of naturally

occurring radioactivity, electroretinography, ultrasound, diagnostic infrared imaging, Doppler blood flow, and echocardiography; (e) moderate exercise, muscular strength testing, body composition assessment, and flexibility testing where appropriate given the age, weight, and health of the individual.

- Data from voice, video, digital, or image recordings made for research purposes.
- Data on individual or group characteristics or behavior (such as research on perception, cognition, motivation, identity, language, communication, cultural beliefs or practices, and social behavior) or research employing survey, interview, oral history, focus group, program evaluation, human factors evaluation, or quality assurance methodologies.

Human Experimentation Involving Special Populations

Incompetent patients

It is often assumed that those with mental illness or those who are not able to provide informed consent must be excluded from all studies. This is not the case. Consent must be provided by the legally responsible person, and the study must be designed in such a way that adequate safeguards exist for the participants. It would seem unfair to deprive these people of the right to participate in potentially therapeutic studies or to prevent information from being gained to help people with mental disorders. Clearly, the IRB and the researchers must ensure that individual rights are respected. They must also take into account that participation in arduous programs without being able to understand the reason for the treatments makes such programs much more difficult to endure. This type of research (certain chemotherapy trials, for example) may therefore be inappropriate for certain populations. Psychiatric patients may be particularly vulnerable emotionally. Particular attention must be paid to avoid covert (and likely unintentional) coercion. Furthermore, it has been suggested that research personnel should use the medical definitions of informed consent for certain studies in this patient population rather than the more comprehensive information usually required (4) in an effort to reduce patient anxiety. Thus, the IRB has special responsibilities for protocols involving these patients.

Prisoners

Prisoners constitute an excellent example of a population that requires additional safeguards for consent for scientific study. The nature of incarceration affords numerous potential coercions, and thus federal regulations specifically offer additional safeguards for this population. Only certain types of federally sponsored research can be performed on prisoners.

These include:

- Study of possible causes, effects, and processes of incarceration or criminal behavior that present no more than minimal risk or inconvenience to the prisoner.
- Studies of prisons as institutional structures or of prisoners as incarcerated persons.
- Research on conditions affecting prisoners as a class, such as vaccine studies on hepatitis due to the increased incidence of hepatitis in prisons, or social or psychological problems such as alcoholism or drug addiction. The Secretary of Health and Human Services must consult with experts in penology, medicine, and ethics and give notice in the *Federal Register* of intent to approve such research.
- Research on both innovative and accepted practices that have the intent to improve the health or well-being of the subject. If control groups will be used in the protocol, the Secretary must again consult with experts and give notice as above.

There are very specific requirements for the IRB, including the requirement that a prisoner or a prisoner representative must be a member of the IRB. A prisoner representative must have the appropriate background and experience to serve as a true representative of the prisoners. Another requirement is that a majority of the IRB (exclusive of prisoner members) must have no association with the prisons involved. There is no requirement that the prisoner or prisoner representative must vote for a given proposal for it to be enacted.

The IRB must further determine that any advantages gained by the prisoner by participating are not of such magnitude that the prisoner's ability to weigh the risks of participation are impaired. These would include advantages in living conditions, medical care, food quality, amenities, potential earnings, and outside contacts. The risks involved must also be risks that would be accepted by nonprisoner volunteers. Study information must be presented in a manner the population can understand.

Selection of subjects in prison must be fair to all prisoners and cannot be arbitrarily used or influenced by prisoners or prison authorities. Studies must not be used as a reward or method to control the inmate population.

Participation in scientific or medical studies cannot be taken into account by parole boards in determining eligibility for parole. The prisoner must be specifically informed that parole considerations will not be affected. Allowing participation to affect parole would be an example of undue influence or coercion to participate.

Where follow-up is required, arrangements must be made for the various lengths of sentence of the prisoners. The researchers should also consider the likelihood of noncompliance after the sentence is over. The potential import of these arrangements is illustrated by the case of a

35-year-old prisoner who developed testicular cancer while incarcerated. The prisoner was placed on a standard, noninvestigation therapy with his consent. With aggressive chemotherapy, testicular cancer is largely curable. After the first course of chemotherapy resulted in a good response, the court, at the county's request, paroled the prisoner. The reason for parole was not made clear to the medical staff, but it was suspected that either it was a compassionate parole (which seemed strange for a largely curable, as opposed to terminal, cancer) or the county did not wish to pay the costly medical bills for the therapy. The prisoner, who had tolerated the chemotherapy well, left the hospital against medical advice in the middle of a treatment saying he had "things to do." He never returned for the needed therapy and was lost to follow-up. While it was clearly his right to leave, it is also likely that the cancer recurred. Recurrent cancer has a diminished prognosis and, if left untreated, would undoubtedly be fatal. If the prisoner had been on a study, it is certain he would not have continued with it. In this particular case, some of the medical staff thought that the county, by paroling the prisoner, had converted his sentence to a death sentence (albeit with the prisoner's unintentional collaboration).

Children

For children under the age of 18 years, parents or guardians must give consent. For research that involves significant risk, both parents must consent when available, unless only one parent has legal responsibility or custody. In addition, assent or agreement of the child is required when the IRB deems that he or she is capable. In making this determination, the IRB must consider the age, maturity, and psychological state of the children involved. This can be done for all children involved in a given protocol or individually. If the IRB determines that the capacity of the child is too limited or if the research may offer benefits important to the health or well-being of the child, assent is not required.

Recently, it has been pointed out that certain behaviors commonly accepted in society put children at much greater risk than do most research studies. Koren et al. calculated the risk of a babysitter's having to deal with a severe medical emergency in Canada (9). They calculated that each year at least 900 Canadian babysitters would have to deal with an acute asthmatic attack in one of their charges and that 26 would likely have a child who experiences sudden infant death syndrome while under their care. These situations would place the babysitters, often between the ages of 10 and 15, at risk of emotional trauma far greater than would most research studies. The work of Koren et al. suggests that if a child is deemed mature enough to supervise younger children in potentially extremely dangerous situations, he or she should be able to consent to most research studies.

Children who are wards of the state or any other agency can be included in research only if the research is either related to their status as ward or conducted in institutions in which the majority of children involved are not wards. In such cases, the IRB shall require appointment of an advocate — not associated with the research, the investigator, or the guardian organization — who agrees to act in the best interests of the child for the duration of the child's participation in the research.

Additional restrictions are imposed for research with greater than minimal risk. However, when there is greater than minimal risk but also the possibility of direct benefit to the child, the IRB must determine that the risk is justified by the anticipated benefits. The risk-to-benefit ratio must also be at least as good as that of all alternative approaches. When there is no prospect of direct benefit, but the research is likely to yield important knowledge about a disorder, the risk must represent a minor increase over minimal risks. The interventions must be comparable to those inherent in the actual or expected medical, dental, social, or educational situations. The information to be obtained must be of vital import for the understanding or amelioration of the subject's disorder or condition. To bypass these restrictions, there must be a reasonable opportunity to achieve further understanding, prevention, or alleviation of a serious problem affecting the health or welfare of children. Nevertheless, the Secretary of Health and Human Services must consult with a panel of experts and ensure that such a condition exists and that the research will be ethically conducted.

These restrictions may seem excessive and may indeed slow research in some areas. It must be remembered, however, that for children who are not old enough to consent, the parents and the IRB remain their sole advocates. There is even some indication that parents who volunteer their children for studies may be psychologically different from those who do not, making the issue of study regulation and control even more important (5).

More recently, efforts have been made to ensure that children are incorporated into studies of most new medications. This is part of an effort to include all underrepresented groups in research studies to ensure widespread applicability of the results. Efforts to include women and minorities are also under way. Many medications routinely used in pediatrics have not been studied in children but merely used after approval for adults. By requiring pediatric studies (i.e., persons less than 21 years old) for most medications, it is hoped that this situation can be reversed.

Fetal Tissue Research

There has been a good deal of controversy surrounding the use of fetal tissue in research, specifically in transplantation research (3, 13). A national Commission for the Protection of Human Subjects was established in 1974, and a moratorium was placed on fetal research until it set up appropriate regulations. Its findings are now part of Department of Health and Human Services regulations. Part of the regulations mandate that applications involving human in vitro fertilization must be reviewed by an Ethics Advisory Board before they can be federally funded. The Board's charter expired in 1980, and no federal administration since has renewed the charter, imposing a de facto moratorium on such federally funded research. Fetal research was also put on hold for years, but on February 1, 1993, the Secretary of Health and Human Services ended the moratorium on funding of fetal research. Criteria in several categories have now been promulgated by multiple sources regarding the conduct of fetal research.

It is believed important to separate the abortion from the research. This includes issues such as the decision to terminate a pregnancy, the timing of the abortion, and which abortion procedures to use. Payments and other inducements to participate in research on fetal tissues are prohibited. Directed donations are prohibited, including the use of related fetal tissue transplants. Anonymity between donor and recipient must be maintained. The donor will not know who will receive the tissue nor will the recipient or transplant team know the donor.

Consent of the pregnant woman is required and is sufficient unless the father objects (except in cases of incest or rape). The decision and consent to abort must precede discussion of the possible use of fetal tissue and any request for such consent that might be required for such use. Recipients of such tissues, researchers, and health care participants must also be properly informed about the source of the tissue in question.

The guidelines may well undergo continued revision. Some suggest that the person performing the abortion or any physician supplying fetal tissue not be allowed to be a coauthor or receive support from the study. Others believe that the consent of the mother is not appropriate and that an external consent should be sought.

Martin (10) argues that, contrary to National Institutes of Health guidelines, it is unethical to withhold information on fetal tissue donation because it may be important information for women making the decision to abort. He also argues that in time women will know of the option anyway. Attempts to make the abortion decision in the absence of knowledge of tissue donation options will become futile as the information about such transplants is disseminated. It should be emphasized that this is currently a minority opinion.

Research directed toward the fetus in utero can be approved by an IRB if (i) the purpose of the research is to meet the health needs of the fetus and is conducted in a way that will minimize risk or (ii) the research poses no more than minimal risk and the purpose is to obtain important biomedical knowledge that is unobtainable by other means. Risk-to-benefit ratios need to be carefully considered under the first category, especially as medical and surgical intervention in utero becomes more prevalent.

Research directed toward the fetus ex utero depends on viability. If the fetus is judged viable, it is then an infant and is covered by standard pediatric regulations and policies. If it is nonviable (i.e., cannot possibly survive to the point of sustaining life independently despite medical care), then research cannot either artificially maintain vital functions or hasten their failure. Researchers must maintain the dignity of the dying human and avoid unseemly intrusions in the process of dying for research purposes. Research with dead fetal material, cells, and placenta is regulated by the states.

Use of fetal tissue for transplantation, particularly for the treatment of Parkinson disease and juvenile diabetes, has been particularly controversial. A moratorium was placed on federally funded fetal transplant research from 1988 to 1993. Interim guidelines require adherence to all fetal research conditions listed above; in addition, there must be sufficient evidence from animal experimentation to justify the human risk.

The increased inclusion of women in research studies raises the issue of pregnancy. In research directed primarily toward the health of the mother, her needs generally take precedence over those of the fetus. For example, if a new therapeutic agent is considered necessary to improve a pregnant woman's condition, her consent alone is sufficient even if the treatment poses greater than minimal risk to the fetus. The study must, however, try to minimize the risk to the fetus consistent with achieving the research objective. When there is no health benefit to the mother, research on non-pregnant participants must be used as a guide to the level of risk to the fetus. If there is greater than minimal risk, the research cannot currently proceed, as it requires review by the Ethics Advisory Board (which currently has no charter) before going to the Secretary of Health and Human Services, who could approve the research.

Surprisingly, there are no regulations for studies on lactating women, enhancing conception or contraception, or on abortion techniques. However, many of the above considerations will apply to IRB deliberations of such research.

Conclusion

In contrast to most areas of biomedical research, human subject experimentation is governed stringently by policies and regulations that have their underpinnings in federal law. Although this history of formal regulation dates back just over 50 years, the regulatory network that applies to human subject experimentation increasingly spans research efforts worldwide. Biomedical researchers thus have both ethical and legal obligations. Research using human subjects demands careful planning that will pass rigorous peer review before the performance of any experimentation. Scientists wishing to do human subject research must be conversant with the applicable policies and regulations. Other topic areas discussed in this book also have strong implications for human subject research. Record keeping (chapter 11) plays heavily into clinical research with humans in terms of maintenance, form, storage, retention, and confidentiality of results. Conflicts of interest (chapter 7) must be frequently dealt with in clinical research. For example, investigators need to disclose industrial support or commercial affiliations at various stages in the project, e.g., to IRBs, patients, editors, and reviewers. Finally, issues relating to collaborative research (chapter 8) and authorship (chapter 4) are common in human subject experimentation owing to the frequent interdisciplinary nature of this research.

Case Studies

5.1 First-year dental students have cultivated oral bacteria from their own mouths as part of a microbiology laboratory class. Organisms recovered are identified to the species level as part of the laboratory exercise. At its conclusion, the instructor informs the class that he wants to use all of their cultures in his research project. He is doing epidemiologic studies on the occurrence of antibiotic-resistant oral bacteria. The students comply and submit the culture dishes containing their bacterial isolates. Are the instructor's actions appropriate? Is an IRB-approved protocol needed? Do the students need to give informed consent?

5.2 An investigator wants to screen a recombinant bacteriophage library with antiserum to a specific protein made by a common human oral bacterium. Her initial intent is to raise polyclonal antibodies in rabbits using the purified protein. However, following discussions with colleagues, the investigator learns that many adults produce antibodies to a number of proteins associated with this bacterium. One of her colleagues indicates that he has dozens of serum samples from individual patients stored in his freezer. He suggests that she pool these sera and use the pooled material to screen her library. Is this strategy appropriate in terms of proper use of human materials in research? What if any paperwork should be filed?

5.3 A researcher wishes to study the metabolism of alveolar macrophages in patients with severe pneumonia and to correlate these findings with the clinical outcome of the patients. She proposes to take small samples of pulmonary lavage fluid to obtain the required cells. The fluid would be obtained only from those patients who would be undergoing pulmonary lavage for medical reasons. Pulmonary lavage is a standard medical procedure in which a flexible fiberoptic bronchoscope is passed into the lungs of a sedated patient. Fluid is then used to wash out the contents of the alveoli in lung and is then withdrawn. The procedure is used clinically to help diagnose difficult pulmonary problems. Does she need to include the risks of pulmonary lavage in the consent form? Does she need to worry that her pulmonologist collaborator will perform lavage on patients not needing the procedure to provide additional study samples?

5.4 You are sitting as a member of an IRB that examines proposals for the use of humans in medical experiments. A proposal currently under consideration involves the administration of fluorescently labeled, mouse-derived monoclonal antibodies to patients. These immunologic reagents would be used to test their ability to localize and diagnose tumors. The committee discusses the informed consent form proposed for use in these experiments. Specifically, one member of the committee argues that the consent form fails to reveal that participation in this study could preclude the future use of antitumor, mouse-derived monoclonal antibody therapy in these patients. This argument is based on the possibility that such patients could mount an anti-mouse antibody response. Considerable disagreement among the committee members erupts as a result of this issue. Where do you stand?

5.5 You ask a fellow graduate student and her husband, whom you run into in the hall, to donate 10 ml of blood each for one of your established protocols. While enumerating the T-cell subsets, you find very low values in the husband. This is worrisome, as AIDS can cause such low values. You ask your mentor what should be done. She says she will take care of the situation. Two days later you see the husband in your mentor's office, and she later informs you that he was told. Two weeks later it becomes apparent that your fellow graduate student, the wife of the man with the low T-cell subsets, still doesn't know about the problem. Do you tell her?

5.6 A researcher has developed a new drug for malaria and wishes to test it in humans. Since there are too few cases in the United States, he wishes to test it in Africa on a population at significant risk for malaria. Unfortunately, the drug is quite expensive to manufacture and is unlikely to be made available to the population of the country chosen for testing. Is it reasonable to use this population for drug study when they will not be likely to benefit from the research? What are the considerations regarding informed consent? Should local standards, which may be lax or nonexistent, be used? Should the researchers impose outside standards on the research? Are there obligations of the research team toward the medical community in the country where the testing occurs? (See reference 1.)

5.7 A 64-year-old man presents with advanced lung cancer. He has a long history of depression, resulting in several suicide attempts and several hospitalizations for clinically severe depression. He often skips taking his antidepressant medications. No standard therapy has been shown to statistically prolong life with his cancer, but some patients do seem to respond to aggressive chemotherapy and may live longer. An experimental protocol is available for a new therapy that has been tested in small trials and is now being compared with the aggressive chemotherapy. The family is most interested in treatment, including the patient's brother, who has been appointed his legal guardian. The patient is clinically depressed and states he is not interested in therapy. His psychiatrist is concerned that this may be a subtle way of committing suicide. Should the researcher attempt to place the patient in the study? This would require the consent of the guardian and probably also a court ruling, since the patient is opposed to the therapy. If not placed in the study, should chemotherapy be given anyway? What would be the effect on the patient of complications of chemotherapy given that the therapy is not wanted in the first place?

5.8 Before the emergence of AIDS, a noted hematologist wished to study if idiopathic thrombocytopenic purpura (ITP), a disease in which the platelet count drops dangerously low despite increased platelet production, were due to the presence of an autoantibody directly against the person's own platelets. He drew blood from one of these patients, isolated serum, which contains the antibodies in the blood, and injected the serum into his own vein. He then monitored his platelet count and documented an abrupt drop. Indeed, the platelet count dropped so far that he had a nearly fatal bleed and required hospitalization. After recovery, he drew blood from another patient and injected a smaller amount into his arm to document that antibodies existed in this patient as well. This time the platelet count did not drop as far, and he did not bleed. Subsequently, a paper was published that documented similar results from a larger number of patients whose serum was injected into an undisclosed number of volunteers. This information was a major contribution to the understanding of ITP. Is such research reasonable? Should it be regulated by the IRB? Is self-coercion involved? Is the researcher so committed to the science that he or she has no perspective on potential self-injury? In sum, should researchers be allowed to experiment on themselves?

5.9 You have been attending a meeting on eukaryotic growth factor biology and have just finished listening to Dr. Roman give his keynote address. His overview involved some clinical studies, and he showed slides of patients undergoing procedures as part of an institutionally approved clinical trial. In all instances the faces of the patients were clearly visible. On two other slides there were clinical materials depicted, and these were labeled with the patient's name. One tissue sample was clearly labeled with a tag that read "Mrs. MacDonald." After the lecture you leave to make a phone call. As you return to the lecture you are intercepted by Susan Jeris, a colleague you know casually from another institution. Susan confides in you that one of the slides shown by Dr. Roman was a picture of her step-mother, Shirley MacDonald. She is agitated and claims that Roman's use of the picture and disclosure of her stepmother's name is a violation of her stepmother's privacy and in violation of accepted standards of clinical re-search. She claims that Dr. Roman's presentation is an egregious violation of human subject research practices and thinks he should be punished. She asks you what she should do about this situation.

5.10 The frequencies of hospital-acquired infections in both the medi-cal intensive care unit (MICU) and the surgical intensive care unit (SICU) of a university medical center have reached alarming proportions. In response to this crisis, a research team implements a clinical study designed to reduce the frequency of occurrence of hospital-acquired infections in the MICU. This study involves the use of a series of aggressive strategies, which include (i) the use of experimental antibacterial towelettes for hand cleansing, (ii) controlled use of antibiotics to counter the emergence of antibiotic-resistant bacteria, and (iii) daily environmental monitoring for potential pathogenic bacteria. The researchers seek and receive IRB ap-proval for this work, and every patient in the MICU is enrolled. Of course, MICU patients are required to sign informed consent forms, approved by the IRB. Over the next 4 months, a dramatic decrease in hospital-acquired infections is seen in the MICU. During the same period, the infections in the SICU remain at high levels, and one patient in this unit dies owing to an infection caused by a multiply antibiotic-resistant bacterium. In preparing their results for presentation at a national meeting, the research team com-pares the frequency, type, and seriousness of infections between the MICU and the SICU. Comment on the ethical implications of this study. Should the SICU patients have been required to sign informed consent forms?

5.11 Professor Roger Fred is the course director of a physiology lab taught to medical students. One of the laboratory exercises involves students' drawing blood from one another (under supervision) and using the serum to perform a variety of chemical and cellular analyses. The lab ex-ercise is carried out successfully. At its conclusion Professor Fred announces to the class of 100 students that he would like to retain their leftover blood sera. He informs them that some of the sera will be used individually while some will be pooled. In all cases these sera will be used to gather baseline control data in a number of research projects. He asks if anyone wants to refuse having his or her serum used for research but receives no objections. Are Fred's actions appropriate? Is an IRB-approved protocol needed? Do the students need to give informed consent?

5.12 Dr. Mike Berle, an assistant professor at Vision University, has a seed grant from the University Foundation to study student attitudes related to academic honesty. As part of his research, Berle prepares a survey that he administers to 200 undergraduates taking his course in introductory psychology. One survey question asks the students to disclose if they have ever cheated on tests or assignments. Another asks the students to reveal if they have witnessed cheating at Vision University. Follow-up questions allow the students to indicate what they did about any cheating they witnessed. The results of the survey reveal that there is a baseline level of cheating that occurs at the university. Dr. Berle repeats the survey the following semester with a different class and gets similar results. He writes a manuscript, which he plans to submit to the *Journal of Academic Standards*. Dr. Berle discusses his survey results with some of his departmental colleagues, and the topic of his research rapidly becomes known in faculty and administrative circles. Shortly before he plans to mail the manuscript, he receives a phone call from the Vice President for Research and Graduate Affairs, Shirley Forest. She pointedly tells Berle that he cannot submit the manuscript unless he can certify that he has filed a human subject use protocol for the work and received approval from Vision's IRB. Berle, of course, thinks this is ludicrous, but Dr. Forest reminds him that his research used human subjects. He never considered asking for IRB approval, and he ends his conversation with Dr. Forest abruptly and without resolution. In subsequent discussions with some of his colleagues, Berle learns of rumors implying that the Vice President's position is an attempt to prevent the publication of results that may be embarrassing to Vision University. An angered and perplexed Berle comes to you for advice. What do you tell him?

Declaration of Helsinki

WORLD MEDICAL ASSOCIATION DECLARATION OF HELSINKI[a]

Adopted by the 18th World Medical Assembly, Helsinki, Finland, June 1964
and amended by the
29th World Medical Assembly, Tokyo, Japan, October 1975
35th World Medical Assembly, Venice, Italy, October 1983
41st World Medical Assembly, Hong Kong, September 1989
and the
48th General Assembly, Somerset West, Republic of South Africa,
October 1996

Introduction

It is the mission of the physician to safeguard the health of the people. His or her knowledge and conscience are dedicated to the fulfillment of this mission.

[a] © World Medical Association, Used with permission.

The Declaration of Geneva of the World Medical Association binds the physician with the words, "The health of my patient will be my first consideration," and the International Code of Medical Ethics declares that, "A physician shall act only in the patient's interest when providing medical care which might have the effect of weakening the physical and mental condition of the patient."

The purpose of biomedical research involving human subjects must be to improve diagnostic, therapeutic and prophylactic procedures and the understanding of the aetiology and pathogenesis of disease.

In current medical practice most diagnostic, therapeutic or prophylactic procedures involve hazards. This applies especially to biomedical research.

Medical progress is based on research which ultimately must rest in part on experimentation involving human subjects.

In the field of biomedical research a fundamental distinction must be recognized between medical research in which the aim is essentially diagnostic or therapeutic for a patient, and medical research, the essential object of which is purely scientific and without implying direct diagnostic or therapeutic value to the person subjected to the research.

Special caution must be exercised in the conduct of research which may affect the environment, and the welfare of animals used for research must be respected.

Because it is essential that the results of laboratory experiments be applied to human beings to further scientific knowledge and to help suffering humanity, the World Medical Association has prepared the following recommendations as a guide to every physician in biomedical research involving human subjects. They should be kept under review in the future. It must be stressed that the standards as drafted are only a guide to physicians all over the world. Physicians are not relieved from criminal, civil and ethical responsibilities under the laws of their own countries.

I. Basic Principles

1. Biomedical research involving human subjects must conform to generally accepted scientific principles and should be based on adequately performed laboratory and animal experimentation and on a thorough knowledge of the scientific literature.

2. The design and performance of each experimental procedure involving human subjects should be clearly formulated in an experimental protocol which should be transmitted for consideration, comment and guidance to a specially appointed committee independent of the investigator and the sponsor provided that this independent committee is in conformity with the laws and regulations of the country in which the research experiment is performed.

3. Biomedical research involving human subjects should be conducted only by scientifically qualified persons and under the supervision of a clinically competent medical person. The responsibility for the human subject must always rest with a medically qualified person and never rest on the subject of the research, even though the subject has given his or her consent.

4. Biomedical research involving human subjects cannot legitimately be carried out unless the importance of the objective is in proportion to the inherent risk to the subject.

5. Every biomedical research project involving human subjects should be preceded by careful assessment of predictable risks in comparison with foreseeable benefits to the subject or to others. Concern for the interests of the subject must always prevail over the interests of science and society.

6. The right of the research subject to safeguard his or her integrity must always be respected. Every precaution should be taken to respect the privacy of the subject and to minimize the impact of the study on the subject's physical and mental integrity and on the personality of the subject.

7. Physicians should abstain from engaging in research projects involving human subjects unless they are satisfied that the hazards involved are believed to be predictable. Physicians should cease any investigation if the hazards are found to outweigh the potential benefits.

8. In publication of the results of his or her research, the physician is obliged to preserve the accuracy of the results. Reports of experimentation not in accordance with the principles laid down in this Declaration should not be accepted for publication.

9. In any research on human beings, each potential subject must be adequately informed of the aims, methods, anticipated benefits and potential hazards of the study and the discomfort it may entail. He or she should be informed that he or she is at liberty to abstain from participation in the study and that he or she is free to withdraw his or her consent to participation at any time. The physician should then obtain the subject's freely-given informed consent, preferably in writing.

10. When obtaining informed consent for the research project the physician should be particularly cautious if the subject is in a dependent relationship to him or her or may consent under duress. In that case the informed consent should be obtained by a physician who is not engaged in the investigation and who is completely independent of this official relationship.

11. In case of legal incompetence, informed consent should be obtained from the legal guardian in accordance with national legislation. Where

physical or mental incapacity makes it impossible to obtain informed consent, or when the subject is a minor, permission from the responsible relative replaces that of the subject in accordance with national legislation.

Whenever the minor child is in fact able to give a consent, the minor's consent must be obtained in addition to the consent of the minor's legal guardian.

12. The research protocol should always contain a statement of the ethical considerations involved and should indicate that the principles enunciated in the present Declaration are complied with.

II. Medical Research Combined with Professional Care (Clinical Research)

1. In the treatment of the sick person, the physician must be free to use a new diagnostic and therapeutic measure, if in his or her judgment it offers hope of saving life, reestablishing health or alleviating suffering.

2. The potential benefits, hazards and discomfort of a new method should be weighed against the advantages of the best current diagnostic and therapeutic methods.

3. In any medical study, every patient — including those of a control group, if any — should be assured of the best proven diagnostic and therapeutic method. This does not exclude the use of inert placebo in studies where no proven diagnostic or therapeutic method exists.

4. The refusal of the patient to participate in a study must never interfere with the physician-patient relationship.

5. If the physician considers it essential not to obtain informed consent, the specific reasons for this proposal should be stated in the experimental protocol for transmission to the independent committee (1, 2).

6. The physician can combine medical research with professional care, the objective being the acquisition of new medical knowledge, only to the extent that medical research is justified by its potential diagnostic or therapeutic value for the patient.

III. Non-Therapeutic Biomedical Research Involving Human Subjects (Non-Clinical Biomedical Research)

1. In the purely scientific application of medical research carried out on a human being, it is the duty of the physician to remain the protector of the life and health of that person on whom biomedical research is being carried out.

2. The subjects should be volunteers — either healthy persons or patients for whom the experimental design is not related to the patient's illness.

3. The investigator or the investigating team should discontinue the research if in his/her or their judgment it may, if continued, be harmful to the individual.

4. In research on man, the interest of science and society should never take precedence over considerations related to the well being of the subject.

References

1. **Barry, M., and M. Molyneux.** 1992. Ethical dilemmas in malaria drug and vaccine trials: a bioethical perspective. *J. Med. Ethics* **18:**189–192.

2. **Beecher, H. K.** 1966. Ethics and clinical research. *N. Engl. J. Med.* **274:**1354–1360.

3. **Begley, S., M. Hager, D. Glick, and J. Foote.** 1993. Cures from the womb, p. 49–51. *Newsweek*, February 22.

4. **Fulford, K. W. M., and K. Howse.** 1993. Ethics of research with psychiatric patients: principles, problems and the primary responsibilities of researchers. *J. Med. Ethics* **19:**85–91.

5. **Harth, S. C., R. R. Johnstone, and Y. H. Thong.** 1992. The psychological profile of parents who volunteer their children for clinical research: a controlled study. *J. Med. Ethics* **18:**86–93.

6. **Jones, J. H.** 1993. *Bad Blood, the Tuskegee Syphilis Experiment.* The Free Press, New York, N.Y.

7. **Kolata, G.** 1999. Drug trials hide conflicts for doctors, p. 1. *New York Times*, May 16.

8. **Kolata, G.** 1999. A doctor's drug trials turn into fraud, p. 1. *New York Times*, May 17.

9. **Koren, G., D. B. Carmeli, Y. S. Carmeli, and R. Haslam.** 1993. Maturity of children to consent to medical research: the babysitter test. *J. Med. Ethics* **19:**142–147.

10. **Martin, D. K.** 1993. Abortion and fetal tissue transplantation, p. 1–3. *In IRB — A Review of Human Subjects Research*, vol. 15. Hastings Center, Briarcliff Manor, N.Y.

11. **Riis, P.** 1993. Medical ethics in the European Community. *J. Med. Ethics* **19:**7–12.

12. **Sugarman, J., A. C. Mastroianni, and J. P. Kahn (ed.).** 1998. *Ethics of Research with Human Subjects: Selected Policies and Resources.* University Publishing Group, Frederick, Md.

13. **Woodward, K., M. Hager, and D. Glick.** 1993. A search for limit, p. 52–52. *Newsweek*, February 22.

Resources

Internet

The general reference used in preparation of this chapter was the Code of Federal Regulations, Title 45 Part 46, Protection of Human Subjects. This is available on-line at the Web site of the U.S. Department of Health and Human Services Office of Protection from Research Risks (OPRR):

http://grants.nih.gov/grants/oprr/library_human.htm

The IRB Guidebook may also be found on this Web site, along with other relevant regulations, educational materials, and news on issues of importance to human subject research.

IRB — A Review of Human Subjects Research also provides a wealth of practical and useful information for those interested in human research. This periodical is published by the Hastings Center, Briarchiff Manor, N.Y., and is available in most university and medical center libraries.

A powerful bibliography of books, audiovisual materials, and journal articles relevant to ethical issues in human subject experimentation may be found on-line at

http://www.nlm.nih.gov/pubs/cbm/hum_exp.html

The U.S. Department of Energy maintains a Web site devoted to protecting human subjects at

http://www.er.doe.gov/production/ober/humsubj/

General reading

Altman, L. K. 1998. *Who Goes First? The Story of Self-Experimentation in Medicine.* University of California Press, Berkeley.

Kahn, J. P., A. C. Mastroianni, and J. Sugarman (ed.). 1998. *Beyond Consent: Seeking Justice in Research.* Oxford University Press, New York, N.Y.

Sugarman, J., A. C. Mastroianni, and J. P. Kahn (ed.). 1998. *Ethics of Research with Human Subjects: Selected Policies and Resources.* University Publishing Group, Frederick, Md.

Vanderpool, H. Y. (ed.). 1996. *The Ethics of Research Involving Human Subjects: Facing the 21st Century.* University Publishing Group, Frederick, Md.

Text appendix material

Appendix IV of this book contains the full text of an actual human subject protocol and the corresponding informed consent form used in that study.

Use of Animals in Biomedical Experimentation

Bruce A. Fuchs

Introduction • Ethical Challenges to the Use of Animals in Research
• Practical Matters: Constraints on the Behavior of Scientists
• Political Realities: Then and Now • Case Studies • References
• Resources

Introduction

A consensus challenged

ANIMAL EXPERIMENTATION HAS BEEN AN IMPORTANT RESEARCH tool for more than 100 years. At the dawn of the 19th century, scientific medicine was beginning to challenge medical traditions more than 1,000 years old. Physiological research involving animals was one of the key technologies that spurred this transition and led to an understanding of bodily functions and the physical basis of disease. However, the new approach was resisted by traditionalists who employed as one of their foremost criticisms the cruel nature of animal research. Present-day scientists should not delude themselves; early animal experiments could be exceedingly brutal. Fully conscious dogs were nailed to boards by their four paws, before being cut open, so that the beating of a heart might be observed. While the advent of anesthesia in the mid-1800s addressed some concerns, it by no means ended the debate over the fundamental morality of animal research. Numerous groups formed in the late 1800s to challenge the existing social order with regard to animals. These "antivivisectionists" were the antecedents of the contemporary "animal rights" movement.

In 1975 Peter Singer, an Australian philosopher who is now on the faculty at Princeton University, published the book *Animal Liberation* (26), which many believe was the seminal event in the rebirth of modern antivivisectionism. Since that time, animal rights activists have assiduously set about achieving their ultimate goal — the abolition of the use of animals for biomedical research, for food and clothing, and for entertainment. The most extreme activists even question the morality of pet ownership. The animal rights movement is viewed by many scientists as a threat to scientific progress and, ultimately, to the health and well-being of humankind. But the majority of scientists have not actively participated in the debate by responding to the charges of the animal rights activists at the local level, preferring instead to allow a defense to be mounted by national scientific organizations. This is arguably a serious mistake. The animal rights organizations have been quite successful in carrying their message to the general public. While the majority of the population still expresses support for the use of animals in biomedical research, the efforts of animal rights activists have clearly eroded this support, especially among young people. Additionally, the animal rights movement has sought to link its agenda with that of other popular causes, such as environmentalism, saying in essence, "If you care about our environment, you must support animal rights."

It is important that individual scientists take the time to become better educated about the moral and political controversies that surround the use of animals in biomedical research. Scientists often have a tendency to dismiss the animal rights philosophy as irrational. Yet the movement's leading philosophers, people like Peter Singer and Tom Regan, are respected scholars who present eloquently argued, and intensely rational, cases for their belief in animal rights. Inadequately prepared scientists can embarrass themselves and the larger scientific community when trying to debate some of the articulate, well-prepared leaders of the animal rights community. One will not catch these individuals in trivial moral blunders — they do not eat meat, wear leather shoes, or frequent the circus. Many of them struggle to live an ethically consistent (and difficult to maintain) lifestyle because of the moral status that they ascribe to animals. The fact that scientists are often unfamiliar with the ethical theories of the leading animal rights philosophers is bound to reduce their effectiveness in any public debate of the issues.

Scientists occasionally have a tendency to dismiss all animal rights activists as the members of a "lunatic fringe." This view is untenable. The vast majority of people in attendance at animal rights meetings are not lunatics, but rather people just like our neighbors. It is important to realize that most of the people at such meetings are not fervent animal rights activists. They continue to eat meat, value the benefits of medical research, and own pets, no matter what their leadership might have to say about these practices. These people are, however, *extremely* concerned about how the animals used in biomedical research are being treated. And unfortunately, their major source of information is often the animal rights groups themselves. Because of this, they are often inherently distrustful of the scientific establishment. It is not likely that any impersonal scientific organization is going to be able to quiet their fears without the help of large numbers of individual scientists explaining to their own neighbors exactly how they do biomedical research.

"Rights" for animals?

While most scientists would probably not claim that animals have rights, it is important to realize that we nevertheless act as though animals do have something *like* rights. It is worth spending a moment to consider why most working scientists support the use of animals in biomedical research and are also concerned that such research be conducted humanely. Likewise, while fairly large percentages of the general public (especially young people) express support for the concept of animal rights, they simultaneously eat animals and support the use of animals in biomedical research. Therefore, while it is apparent that nearly all of us perceive animals to be objects of moral concern, the exact nature and extent of our moral obligations are not entirely clear.

If asked to describe the difference between a test tube and a mouse, none of you would have any problem in doing so. Precisely how you choose to reply might well depend on whether you have been trained in biology, chemistry, genetics, etc. However, it seems likely that your initial answer would focus on the most compelling distinction between the test tube and the mouse — the fact that the mouse is a living creature. Now let's suppose that someone enters your laboratory with a hammer and smashes one of your test tubes. Clearly it would be wrong for them to do so. They would have intentionally, and senselessly, destroyed your property. To be sure, in these days of disposable culture tubes, the actual loss to you would be a small one. But now let us change the scenario and suppose that instead of destroying one of your test tubes, the person enters your laboratory to smash one of your mice. This act, too, would be wrong. But is it wrong for precisely the same reasons as the previous destruction of the test tube? The person has once again destroyed your property, and it is also true that the mouse is undoubtedly worth more in purely monetary terms. But is this the full measure of the difference between these acts? Few of us would equate the senseless destruction of a whole shelf pack of test tubes (to equalize the monetary value) with that of a single laboratory mouse.

It is important to understand that ownership of property is not the key issue. What if, instead of using the hammer to smash *your* mouse, the person in question used it to smash one of his own? How many would feel significantly better about the event? So what is the fundamental difference in the destruction of these two objects? Is it only the fact that the mouse is alive while the test tube is not? Then let's suppose that it is not a test tube that is about to be destroyed, but rather, a tissue culture flask full of living animal cells. Clearly, the senseless destruction of a mouse is more troubling than that of a flask of cells. Therefore, it is not the mere fact that the mouse is alive that we are responding to — it must be something else.

Moral judgments

At some level, many scientists are abolitionists. That is, if we were able to acquire the information needed to adequately answer compelling research questions without the use of animals, who among us would not gladly do so? Nevertheless, one of the best methods we have developed to advance biomedical knowledge involves the use of animals which, unlike the test tube, have interests. They have interests in obtaining sufficient food, in remaining free from pain, in reproducing themselves, and perhaps in living a normal life span. Experiments can frustrate the interests of laboratory animals, and most scientists recognize this both in their concern for the humane treatment of animals and in their belief that research should be directed at important problems. The fact that animals have interests does not necessarily mean that we should never use them in biomedical experiments; however, it does mean that any such use should be preceded by a moral judgment. Do the benefits derived from the biomedical research that is being considered offset the associated moral costs?

Animal rights groups are challenging the existing societal consensus on many questions involving animals. Their actions will undoubtedly have an influence on public policy — decisions that will be made whether or not scientists choose to participate in the ongoing debate over the issues.

Ethical Challenges to the Use of Animals in Research

Peter Singer and Tom Regan are the two most influential animal rights philosophers currently working. Each argues that society should radically restructure the moral status it grants to animals from his own ethical perspective, utilitarianism (Singer) or deontology (Regan). While chapter 2 provided an introduction to the utilitarian and deontological approaches to ethical decision making, we will now briefly consider how these well-known opponents of animal research apply them.

Singer's utilitarianism and animal "rights"

Peter Singer's book *Animal Liberation* (26) is credited with the modern revival of the animal rights movements when it was first published in 1975. There is a small irony in this because Singer, like utilitarianism's founder Jeremy Bentham before him, does not believe in the philosophical concept of rights. Although Singer uses the term "rights," he does not consider it to have philosophical meaning but instead to be a "convenient political shorthand" (28). Singer echoes an assertion made by Bentham that the key moral question related to animals is not whether they can reason but whether they suffer. For Singer, sentience — the ability to feel pleasure or pain — is the key characteristic required for admittance into the moral universe. Singer concludes that many animals can suffer from physical pain, deprivation, loneliness, etc., while fully acknowledging that humans can suffer in ways that animals cannot (e.g., the fear of a future catastrophe). Singer, again drawing from Bentham, proposes that a principle of equality requires that we give equal consideration to the suffering of individuals, regardless of their species. Failure to do so amounts to "speciesism," an offense that Singer finds analogous to racism or sexism (26). It is important to realize that Singer is not claiming that there are no relevant moral differences between humans and animals. Human children have an interest in learning to read. Therefore, it would be immoral for us to raise a child and intentionally prevent him or her from acquiring this skill. Clearly, such disapprobation is meaningless for animals, which have no interest (or capability) in reading. Nevertheless, Singer argues that both animals and humans have an equal interest in being free from torment. Because of this, he maintains that it is just as wrong to torture an animal as it is to torture a human being. But once again, this does not mean that Singer believes that all lives are of equal moral worth. He plainly states that if one is required to decide between the life of a human being and the life of an animal, that one should choose to save the life of the human (28). Singer can envision circumstances that might alter this decision. If the life of a normal animal is placed in the balance with that of a severely impaired human, the normal decision might be reversed and the life of the animal saved.

Thus, Singer does not say it is never appropriate to use animals in scientific research. As a utilitarian he *must* support such use if the benefits obtained outweigh the harm done. But Singer places an enormous barrier in the way of such research, one he believes will forbid essentially all of it. Since pain in animals and humans is viewed as exacting an equivalent moral cost, no animal experiment should be conducted unless it would also be permitted on a human.

> We have seen that experimenters reveal a bias in favor of their own species whenever they carry out experiments on nonhumans for purposes that they would not think justified them in using human beings, even brain damaged ones. This principle gives us a guide toward an answer to our question. Since a speciesist bias, like a racist bias, is unjustifiable, an experiment cannot be justifiable unless the experiment is so important that the use of a brain-damaged human would also be justifiable. (27)

It is clear that Singer does not believe that very much animal research would be able to overcome this obstacle. It is also clear that he does not believe this loss to be a serious one. He believes that "animal experimentation has made at best a very small contribution to our increased lifespan" (27). For Singer the benefits of animal research (or of meat eating) are not worth the moral costs.

Fellow utilitarian R. G. Frey of Bowling Green State University (7, 8) has criticized Singer's philosophy. Frey defends the use of animals in medical research using essentially the same utilitarian ethic as does Singer! In some of their writings it is difficult to understand where Frey and Singer differ in method, even though they differ radically in their conclusions. Frey, too, believes that animal research must pass a test similar to the one described by Singer. Frey believes that it would be wrong to perform an experiment on an animal if we were not willing to perform it on a human with an even lower quality of life (e.g., an orphaned infant born without a brain). However, Frey recognizes the benefits that flow from animal research and seems intent on preserving them. Therefore, while he maintains that we should be willing to perform such human experiments, he also recognizes reasons why we might choose not to. The side effects of such human research (e.g., societal uproar, outraged relatives) may outweigh the benefits derived and thereby cause us to refrain from conducting them in the first place.

Singer's claim that speciesism is analogous to racism has also been criticized. Peter Carruthers, a British philosopher and supporter of animal research, believes that species membership is a morally relevant characteristic (2), as do Stephen Post of Case Western Reserve University (22) and Carl Cohen of the University of Michigan (3). Animal rights philosopher Mary Midgley, who is clearly willing to demand limitations on the use of animals in research (15), also rejects the speciesism-racism analogy. She argues that "race in humans is not a significant grouping at all, but species in animals certainly is. It is never true that, in order to know how to treat a human being, you must first find out what race he belongs to. . . . But with an animal, to know the species is absolutely essential" (16). For Midgley there are morally significant bonds between species members just as between the members of a family. However, these species bonds are not absolute, and it is important to realize that we also form significant bonds with members of other species.

Regan's deontology and animal rights

Tom Regan, a professor of philosophy at North Carolina State University, is the author of *The Case for Animal Rights* (23). Whereas Singer rejects the philosophical concept of rights, Regan embraces it. He describes "the rights view" as a type of deontological theory distinct from that articulated by Kant.

> According to this theory, certain individuals have moral rights (e.g., the right to life) and they have these rights independently of considerations about the value of the consequences that would flow from recognizing that they have them. For the rights view in other words, rights are more basic than utility

and independent of it, so that the principal reason why, say, murder is wrong, if and when it is, lies in the violation of the victim's moral right to life, and not in considerations about who will or will not receive pleasure or pain or have their preferences satisfied or frustrated, as a result of the deed. Those who subscribe to the rights view need not hold that all moral rights are absolute in the sense that they can never be overridden by other moral considerations. For example, one could hold that when the only realistic way to respect the rights of the many is to override the moral rights of the few, then overriding these rights is justified. (23)

In his rejection of utilitarian ethics, Regan charges that the consequentialist philosophies make a mistake in viewing individuals as little more than a receptacle to be filled with pleasure or displeasure. Regan's analogy is that of a cup filled with a sweet liquid, a bitter liquid, or some combination of the two. He maintains that utilitarians ignore the value of the cup (the individual) and only concentrate on the liquid within it (pleasure or displeasure). Regan argues that individuals themselves possess a property that he calls "inherent value." Inherent value, according to Regan, is not dependent on the race, sex, religion, or birthplace of an individual. Further, it does not depend on the intelligence, talents, skills, or importance of a person. "The genius and the retarded child, the prince and the pauper, the brain surgeon and the fruit vendor, Mother Teresa and the most unscrupulous used car salesman — all have inherent value, all possess it equally, and all have an equal right to be treated with respect, to be treated in ways that do not reduce them to the status of things, as if they existed as resources for others" (24). Regan also claims that it would be blatant speciesism to insist that only humans have inherent value. He argues that many animals also possess it. But how does he decide which animals possess inherent value and which animals do not? Regan's test for the possession of inherent value is something he terms the "subject of a life criterion." This does not require that one merely be alive but also that one "have beliefs and desires; perception, memory, and a sense of the future, including their own future, an emotional life together with feeling of pleasure and pain; preference- and welfare-interests; the ability to initiate action in pursuit of their desires and goals; a psychophysical identity over time; and an individual welfare in the sense that their experiential life fares well or ill for them. . . ." (23). At the time he wrote *The Case for Animal Rights*, Regan seemed to think that all mammals over the age of 1 year possess inherent value. In more recent statements, he seems to believe that this range should be expanded considerably.

The claim that animals have inherent value seems to agree with our sense of moral intuition, and up to this point many of you may have found little to argue with. However, Regan's insistence that inherent value is a "categorical concept" is likely to prove more controversial. By this, Regan means that humans cannot be said to possess any more inherent value than any other animal. Either animals are in the category of beings that possess inherent value or they are not. "One either has it, or one does not. There are no in-betweens. Moreover, all those who have it, have it equally. It does not come in degrees" (23). When pressed to delineate the exact point of demarcation between those beings said to possess inherent value and those who do not, Regan deflects the question as essentially moot. "Whether it belongs to others — to rocks and rivers, trees and glaciers, for example — we do not know and may never know. But neither do we need to know, if we are to make the case for animal rights. We do not need to know, for example, how many people are eligible to vote in the next presidential election before we can know whether I am" (24). But Regan's position does not imply that he believes that there are no moral differences between animals and humans. If there are five individuals (four humans and a dog) who seek sanctuary in a lifeboat that can hold only four of them, what should be done? Regan believes that it is the dog that should be thrown overboard to die. He argues that while the inherent value of each of these beings is equivalent, the harm that would be done to them through their deaths is not. Humans have a much greater range of possibilities open to them in their lives than do dogs. Humans can experience joys and satisfactions that no dog will ever experience. Because of this, death forecloses far more potential opportunities for satisfaction in the human than it will in the dog. Regan argues that it would be allowable to throw even 1 million dogs overboard to save the humans because each dog's death, when considered one at a time, is less harmful than the death of a human considered one at a time.

One might imagine, from such a position, that Regan would be disposed to permit animal research that could save the lives of humans. However, Regan's position is, if anything, more severe than Singer's on the question of animal research. Regan states that his ethic requires the immediate abolition of all such research. Why isn't medical research seen as analogous to the lifeboat ethics described above? In the lifeboat example, all (including the dog) would have perished if one individual were not sacrificed. A decision had to be made as to whether a human or a dog had to die so that the others could live. Regan does not see that choice as analogous to using animals in research on human disease. The animals are not in the lifeboat, because they are not sick. No decision has to be made to sacrifice one or the other. In Kantian terms one can imagine that Regan believes that medical research uses animals merely as a means and not also as an end.

While Regan is quite comfortable with his abolitionist position, it should be noted that he, like Singer, does not seem to view the loss of the ability to use animals in research as having grave consequences for medical advances. Regan writes, "Like Galileo's contemporaries, who would not look through the telescope because they had already convinced themselves of what they would see and thus saw no need to look, those scientists who have convinced themselves that there can't be viable scientific alternatives to the use of whole animals in research (or toxicity tests, etc.) are captives of mental habits that true science abhors" (23).

Regan's views have been extensively criticized. Frey, the utilitarian philosopher and cautious supporter of animal research, questions the claim that animals have moral rights. As a utilitarian, Frey doubts the existence of moral rights in the first place, but his criticism extends beyond his philosophical viewpoint. (It can be hypothesized that Singer would largely echo Frey's criticism of the philosophical concept of rights.) Frey notes that the concept of moral rights is especially popular in the United States and that, in this country, the position in contentious social issues is often stated using rights language (women's rights, gay rights, children's rights). Often the opposing sides in a debate will each make appeals using rights language — the "right to life" versus "a woman's right to choose." Frey argues that, in the United States, for a group "to fail to cast its wants in terms of rights . . . is to disadvantage itself in this debate. . ." (10). In contrast, he observes that debates over the moral treatment of animals have proceeded in Britain and Australia with relatively little mention of rights.

Carl Cohen (3) has argued that animals are not the *kind* of creatures capable of possessing rights. He states that rights can only be accorded to "beings who actually do, or can, make moral claims against one another." Peter Carruthers criticizes Regan on a much more fundamental level (2). He claims that Regan has not adequately provided groundwork for his moral theory. Where are the rights he argues for supposed to have come from? What exactly is the inherent value which Regan claims is possessed (at least) by all mammals of 1 year of age or older? How do we detect inherent value; that is, how are we to determine which life forms have it and which do not? Carruthers accuses Regan of altogether failing to provide the kind of "governing conception" necessary to explain his moral theory.

Practical Matters: Constraints on the Behavior of Scientists

Overview

We have seen that there is no unanimity among those philosophers critical of the use of animals in biomedical research. Likewise, there is no unanimity among the philosophers who support such use. Frey (7–10), Carruthers (2), Leahy (13), and Cohen (3) all argue from their own philosophical perspectives. So while these readings can provide us with useful frameworks for thinking about ethical problems, those hoping for a simple consensus view on why it is morally permissible to experiment on animals will be just as disappointed as those hoping for a consensus supporting the opposite view. But we do not require a confluence of philosophical opinion to recognize that the use of animals in research entails a moral responsibility.

Table 6.1 Brief U.S. legislative and regulatory history

1960	Animal Welfare Institute initiatives lead to proposed federal legislation that would require individual animal researchers to be licensed. No legislation enacted.
1963	NIH publishes first voluntary *Guide for the Care and Use of Laboratory Animals*. The *Guide* was revised in 1965, 1968, 1972, 1978, 1985, and 1996.
1966	Congress enacts the Laboratory Animal Welfare Act in response to public outcry over a *Life* magazine article. Amended and strengthened in 1970, 1976, and 1985. The legislation is now called the Animal Welfare Act.
1985	Health Research Extension Act of 1985 requires the NIH to establish guidelines for the use of animals in biomedical and behavioral research. First animal law covering the USPHS.
1986	NIH Office of Protection from Research Risks (OPRR) publishes the *Public Health Service Policy on the Humane Care and Use of Laboratory Animals*. USPHS laboratories and any institutions wishing to receive USPHS funding must agree to comply with the *PHS Policy* and the *Guide*.
1996	Publication of the *Guide for the Care and Use of Laboratory Animals* by the National Academy of Sciences indicates the broad acceptance of the *Guide* within the U.S. and international animal research communities.

Legislation

Scientists no longer have the luxury, or burden, of being the sole arbiter of the acceptability of their own experiments. In the early days of animal research, there was little to restrict scientists' use of animals other than their own individual consciences. This is not the case today. Discuss animal care with any of the older scientists or animal care technicians with whom you work, and you will discover how dramatically the definition of acceptable treatment has changed over the years. Scientists work under a number of restrictions — legal, institutional, and moral — that constrain how animals may be used in experiments.

Table 6.1 presents a brief history of legislation and regulations pertaining to animal care and use. In 1963, the National Institutes of Health (NIH) published the first edition of its *Guide for the Care and Use of Laboratory Animals* (17). At first, compliance with the recommendations set out in the guide was voluntary. The movement to pass restrictive legislation on animal use gained momentum in early 1966 when an article in *Life* (1) caused public outrage by chronicling the despicable conditions under which many animal dealers maintained their dogs. In August 1966, Congress passed the Laboratory Animal Welfare Act. A major goal of this legislation was to require the registration of research facilities and dog dealers with the U.S. Department of Agriculture. A clear intent of the bill was to minimize the number of instances of people's cats and dogs being stolen and sold to research institutions. These institutions were now required to buy their cats and dogs from licensed dealers. This legislation was amended in 1970, 1976, and 1985 and is now referred to as the Animal Welfare Act. The legislation mandated humane care and treatment for dogs, cats, rabbits, hamsters, guinea pigs, and nonhuman primates. However, it provided no protection for rats and mice, the two species that account for the vast majority of all animals used in research.

Shortly before the last set of amendments to the Animal Welfare Act was instituted, Congress passed the Health Research Extension Act of 1985 (Public Law 99-158). This was the first law concerning animals under which the U.S. Public Health Service (USPHS), now the U.S. Department of Health and Human Services (DHHS), was required to operate. This law, in effect, caused the heretofore voluntary *Public Health Service Policy on the Humane Care and Use of Laboratory Animals* (19) to become mandatory for both USPHS/DHHS research labs and any nongovernmental institutions that received funding from any DHHS agency. The DHHS policy includes a number of key elements, one of which is an assurance obtained from research institutions stating that they are committed to following the DHHS policy and the *Guide for the Care and Use of Laboratory Animals* (17).

The *Guide for the Care and Use of Laboratory Animals*, often referred to simply as the *Guide*, is an important document for scientists and animal care personnel. While previous versions of the *Guide* were supported solely by the NIH and published by the Government Printing Office, the 1996 edition received support from NIH, the U.S. Department of Agriculture, and the U.S. Department of Veterans Affairs. The current edition was revised by an ad hoc committee of the Institute of Laboratory Animal Resources within the National Research Council and published by the National Academy Press. (The National Research Council is the operational arm of the nongovernmental National Academy of Sciences.) The broader financial support of this new edition, as well as its publication by the National Academy Press, gives some indication as to how widely the *Guide* is used by the animal research community.

The *Guide* describes details on how animal research should be carried out within an institution. Although the Animal Welfare Act itself does not address standards in regard to rats and mice, the *Guide* does include these species. It details a number of institutional policies that should be put into place concerning issues such as the qualifications and training of the professional animal care staff and the establishment of an occupational health program to protect personnel who come into contact with the animals. Other special considerations include policies discouraging the prolonged physical restraint of animals and the use of multiple major surgical procedures on a single animal. The *Guide* also addresses issues surrounding the animal facilities and housing requirements for laboratory animals. Minimum space recommendations are given in detail for a number of different species. (For example, it is suggested that a 20-gram mouse be allotted at least 12 square inches of floor space in a cage that is at least 5 inches high.)

Further, it is recommended that attention be given to the particular social requirements of the animal species in question. Communal animals should be housed in groups whenever appropriate, while taking into account population density, familiarity of individuals, social rank, etc. For highly social animals (such as dogs and nonhuman primates), it is suggested that group composition be held as stable as possible. It is also suggested that the environment of the animals be enriched to prevent boredom, especially when animals are to be held for a long period of time.

The physical environment under which the animals are maintained is also addressed in the *Guide*. Temperature and humidity ranges are given for a number of species, as well as suggestions for ventilating the animals' rooms (10 to 15 room air changes per hour). Levels of illumination are suggested because light that is within the comfortable range for humans can actually be so bright that it damages the retina of albino mice. In addition, the *Guide* discusses noise levels and requirements for bedding, water, sanitation, waste disposal, and vermin control. Veterinary care issues such as quarantine, separation by species, and disease control are discussed, as are anesthesia, surgical and postsurgical care, and recommended means of euthanasia. The *Guide* also addresses many aspects of the actual physical plant in which animals are housed and experimented upon. Recommendations are given for corridor sizes, animal room door sizes, ceiling heights, placement of floor drains, the surface material from which the walls should be constructed, and suggested locations of storage areas for food and bedding.

Institutional animal care and use committees

Both the Animal Welfare Act and the DHHS policy mandate the establishment of an institutional animal care and use committee (IACUC), which oversees the animal care and use program for each institution. The Animal Welfare Act and the DHHS policy differ somewhat in their minimal requirements for the committee. The Animal Welfare Act requires a committee of at least three people. The members of the committee are to possess "sufficient ability to assess animal care, treatment, and practices in experimental research . . . and shall represent society's concerns regarding the welfare of animal subjects." At least one of the committee members is to be a doctor of veterinary medicine and one member is not to be affiliated with the institution in any way (other than as a member of the IACUC). The nonaffiliated member is supposed to represent the interests of the general community in the proper care and treatment of animals. The nonaffiliated member cannot be an immediate family member of a person affiliated with the institution.

The DHHS policy requires a committee of at least five people. One of the members must be a doctor of veterinary medicine with training or experience in laboratory animal science and medicine. This individual must have direct or delegated authority and responsibility for the research activities involving animals at the institution. The committee must also include one practicing scientist with experience in animal research, one individual whose primary concerns are in a nonscientific area (e.g., clergy member, lawyer, ethicist), and one individual who is not affiliated with the institution in any way (other than as an IACUC member).

The *Guide* does not specify a minimum number of members for an IACUC (and so is compatible with the policies of institutions operating under the Animal Welfare Act or the DHHS policy) but suggests that the number should be determined by the size of the institution and the extent of the program. The *Guide* uses slightly different wording to describe the requirements for the members of the committee. This difference is most significant in the requirements for the nonaffiliated or public member. As in the other policies, the public member is not to be affiliated with the institution or a member of the immediate family of a person affiliated with the institution. Again, the public member is to "represent the general community interests in the proper care and use of animals." However, the *Guide* adds the requirement that the public member not be a user of laboratory animals. This requirement prevents an animal research scientist from one institution from serving as the public member on the IACUC of another institution.

IACUCs are often larger than the minimum size required and may have 10 or more members. The IACUC is charged with evaluating the institution's animal and care use program and animal facilities every 6 months and preparing a report on its findings. The IACUC also evaluates and makes recommendations regarding all aspects of an institution's animal program, including training of the personnel. The IACUC has the authority to suspend any activity that involves animals should it determine that the activity is not being conducted in accordance with the Animal Welfare Act or, if applicable, the *Guide*. Table 6.2 presents a comparison of the Animal Welfare Act, DHHS policy, and the *Guide* in their requirements for an IACUC.

Most scientists will interact directly with the IACUC when they submit a research protocol for approval. An approved protocol is required before any experiments involving animals, even pilot projects, are conducted. The NIH will not fund a grant that has not had its animal research protocol reviewed and approved. Graduate students, postdoctoral students, and technicians who work with animals must be operating under an approved protocol submitted by the laboratory's principal investigator. It is important that persons working under an approved animal protocol be familiar with that protocol to prevent accidental deviations from existing techniques that might require new approval before being adopted.

When preparing a protocol for submission, the investigator should use clear language and avoid the use of unnecessary jargon. The nonaffiliated

Table 6.2 Comparison of the Animal Welfare Act, DHHS policy, and the *Guide for the Care and Use of Laboratory Animals* in their requirements for IACUCs

Requirement	Animal Welfare Act	DHHS policy	The *Guide*
Minimum no. of members	3	5	Not specified (but minimum of three because of special requirements)
Special requirements for members	• 1 DVM[a] • 1 nonaffiliated	• 1 DVM • 1 practicing scientist • 1 nonscientist • 1 nonaffiliated	• 1 DVM • 1 practicing scientist • 1 nonaffiliated, non-animal researcher
Applies to rodent use	No	Yes, through reference to the *Guide*	Yes

[a]DVM, doctor of veterinary medicine.

public member of the IACUC should be able to understand what types of procedures are being proposed and why the research is important. The investigator should be careful to address the same topics that the IACUC will consider in its review. Most institutions have a form that will act as a guide for the process. The submission should discuss the rationale of the experiments, and the selection of the species should be justified. Alternatives to the use of animals (cell cultures, computer models, etc.) that were considered should be discussed. The investigator should explain why the use of these alternatives was rejected for the proposed study. Any steps that were taken to make the proposed experiments less invasive, or to make use of a species lower on the phylogenetic tree, should be explained for the committee. The investigator should justify the number of animals requested for the series of experiments planned. Whenever possible, this justification should include a statistical analysis to demonstrate that appropriate numbers of animals (neither too many, nor too few) will be used.

Along with a detailed explanation of the experimental procedures to be performed on the animals, the use of appropriate anesthetics, analgesics, or sedatives should be described. Description of the drugs used for these purposes, as well as the dosages and frequency of administration, should be detailed enough that the committee can determine that they are appropriate for the species and experimental procedures involved. An assessment of pain and distress anticipated can be useful for the committee. (Procedures that are painful in humans must be considered to be painful in animals unless evidence to the contrary is supplied.) The investigator must also describe the criteria and process that will be used to remove animals from a study, or euthanize them, if painful or stressful outcomes may be anticipated.

Postprocedure care of the animals should be described as well as the method of euthanasia or ultimate disposition.

The investigator should assure the IACUC that the experiments proposed do not unnecessarily duplicate previous work. The training and experience of the laboratory personnel in the specific procedures proposed should be discussed. The safety of the work environment and any precautions taken to protect laboratory personnel should be described.

Protocol review by the IACUC

When reviewing an investigator's research protocol, the IACUC must determine whether the proposed experiments are being conducted in accordance with the Animal Welfare Act and, if applicable, the *Guide for the Care and Use of Laboratory Animals*. The scientist must justify any departures from these guidelines to the satisfaction of the committee. The committee must ensure that protocols are designed to avoid or minimize discomfort, distress, and pain to animals consistent with sound research design. Any procedure that is judged to cause more than a "momentary or slight pain or distress" should be performed with the appropriate sedation, analgesia, or anesthesia unless the investigator can convince the committee that withholding such treatment is justified for scientific reasons. Animals that would suffer severe or chronic pain and distress that cannot be relieved must be euthanized. The committee must also ensure that the laboratory animals covered by a particular protocol will be housed under conditions that are appropriate for the species and will contribute to "their health and comfort" (19). A veterinarian, or other scientist trained and experienced in the care of the species being used, must direct the housing, feeding, and nonmedical care of the animals. A qualified veterinarian must provide medical care for the animals. Any means of euthanasia employed must be consistent with the recommendations of the American Veterinary Medical Association Panel on Euthanasia unless the investigator is able to justify any deviation on scientific grounds to the satisfaction of the IACUC.

In recent years, increasing thought has been given to how to assign ethical scores to animal protocols (20, 21). While this may be difficult to do with precision, it is clear that the committee can usually agree with relative ease on those proposals that are the most problematic. The IACUC is not restricted to simply accepting or rejecting the investigator's protocol. Often the IACUC will suggest alterations to a protocol that would make it acceptable. The committee may suggest a different anesthetic, or perhaps an alternative dose or schedule of treatment. The IACUC can draw upon the expertise of its various members in order to work with the investigator to see that both scientific and animal welfare concerns are met. Occasionally, investigators feel that the suggestions of the IACUC are intrusions into their scientific experimental design. This is unfortunate, but it is nonetheless the responsibility of the committee to ensure that all animal welfare concerns are satisfied. An investigator's attempt to justify a particular technique by using the argument that "this is the way that we have always done it" is not a sufficient rationale for an IACUC to approve a protocol that might otherwise be questionable. Likewise, it is not a sufficient rationale to claim that similar (or identical) techniques have been approved for use at other institutions. Each IACUC is responsible for making decisions on the protocols that come before it, and differences of opinion from one institution to another as to the acceptability of a specific technique are bound to occur.

Another important element of the Animal Welfare Act and the DHHS policy is the requirement that the institution provide training for those staff members involved in the care and/or research use of animals. This training is to include a discussion of humane methods of animal care and experimentation, techniques available to minimize the use of animals and animal distress, the proper use of anesthetics and analgesics, methods by which deficient animal care procedures may be reported, and how to use available services to learn more about appropriate animal care and alternatives to animal techniques. The National Research Council has prepared a book to assist in the development of such institutional programs (18).

Although the Animal Welfare Act has not been legislatively amended since 1985, some animal welfare or animal rights organizations have attempted to alter the scope and specifics of the act through judicial action. Given the current legal climate, it is impossible for a textbook to present a current assessment of the laws regulating the care and use of laboratory animals. The Animal Welfare Act has been amended in the past, and it is certain to be modified again in the future, whether by legislation or lawsuit. For current information, scientists will have to depend on the Division of Animal Care within their own research institutions. As we will discuss below, the relationship between animal care professionals and scientists will become an increasingly important one.

Beyond legislation

While laws define the minimum requirements scientists must follow in their care and use of animals, most scientists will want to strive for levels of care that exceed these minimums. The scientist's primary ally in this goal is the institution's Division of Animal Care or equivalent body. The veterinarians and animal care professionals employed by this department serve as a powerful resource to scientists. Using their knowledge can lead to both better animal care and better science.

In most instances, it will be these professionals who provide the training that is now mandated by law for those who are going to use animals in their research. New graduate students should be sure that they attend these training sessions as early as possible. Traditionally, the training in animal procedures for new graduate students has taken place within the laboratory of their chosen advisor. However, animal care professionals are better able to provide a comprehensive training experience than the old ad hoc system in place in most laboratories. In addition to this formal training experience, students should realize that their institution's animal care professionals could also be an invaluable resource when they are seeking to learn a new procedure or technique. In addition to being able to advise students as to what the law requires when, for example, performing rodent surgery, they will also be able to advise them on the appropriate surgical techniques, use of anesthetics, and postoperative care. This advice can ensure both that the animal does not suffer any unnecessary pain or distress and that the students obtain the best data possible from their experimental efforts.

Although it is the legal responsibility of the faculty advisor (principal investigator) to submit protocols to the IACUC, students would be well advised to look at the protocols under which they are conducting their research. Laboratory techniques often drift over time as personnel and experience change. Graduate students are likely to be in a better position than their advisors to see this happening and realize that it is time to submit an amended protocol to the IACUC. Additionally, scientists are required to consider the use of nonanimal alternative techniques before resorting to the use of animals for any procedure likely to cause pain or distress. Senior graduate and postdoctoral students are often on the cutting edge of technology and thus in an excellent position to make suggestions to their advisor for improving laboratory procedures. In 1959 Russell and Burch (25) enumerated three principles that should act as a guide for the humane use of animals in research. These are commonly referred to as the 3 R's: Replacement, Reduction, and Refinement.

- *Replacement* refers to the attempt to substitute insentient materials, or if this is not possible, a lower species that might be less susceptible to pain and distress than a higher species. Why sacrifice the life of a monkey for an experiment in which a dog would suffice? Why use a dog where a mouse would do? Why use a mouse if the research question could be answered using a cell culture?
- *Reduction* refers to the attempt to use the minimum number of animal lives necessary to answer the research question. To design an experiment in which the n of a treatment group is 25 in a situation where statistical significance could be achieved with an $n = 8$ is both economically wasteful and morally troubling. However, it is equally troubling to see an experimental design in which too few animals are used. If the group size is too small to permit any reasonable chance of

demonstrating a statistically significant difference, then the entire experiment is a wasted effort. There are techniques available to assist in the estimation of the appropriate numbers of animals to be used in an experiment (14). Additionally, one can seek the advice of a professional statistician before conducting a series of experiments, both to prevent the waste of animal lives and to ensure a more rigorous scientific study.

- *Refinement* refers to the attempt to reduce the incidence or severity of pain and distress experienced by laboratory animals. Use of anesthetics and analgesics that are appropriate for the species, as well as appropriate doses and intervals of administration, are all important. Additionally, use of trained personnel to perform experimental or surgical manipulations and effective postoperative procedures will improve both animal welfare and scientific validity. (Who would want pain introduced as an uncontrolled variable into their experimental design?)

Finally, it should be recognized that animal care professionals play something of a dual role within the institution. As we have discussed, they can serve as an invaluable resource to the research scientist. However, they also must ensure the welfare of the animals under their care. This role could potentially put them at odds with the research scientist. The animal care staff is also there to protect the animals from any researcher who refuses to observe the rules. This dual role can be stressful; they are at the same time advocates for both scientific research and animal welfare.

We should recognize that while the work we do is important and morally justified in the minds of most people, our system is not perfect and there are ways in which we can contribute to improved animal care. Each of us should be on the lookout for animals that are suffering, either from neglect or from abuse at the hands of a careless or poorly trained scientist. In some instances, the situation might be resolved by talking to the person involved. In other situations, a report might have to be made (formally or anonymously) to the head veterinarian of the animal care staff.

It is also beneficial to realize that there are moral inconsistencies in the way we relate to animals. Harold Herzog has written provocatively on this matter. He wonders why it is that we have strict rules for how we may use and euthanize laboratory mice, and yet we are allowed to catch and kill escaped mice in inhumane "sticky traps" (11). After once being accused (unjustly) by an animal activist of obtaining kittens from a local animal pound in order to feed his son's boa constrictor, Herzog began to think about the ethics of pet food. Is it more moral to raise a rat to feed to a boa than it is to use a kitten that is about to die anyway? For that matter is it any more moral to keep a kitten (an obligate carnivore) than it is a boa?

Political Realities: Then and Now

The political realities are such that there is no chance that scientists will be left to decide by themselves how laboratory animal welfare may be improved. Recent history has seen the rise of a number of well-funded animal rights groups that can be expected to press for legislative and judicial mandates to alter the existing procedures. While some of these initiatives will originate from a genuine concern to improve the treatment of laboratory animals, others seek to harass animal researchers until such time when the groups believe that they will amass the political might to see such research abolished. While scientists often like to believe that the animal rights movement consists of a lunatic fringe, such an assertion is not true and carries with it great danger. (For who seriously worries about the demented rambling of a group of lunatics?) Apart from a few apparently irrational statements made by the leaders of some animal rights groups, there is no evidence that the membership is anything other than a group of highly concerned citizens. It is important that we set aside the easy (and erroneous) explanations of the animal rights phenomenon and seriously consider who is involved in the movement and why.

A survey of animal rights activists attending a march in Washington, D.C., revealed a level of political activity that was termed "truly extraordinary" (12). The facts that 74% percent of those surveyed had contacted their elected representatives about animal rights and that 38% had made political donations to candidates supportive of such rights suggest a highly motivated and politically sophisticated activist group. The facts that nearly 14% of the activists reported having incomes in excess of $70,000 per year and more than 30% had in excess of $50,000 per year help to explain why the animal rights movement is so well funded. It is also important to realize that the typical animal activist is well educated — nearly 79% reported some college education, 47% a bachelor's degree, and nearly 19% a graduate or professional degree. An important distinction can be made between those organizations that are concerned primarily with the humane treatment of animals (animal welfare) and those that press for radical alterations in the predominant world view (animal rights). It is not always possible to identify with certainty a particular organization as being one or the other. It is not unusual for both sentiments to coexist within an organization. Not surprisingly, the more radical beliefs sometimes lead to internal inconsistencies between the leaders of the movement and the rank-and-file membership over issues such as the morality of pet ownership (12).

Many in the animal rights movement also display profound doubts about scientific enterprise. Fifty-two percent of the animal rights activists surveyed felt that science does "more harm than good." This opinion sets them dramatically apart from the general public, only 5% of whom express this belief. The activists view scientists in the same suspicious light reserved for other traditional authority figures, such as politicians and businessmen (12).

Further, it is a mistake to believe that this skepticism is limited to the benefits derived from scientific research or to the character of the scientists performing such work. Gary Francione, professor of law at Rutgers University and former legal advisor to People for the Ethical Treatment of Animals (PETA), has expressed mistrust of the scientific process itself.

> ... science no longer enjoys a position as epistemologically superior to other forms of knowledge. Despite the seductive simplicity of the traditional empiricist point of view — that science represents "objective" truth, the assumptions supporting this traditional view have been challenged effectively in recent years. Philosophers and sociologists of science have argued persuasively that factual assertions are completely contingent on theoretical assumptions, and that observation itself is subject to interpretation . . . .
>
> This recognition is slowly eroding the pedestal upon which science has presided for many years. More and more people in the animal rights movement, the environmental movement, and the alternative health care movement recognize that science is as value-based as any other activity. Indeed, there is increasing criticism of the fundamental premises of Western medicine. (6)

Francione has also challenged the "general view" that scientific inquiry is protected under the First Amendment to the United States Constitution. Francione's view is that the First Amendment provides very little protection to the conduct of scientific research although, somewhat paradoxically, the dissemination of the research results themselves is protected. "For example, under this analysis, the government could . . . prohibit all research involving genetic engineering as long as the purpose of the prohibition is not to suppress the dissemination of the information derived from such research" (4).

While it is not clear that Francione would be in favor of prohibiting all genetic engineering research, there is little doubt that he opposes all use of animals in scientific research. Statements made in Francione's concluding paragraph give us some insight as to why he appears to be so opposed to the concept of constitutional protections for research.

> ... It may be the case, however, that the federal government will, at some point, try to impose on all experimentation a risk/benefit regulatory structure similar to the one that Professor Dresser proposes. *Moreover, it is likely that even though experimenters find themselves with the federal (or other) funds to do an experiment, state and local governments may seek to restrict or even to prohibit such experimentation* . . . . (5) (emphasis added)

We may be seeing in such statements a strategy for political action from a movement that has been unable to convince a majority of society as to the legitimacy of its views. Although during the recent past the animal rights movement has succeeded in causing increasing numbers of the public to question both the validity and humanity of animal research, they have at the same time failed to build anything approaching a consensus for animal rights as they conceive of them. Thus it seems possible that in the future they may try to achieve, through targeted political actions in state and local arenas, what they have been unable to win through philosophical and political debate at the national level. Further, given the political savvy of the movement, this would not appear to be an idle threat. (Francione himself clerked for Supreme Court Justice Sandra Day O'Connor after completing law school.) While there is no possibility, in the foreseeable future, of the movement's securing a legal prohibition of animal research at the national level, things seem less certain at the level of local government. Imagine the impact of a local ordinance proscribing animal research within the city limits of a community such as Berkeley, California, or Cambridge, Massachusetts. The ordinance may not even be phrased in the philosophical terms of animal rights, but rather may appear to be primarily concerned with the alleged environmental impact or health risks to citizens that may be associated with animal research.

Given the political, financial, and human resources available to the animal rights movement, it seems unlikely that these activists will abandon their efforts to reform society anytime in the near future. Individual scientists have an important role to play by educating the public, beginning with family and friends, as to why it is sometimes important to use animals in research. The extent to which any scientist decides to become involved in the political and philosophical debate over animal rights is a matter of individual choice. Nevertheless, all scientists have an obligation to educate themselves about this issue, both to ensure ever-increasing standards of animal welfare and to ensure that society will continue to seek their counsel when searching for answers to this ethical dilemma.

Case Studies

6.1 You are beginning a new postdoctoral position at the same time that your mentor is moving her laboratory into a new building. She is obsessive about animal care and wants to ensure that the colony of animals to be established in the new facility is healthy. You are assigned the task of developing a system of "sentinel" animals to monitor the health status of all new incoming shipments of animals as well those in the established animal colony. You establish a system that involves the euthanizing of selected animals on a regular basis and screening for the presence of specific pathogens by a contract laboratory. Because these animals are not being used for research, do you have to submit a protocol to the IACUC to cover these activities?

6.2 A colleague is planning a project to isolate a protein factor that appears in the blood at a very low level. To facilitate the early stages of the project, she plans to make one trip each week to a local slaughterhouse to collect about 15 gallons of bovine blood after the animals are killed in the usual manner. Do you think that it is a good idea to use tissues collected from a slaughterhouse? Do you think that your colleague needs to submit a protocol to the IACUC for review?

6.3 A graduate faculty advisor has a predoctoral trainee in his laboratory who needs to raise antibodies in rabbits in order to analyze a specific protein. No one in the laboratory has ever used animals or antibodies in their research. The advisor instructs the student to do the necessary homework and determine what is needed to prepare a proposed protocol for approval by the IACUC. The predoctoral trainee does as directed and prepares and types the entire application himself. The predoctoral trainee presents the completed document to his advisor, who, in turn, reads and signs the document. He hands it back to the student, instructing him to submit it to the IACUC office for review. The mentor commends the student for the work and comments that this has been an important "learning experience" for the student. Comment on the mentor's responsibilities and actions in this matter.

6.4 You are a graduate student working on your Ph.D. Your advisor asks to meet with you to discuss your research project. Your advisor suggests a new series of experiments that will hopefully clear up a problem you have encountered. The new series of experiments involves surgical manipulations and you know that the IACUC protocol for the project did not contain any reference to surgery. You ask your advisor about submitting an amended protocol before starting these studies. Your advisor says that he does not wish to go through the trouble if the technique is not going to be useful. He suggests that first you try a few experiments, and if the procedure looks like it is going to work, you can submit an amended protocol at that time. What do you do?

6.5 A faculty member is conducting institutionally approved, NIH-sponsored animal research that involves a surgical technique performed on adult rats. Control rats from such experiments are mildly sedated during the procedure but receive no other medicinal treatment. They are euthanized at the end of the experiment and again examined with a minor surgical technique. An owner of a pet store approaches the faculty member and asks if he can have the euthanized control rats to use as food for his boa constrictors. The owner suggests that the faculty member file an addendum to his animal use authorization form to seek permission to give him the rat carcasses. The faculty member comes to you for advice. What do you tell him?

6.6 A colleague of yours is going to appear at a career day held at a local school. She will use mice to illustrate the effects of various psychoactive drugs (including drugs of abuse). The mice will be used in various experiments for which she has IACUC approval to perform in her lab. The experiments involve such endpoints as loss of balance, flicking of the tail out of the path of an uncomfortably warm light beam, and other standard assays. She wonders whether she should seek IACUC approval for this special appearance at the career day. She is worried that the IACUC approval may not come in time, and the submission represents a considerable amount of work. What would you advise?

6.7 You are a member of an NIH study section and are in the process of reviewing grant proposals for an upcoming meeting. One of your primary assignments involves a proposal in which rabbits will be used to raise antibodies. In reading over the protocol, you discover that the investigator will perform repeated bleedings of these animals to recover enough antiserum to do his experiments. The rabbits will not be euthanized and exsanguinated to recover the maximum amount of blood at the end of the experiment. Instead, the investigator states that the rabbits will be given away as pets at the end of the experiment. Is this proposal appropriate? If you are troubled by this proposal, what would you do?

6.8 You are invited as a guest faculty member to judge a local high school science fair. One entry you judge is entitled "Alcohol Addiction in Mice." The student has purchased six mice from a local pet store. One group of three of these mice has been caged and fed standard mouse chow and given drinking water ad libitum. The other group is fed mouse chow but is allowed water only once per day. This group of mice is instead given unlimited access to 20% ethyl alcohol. After 6 weeks, the student notes a significant weight loss in the latter group of mice as compared with the control animals. He also notes abdominal distention and states that the alcohol-fed mice ate significantly less food throughout the study. He concludes that the alcohol mixture depressed the animals' appetites. At the end of the study, he destroys the animals by cervical dislocation. You consult the school guidelines regarding the use of animals in science projects. The guidelines state that the use of animals in science projects is discouraged. However, animals may be used with permission of the science teacher. In this case, the student has sought and received such permission for his project. What comments, if any, will you offer to the student about his use of animals? Likewise, what, if anything, will you say to his teacher?

6.9   You are a member of your institution's IACUC. A protocol is submitted in which a researcher plans to perform footpad injections in mice using an antigen in complete Freund's adjuvant (CFA) to boost the antibody response. The IACUC used to approve protocols using CFA, but in recent years such use had been denied because of the pain and irritation it causes the mice. The IACUC denies the investigator permission to use CFA. The investigator appeals, arguing that she has just arrived from an institution that allows the use of CFA and she has years of data using the adjuvant. She maintains that she must continue its use so that she is able to make valid comparisons between her old and new studies. How would you respond?

6.10   Your colleague, Dr. Jay Mahata, is an NIH-supported investigator who has an established collaboration with a field biologist, Dr. Ellen Yu, in another state. Dr. Yu does not receive any grant support for her research. Dr. Mahata sometimes receives blood and other tissue samples for analysis from the wild rodents that Dr. Yu traps for her research. Dr. Mahata has asked you to read his latest IACUC protocol before its formal submission. You know about his collaboration with Dr. Yu but note that it is not mentioned in the protocol. When you ask Dr. Mahata about this, he says that he "does not have to report this activity to the IACUC because there are not any animal welfare concerns involved." He points out to you that he does not euthanize the rodents or collect the blood and tissues. He maintains that the relevant animal welfare concerns are between Dr. Yu and her institution. Last, he suggests that because the NIH does not support her work, it does not have to conform to the same guidelines to which his own work is subject. What do you do?

6.11   Dr. Martha Washington is very disappointed that the IACUC has rejected her research protocol because it involves the mouse ascites method of monoclonal antibody production. She appeals to the IACUC, citing her long use of this practice, prior approval to use the method at her previous institution in another state, and the loss of time that an immediate switch to in vitro methods would entail. She asks for permission to continue using the ascites method for 3 years while she phases in the in vitro production methods. The IACUC denies the appeal. She then resubmits the protocol, reporting that, since she has found a commercial source for the monoclonal antibody, she no longer needs to produce it herself. The protocol is quickly approved. Dr. John Louis, a member of the IACUC, has a conversation with Martha at a party a few months later. She tells him that her commercial source is a custom contract lab that she has engaged to produce the antibody using her cell lines and to her specifications (i.e., using the mouse ascites method). The next day Dr. Louis comes to you for advice. What do you suggest he do?

6.12 You are a graduate student working on a project that involves administering nerve toxins directly into the cerebrospinal fluid of rats by using a special infuser connected to tubing that you have surgically implanted into the base of each rat's skull. Administering different nerve toxins to block specific effects of different types of drugs will help determine how the drugs work. After surgery, the nerve toxin is given, and a few days later the investigational drug is given to determine whether it will have an effect. This protocol has been approved by the IACUC and is being funded by a grant from the Department of Defense. Over the past few weeks, you have carefully implanted a catheter into the base of each rat's skull, then infused the specified amount of nerve toxin. When you go to the vivarium to bring the rats to the lab to administer the investigational drugs, you find that a number of the rats are paralyzed or dead. You did not expect this. The lab director is currently out of town, so you go to the lab's senior graduate student, Tom, for advice. Tom will be able to complete his dissertation writing when this experiment is done, and he has made it clear that he wants this experiment to run without delay. You ask him whether you should stop the experiment to determine why some of the rats are dead or paralyzed. He responds that stopping the experiment now would waste several weeks of work and delay completion of his dissertation. Stopping now may mean having to start over later and could result in using even more rats. He further explains that the IACUC might even prohibit restarting the experiment, so the rats would have died for nothing because the data would have to be obtained another way. He suggests that the paralysis and death of some of the rats may be due to your inadequate experience performing rat surgery or infusions, so further practice by continuing this experiment may result in better outcomes for the rest of the rats on which you perform surgery. What do you do now? Do you continue performing surgery and infusions on the rats, knowing that more rats may be harmed? Do you stop the experiment and inform the IACUC, which risks earning the disfavor of Tom, with whom you have to work? How would you explain each course of action to the IACUC?

References

1. **Anonymous.** 1966. Concentration camps for dogs. *Life* **60**(5), February 4, p. 22–29.

2. **Carruthers, P.** 1992. *The Animals Issue: Moral Theory in Practice.* Cambridge University Press, Cambridge, U.K.

3. **Cohen, C.** 1986. The case for the use of animals in biomedical research. *N. Engl. J. Med.* **315:**865–870.

4. **Francione, G. L.** 1987. Experimentation and the marketplace theory of the First Amendment. *Univ. Penn. Law Rev.* **136:**417–512.

5. **Francione, G. L.** 1988. The constitutional status of restrictions on experiments involving nonhuman animals: a comment on Professor Dresser's analysis. *Rutgers Law Rev.* **40:**797–818.

6. **Francione, G. L.** 1990. Xenografts and animal rights. *Transplant. Proc.* **22:**1044–1046.

7. **Frey, R. G.** 1980. *Interests and Rights: The Case Against Animals.* Oxford Clarendon Press, Oxford, U.K.

8. **Frey, R. G.** 1983. *Rights, Killing, and Suffering: Moral Vegetarianism and Applied Ethics.* Blackwell, Oxford, U.K.

9. **Frey, R. G.** 1989. The case against animal rights, p. 115–118. *In* T. Regan and P. Singer (ed.), *Animal Rights and Human Obligations*, 2nd ed. Prentice-Hall, Englewood, N.J.

10. **Frey, R. G., and W. Paton.** 1989. Vivisection, morals, and medicine: an exchange, p. 223–236. *In* T. Regan and P. Singer (ed.), *Animal Rights and Human Obligations*, 2nd ed. Prentice-Hall, Englewood Cliffs, N.J.

11. **Herzog, H. A.** 1988. The moral status of mice. *Am. Psychologist* **43:**473–474.

12. **Jamison, W. V., and W. M. Lunch.** 1992. Rights of animals, perceptions of science, and political activism: profile of American animal rights activists. *Sci. Technol. Hum. Values* **17:**438–458.

13. **Leahy, M. P. T.** 1991. *Against Liberation: Putting Animals in Perspective.* Routledge, New York, N.Y.

14. **Mann, M. D., D. A. Crouse, and E. D. Prentice.** 1991. Appropriate animal numbers in biomedical research in light of animal welfare concerns. *Lab. Anim. Sci.* **41:**6.

15. **Midgley, M.** 1989. The case for restricting research using animals, p. 216–222. *In* T. Regan and P. Singer (ed.), *Animal Rights and Human Obligations*, 2nd ed. Prentice-Hall, Englewood Cliffs, N.J.

16. **Midgley, M.** 1992. The significance of species, p. 121–136. *In* E. C. Hargrove (ed.), *The Animal Rights/Environmental Ethics Debate: The Environmental Perspective.* State University of New York Press, Albany, N.Y.

17. **National Institutes of Health.** 1996. *Guide for the Care and Use of Laboratory Animals.* National Academy Press, Washington, D.C.

18. **National Research Council.** 1991. *Education and Training in the Care and Use of Laboratory Animals: A Guide for Developing Institutional Programs.* National Academy Press, Washington, D.C.

19. **Office for Protection from Research Risks.** 1986. *Public Health Service Policy on the Humane Care and Use of Laboratory Animals.* Office for Protection from Research Risks, National Institutes of Health, Bethesda, Md.

20. **Orlans, F. B.** 1993. *In the Name of Science.* Oxford University Press, Oxford, U.K.

21. **Porter, D. G.** 1992. Ethical scores for animal experiments. *Nature* **356:**101–102.

22. **Post, S. G.** 1993. The emergence of species impartiality: a medical critique of biocentrism. *Perspect. Biol. Med.* **36:**289–300.

23. **Regan, T.** 1983. *The Case for Animal Rights.* University of California Press, Berkeley.

24. **Regan, T.** 1985. The case for animal rights, p. 13–26. *In* P. Singer (ed.), *In Defence of Animals.* Basil Blackwell, Inc., Oxford, U.K.

25. **Russell, W. M. S., and R. L. Burch.** 1959. *Principles of Humane Animal Experimentation.* Charles C Thomas, Springfield, Ill.

26. **Singer, P.** 1990. *Animal Liberation,* 2nd ed. Avon Books, New York, N.Y.

27. **Singer, P.** 1990. Tools for research, p. 25–94. *In Animal Liberation,* 2nd ed. Avon Books, New York, N.Y.

28. **Singer, P.** 1990. All animals are equal..., p. 1–23. *In Animal Liberation,* 2nd ed. Avon Books, New York, N.Y.

Resources

Suggested Readings

In a chapter of this length, it has not been possible to consider all aspects of this complicated issue. For example, it has not been possible to place the modern animal rights movement in the proper political context with other modern movements. The animal rights movement finds itself alternately in agreement and at odds with the environmental movement, the right-to-life movement, and various areas of feminist political and philosophical thought. Additionally, no mention has been made of the use of illegal actions such as laboratory break-ins, arson, violence, and threats of violence by some members of the animal rights movement. The following annotated list will help the student locate additional material for in-depth reading on these and other issues related to the use of animals in biomedical experimentation.

Rowan, A. N. 1984. *Of Mice, Models, and Men: A Critical Evaluation of Animal Research.* State University of New York Press, Albany, N.Y.

A good overview written by Andrew Rowan, a biochemist, former director of the Center for Animals and Public Policy at Tufts University School of Veterinary Medicine, and currently Senior Vice President for Research and Education at the Humane Society of the United States. Rowan is neither an abolitionist nor an advocate for all animal research. His book covers the history, attitudes, and treatment of animals used in both research and education.

Orlans, F. B. 1993. *In the Name of Science.* Oxford University Press, Oxford, U.K.

Another middle-of-the-road book. Orlans is on the faculty of the Kennedy Institute of Ethics at Georgetown University and a former animal researcher at NIH. In addition to outlining the current political status of both animal rights and pro-research groups, she devotes chapters to the consideration of the workings of the IACUC, measurement of animal pain, and the controversial area of determining which species are capable of feeling pain.

Singer, P. 1990. *Animal Liberation*, 2nd ed. Avon Books, New York, N.Y.

Still considered by many the "bible" of the animal rights movement. Even those activists who consider Singer's position too moderate continue to use the book to win new converts for the cause. Singer combines easy-to-read philosophy with arguments for vegetarianism and critiques of animal research and factory farming.

Regan, T. 1983. *The Case for Animal Rights*. University of California Press, Berkeley.

Heavy on philosophy and difficult to read. Nevertheless, the book is important because most who call themselves animal rights activists embrace a philosophy much more like Regan's than Singer's.

Leahy, M. P. T. 1991. *Against Liberation: Putting Animals in Perspective*. Routledge, New York, N.Y.

This British philosopher views the ethical direction taken by the animal rights philosophers as incorrect and argues for a more traditional moral view of animals. Leahy critiques positions taken by Singer, Regan, and other animal rights philosophers.

Carruthers, P. 1992. *The Animals Issue: Moral Theory in Practice*. Cambridge University Press, Cambridge, U.K.

The University of Sheffield philosophy professor argues that while most books published recently seem to support the notion of rights for animals, this is not because it represents a consensus view among philosophers. As well as criticizing the position of the major animal rights philosophers, Carruthers describes a contractualist ethic that demands humane treatment for animals but does not grant them rights.

Sharpe, R. 1988. *The Cruel Deception: The Use of Animals in Medical Research*. Thorsons Publishing Group, Wellingborough, U.K.

Sharpe maintains that animal research has contributed little to human health and that it is generally invalid because its results are not applicable to humans.

Paton, W. 1993. *Man and Mouse: Animals in Medical Research*. Oxford University Press, Oxford, U.K.

A defense of the scientific validity and utility of animal research.

Jasper, J. M., and D. Nelkin. 1992. *The Animal Rights Crusade*. The Free Press, New York, N.Y.

An examination of the animal rights movement by sociologists.

Sperling, S. 1988. *Animal Liberators: Research and Morality*. University of California Press, Berkeley.

An anthropologist examines the animal rights movement in the light of its Victorian predecessors, feminism, and other contemporary movements in an extremely interesting and thought-provoking book.

URLs

Links to information about animal research:

The Association for Assessment and Accreditation of Laboratory Animal Care (AAALAC) administers a voluntary program that evaluates and accredits the laboratory animal care programs of various institutions:

> http://www.aaalac.org/home.html

American Association for Laboratory Animal Science (AALAS) is a professional association of veterinarians, technicians, and others dedicated to exchanging information and expertise in the care and use of laboratory animals:

> http://www.aalas.org/

American Society of Mammalogists. Guidelines for the capture, handling, and care of mammals:

> http://asm.wku.edu/committees/animal_care_and_use/ancarecomm.html

Animal Welfare Act and Regulations. Full-text versions of the Animal Welfare Act, amendments, and related documents at the USDA's Animal Welfare Information Center:

> http://www.nal.usda.gov/awic/legislat/usdaleg1.htm

Guide for the Care and Use of Laboratory Animals. An on-line version of the book published by the National Academy Press:

> http://www.nap.edu/readingroom/books/labrats/

National Association for Biomedical Research (NABR) advocates "sound public policy that recognizes the vital role of humane animal use in biomedical research, higher education, and product safety testing." The Foundation for Biomedical Research (FBR) is NABR's educational arm:

> http://www.nabr.org/
>
> http://www.fbresearch.org/

NIH Office of Protection from Research Risks. Laboratory Animal Welfare Page:

> http://nih.gov/grants/oprr/library_animal.htm

DHHS Policy. A full-text version of the 1996 edition of the DHHS Policy on the Humane Care and Use of Laboratory Animals:

> http://www.nih.gov/grants/oprr/phspol.htm

Scientist's Center for Animal Welfare (SCAW) is an association of individuals and institutions that promotes the humane care, use, and management of animals in research, testing, education, and agriculture. SCAW has started an on-line discussion area for members of IACUCs to "voice their opinions, questions and concerns." This discussion group, called IACUC Talk, is accessible at

http://www.scaw.com/

The Research Defense Society (RDS) was founded in the United Kingdom in 1908, as a society of doctors and medical research scientists, to inform the general public about why animals are used in medical research:

http://www.uel.ac.uk/research/rds/

Resources useful for IACUC members:

http://www.nih.gov:80/grants/oprr/tutorial/index.htm
http://www.nal.usda.gov/awic/pubs/oldbib/acuc.htm
http://www.aphis.usda.gov/ac/
http://www.nal.usda.gov/awic/pubs/oldbib/acuc.htm
http://oerweb.uthscsa.edu/iphcula/iacuc101.htm
http://netvet.wustl.edu/iacuc.htm

Managing Conflicting Interests

S. Gaylen Bradley

Introduction • Conflict of Effort • Conflict of Conscience • Conflict of Interest • Managing Conflicts • Conclusion • Case Studies • References • Resources

Introduction

S CIENTISTS, TECHNICIANS, AND TRAINEES are subject to conflicting demands on their time, have preferences on scientific approaches, have beliefs about social values, are competing for recognition in scientific achievement, and may possess information of substantial economic value. All members of the scientific community are faced with balancing conflicting interests. Most conflicts are resolved personally by subscribing to norms of the immediate scientific community, whether academic, industrial, entrepreneurial, or governmental. Some of these conflicts have become the focus of attention by employers, governing boards, research sponsors, government agencies, and the public. Different conflicts are controlled by different strategies, depending on the nature of the conflict and the concerns of the party with oversight responsibility.

The professional life of a scientist involves choices on what problems to study, what methods to use, which literature to cite, how to collect and organize data, how to interpret data, and how results and interpretations are to be communicated and to whom. The scientist also faces choices on how much effort to devote to various research projects, to teaching, to public service, to professional service, to actual research, to identifying new problems, to interpreting data, to publicizing achievements, to managing and coordinating research, and to the search for funds to support the research enterprise. Numerous factors influence the decision on how scientists expend their effort. Some assignments come from the employer. Some decisions are influenced by the reward system, and some reflect personal qualities and background of the individual. The reward system for scientists is varied, including personal income, job security, prestige, funds for research, recognition by the public, power, and a personal sense of accomplishment. Most of the factors that influence the choices and behaviors of scientists are accepted as normal considerations in the decision-making process. The scientist is expected to weigh the merits of rewards that are given for conflicting goals and to arrive at decisions independent of personal interests. In reality, this is an internally incompatible admonition. At the present time, scientists are encouraged to contribute to the economic development of the nation. Proprietary interests call for restricted access to research directed to products of commercial value whereas the scientific tradition calls for openness, free inquiry, and free exchange of ideas (6). A scientist is subject to a range of conflicting pressures that have different implications, including penalties for transgressions. These conflicting pressures may be categorized as (i) conflicts of interest, (ii) conflicts of effort, and (iii) conflicts of conscience.

Universities and their faculties have entered into business relations with the private sector for a number of reasons, many external to the university (20). The public and government have seen commercialization of research as a means to create jobs that contribute to the gross domestic product, thereby generating tax revenues; to attract domestic and international investment; and to restore a favorable balance of trade by decreasing purchases of foreign goods and products and enhancing purchases of domestic goods and products. Universities and faculty scientists have seen partnerships with business as a new source of revenue for research, for university infrastructure, and for discretionary funds. The search for new sources of revenue has been viewed as particularly important during a period when federal funding of research and research training has become increasingly competitive.

Contractual arrangements between industry and a university or an academic investigator not only raise questions about managing conflicts but may also change the overall intellectual climate in which academic researchers work (3). Universities and faculty members with financial interest in commercial ventures may lose objectivity in making decisions. A number of technical journals now require authors to disclose to the editor any financial interest in a company that might be, or could be construed as, causing a conflict of interest. These guidelines call for disclosure of sources of financial interests that could potentially embarrass an author if the interests become known, whether or not there is an actual conflict of interest (15). Academic science has extolled the virtue of free exchange of ideas, sharing of data to accelerate scientific progress, and maintaining the quality of science by critical peer review at all stages of the scientific method. Individual scientists and university administrators may feel less inclined to discuss research at early stages if there is a perceived potential that economically valuable intellectual property may be generated. Secrecy is viewed by many as contrary to academic science, a position taken by many socially conscious scientists as an argument against university-based research funded by military agencies. There are divergent views about the impact of secrecy on the progress of science. There are those who feel that progress is retarded by the failure to have free exchange of ideas and data. Others hold that the added resources for research with commercial value allow more workers to be recruited to the field, and that this accelerates achieving applied goals. There are some data indicating that research teams receiving the majority of their support from industry publish fewer peer-reviewed articles than those receiving modest amounts of industrial support. Moreover, there is evidence that papers published by investigators without any industrial support have greater scientific impact than those published by colleagues receiving support from industry (2).

Many science educators have expressed concern about the effects of industrially sponsored research on research training. One concern is that the attention devoted to scholarship with economic potential will lead research trainees to develop research strategies for short-term goals and modest extension of knowledge rather than formulating truly novel questions leading to major advances and changes in scientific thinking and problem solving. A related concern is that universities and faculty mentors will use research training to subsidize their commitments to industrial sponsors and will give less attention to nurturing curiosity and innovation. The fear is that mentors will prize well-executed routine studies over creative exploration that goes beyond tried and true methodologies. In fact, students, through their tuition and fees, and benefactors of the university may be unwittingly subsidizing commercial ventures.

Finally, there is a concern that this climate of secrecy and economic competition is contributing to a loss in confidence in the integrity of science and scientists, if not an actual deterioration in the quality of science. It should be noted that there is no established correlation between recent incidents of scientific falsification, fabrication, and plagiarism and economic conflict of interest. In fact, many of the procedures demanded in research for industry (for example, careful record keeping and review of results by a colleague) tend to prevent falsification and fabrication of data. Nevertheless, the perception that scientists today are less rigorous and less self-critical is widely held by the public, news media, legislators, and the scientific community itself.

Conflict of Effort

Members of the scientific community enter into research settings with defined expectations. A trainee expects to receive instruction, counseling, and guidance. A supervisor who has many obligations may not provide adequate direction as measured by the amount of time or quality of advising of the trainee. Faculty members are called upon to serve on institutional, professional, and civic committees; they also strive to excel in their scientific scholarship by writing papers and grants and presenting outside seminars and lectures; and they have assigned duties in teaching and administration. Unscheduled responsibilities such as mentoring research trainees often suffer in the face of multiple demands on faculty time. Trainees too are subject to multiple demands on their time; formal course work, examinations, financial obligations, and personal interactions compete with time spent designing, conducting, and analyzing scientific studies. Perhaps the most stressful conflicts of commitment for trainees relate to financial pressures and personal relationships.

Conflict of effort is distinctly different from conflict of interest, although the same set of external circumstances may precipitate both dilemmas. A conflict of effort arises when demands made by parties other than the primary employer interfere with the performance of the employee's assigned duties in teaching, research, and service. In general, scientists are expected to notify their employer of outside responsibilities, to seek permission in advance in most instances, and to report annually on outside professional activities, whether paid or not. Scientists with successful research programs are asked to present seminars and lectures at other institutions, at conferences, and at meetings. They are also asked to serve on editorial boards, research advisory panels, and policy advisory boards. They may be asked to teach in short courses and to offer methods workshops for peers or professionals in related fields. The university employer encourages some participation in these activities and uses them as criteria in evaluation for promotion, salary increases, and tenure. Good things can be carried to excess, however, and virtually every research-intensive university has a number of faculty members spending an unacceptable amount of time away from the campus. A conflict of effort is serious when the scientist is not available for scheduled classes, for student advising, for guidance of research trainees, for oversight of research projects and resource accountability, and for assigned administrative and service duties. Most universities allow 20% of a faculty member's effort or one day per week for consultation and outside professional activity. Some entrepreneurial faculty members try to define this limitation only in terms of paid consultantships and income generated outside professional activity and do not report professional service or speaking engagements that are unpaid or reimbursed for expenses. This is not the intent of policies on outside professional activities, which are more concerned about faculty effort than faculty compensation. There are those who believe that it is the neglect of, or inattention to, assigned duties at the employing institution that has allowed charges of scientific misconduct to come to public attention (1).

To avoid a conflict of effort, scientists ought to review their assigned duties with their supervisors, discussing the effort involved and the value to the department, institution, or profession. In general, the immediate supervisor (for example, a department chair) is responsible for orchestrating the resources of the unit and for the appropriate deployment of personnel. The immediate supervisor, however, is not usually the person with primary responsibility for making decisions on conflict of interest, although immediate supervisors have a role in alerting the administrator responsible for managing conflict of interest of a potential problem. Immediate supervisors may lack the legal knowledge to interpret conflict-of-interest regulations.

Some of the more difficult conflicts of effort also involve conflict of interest. Scientists who establish for-profit companies may experience increasing demands on their time that interfere with their ability to fulfill assigned duties. What makes these decisions difficult is that the faculty members may be on-site, but their effort may be directed to the interests of the private companies rather than toward the needs of the primary employer. In addition, the faculty member may meet scheduled assignments but arrive inadequately prepared. The faculty member may be inattentive to his or her advisory roles for students, staff, and research trainees. The immediate supervisor has the responsibility to counsel the faculty member about his or her concerns. After mutual agreement, if possible, on the extent of the problem, a date for a follow-up review should be set. If the faculty member and the immediate supervisor cannot reach a mutual accord, the matter may have to be considered by a grievance or disciplinary process.

Most conflicts of effort arise from the enthusiastic aspirations of scientists to gain acceptance from their peers and to achieve national and international stature as an investigator, rather than secondary to conflicts of interest. Universities in particular send mixed messages to young faculty, placing a premium on professional recognition. Faculty members usually respond well to discussion on the expected balance of effort among teaching, research, and service. It is too much to expect young scientists to find the proper balance without role models, mentors, and guidance.

Conflict of Conscience

Deeply held personal beliefs are appropriate determinative factors in career choices. Conflicts of conscience arise when scientists with deeply held personal views are asked to sit in judgment of projects whose very nature is unacceptable to the reviewer. A conflict of conscience does not involve financial reward or personal gain. In all likelihood, a conflict of conscience does not interfere with effort in assigned areas of teaching or research. A conflict of conscience arises when the convictions of an individual are allowed to override scientific merit in reaching a decision. A scientist who abhors abortion and the use of fetal tissue may be unable to act dispassionately on any manuscript or grant application that utilizes fetal tissue. A scientist who opposes all research using laboratory animals may be unable to find merit in any study or report that is based upon such use. These very personal views may not be known to colleagues at the same institution or elsewhere. Conflicts of conscience are not invariably viewed in a negative light. Scientists have occasionally refused to work on projects believed to be immoral applications of science; for example, development of infectious agents for biological warfare, or testing toxicity of drugs in uninformed human subjects such as mental patients, other institutionalized or incarcerated persons, or military personnel. To date, there is no agreement on whether or not to, or how to, manage conflicts of conscience (16). As with other biases in reviewing manuscripts and grant applications, it is likely that responsible leadership would try to identify and resolve any behavior that showed a pattern markedly at variance with other members of the deliberative process. Attempts to resolve conflicts of conscience as they relate to academic matters are apt to raise issues of abridgment of academic freedom. In the academic health science center, delivery of patient care is increasingly confronted with changing patterns of medical ethics with respect to premature births, resuscitation, life support systems, pain control, and suicide. Medical ethics committees have been formed in academic health care centers and other large health care systems, but similar committees or procedures to deal with scientific conflicts of conscience are very rare. Public interest groups are increasingly insisting on the right to sit on institutional bodies that review laboratory animal use, human subjects committees, environmental and occupational health and safety committees, and medical ethics committees.

Conflict of Interest

Orientation

Basic research workers have a tradition of free inquiry and free exchange of ideas, united in a shared purpose to create knowledge, to critique existing knowledge, and to disseminate knowledge. The image of the eccentric scientist lacking worldly aspirations and living in a cloistered ivory tower is giving away to that of a greedy entrepreneur, insensitive to the public good. Science and science administrators have promised, and the public has come to expect, products of research and technology that improve the quality of life. The public has called upon scientists to discover means to prevent or cure cancer, heart disease, mental illness, and AIDS and lavishes great rewards upon those who appear to achieve these goals. It is a small wonder then that many scientists have lost their innocence and fallen afoul of conflicts of interest.

Definitions

Conflict of interest is a legal term that encompasses a wide spectrum of behaviors or actions involving personal gain or financial interest. The definition of conflict of interest, including the scope of persons subject to the provisions in a code or set of rules and regulations, varies according to state and federal statutes, case law, contracts of employment, professional standards of conduct, and agreements between affected parties or corporations or both. A conflict of interest exists when an individual exploits, or appears to exploit, his or her position for personal gain or for the profit of a member of his or her immediate family or household. The identification of members of the immediate family and household is in a state of flux, but these individuals include the spouse and minor children living at home. Case law is evolving with respect to dependent parents and "significant others." Another critical component of conflict of interest pertains to the undue use of a position or exercise of power to influence a decision for personal gain. Many conflict-of-interest codes also prohibit activities that create an appearance of a conflict of interest. Full disclosure may be the only means to combat perceptions of undue influence. Conflict of interest is distinctly different from conflict of effort and conflict of conscience. Conflict of interest is also distinctly different from bias in research, which is the inability or unwillingness to consider alternative approaches or interpretations on their merits. Scientists sometimes develop strong preferences for particular research techniques or become deeply vested in a particular working model to the exclusion of alternative explanations. The origin of these prejudices may be subconscious, or at least unrecognized, reflecting past training, cultural background, experience, or group dynamics. Legislative bodies, governing boards, and the public have tended to define and specify penalties for conflict of interest in science by a unitary code. There is little recognition of a hierarchy of injury to the public well-being. Clearly, the public is harmed to a far greater extent when a conflict of interest is allowed to influence a clinical decision to market a drug for human use than when it is allowed to influence the decision to purchase an item of laboratory equipment from a particular vendor or to hire a relative to work in the laboratory of a scientist.

The changing climate

The federal government has taken a number of actions that encourage universities to enter into agreements with the private sector, thereby creating circumstances that ensnare faculty in potential or real conflict of interest. The 1980 Bayh-Dole Act (Patent and Trademark Laws Amendment, Public Law 96-517) allows a federal contractor to take ownership of the property rights for inventions created in the pursuit of a grant or contract. The Bayh-Dole Act specifies that income from the exploitation of these intellectual properties must be shared with the inventor and the remainder must be used for scientific research or educational purposes (5). The Federal Technology Transfer Act of 1986 extended the incentives for collaboration with industry to technology developed in a government laboratory. This act allows government laboratories to enter into cooperative research and economic development agreements with other governmental agencies and with non-governmental profit and nonprofit organizations. Income from inventions developed under such an agreement, or from other royalties negotiated with a commercial entity, are shared with the government inventor, and the remainder is to be used by the participating company for technology transfer (4).

The biomedical research enterprise expanded dramatically from the mid-1950s to the mid-1970s. There followed a period in which funding from federal agencies such as the National Institutes of Health (NIH) and the National Science Foundation remained at the same level when adjusted for inflation (7). The perception that federal funds for basic research were decreasing stimulated academic administrators to encourage research workers to seek funding from industry (14). Indeed, the amount of industrial money invested in academic research has increased from about $5 million in 1974 to hundreds of millions of dollars at the end of the 20th century (11). During the 1990s, health care costs and other demands on state and federal funds have increased sharply, forcing scientists and research administrators to look for alternative sources of funding. Concurrently, biotechnology has emerged as a significant economic force, with the potential to contribute substantially to the gross domestic product and to the international balance of trade. Scientists have been encouraged by government and academic employers to enter into university-industry ventures and to be entrepreneurs in commercializing new technologies. University employers have seen technology transfer as a new revenue stream to replace decreasing support from state and federal agencies. Local communities have developed economic plans in which research parks are means to provide jobs, tax revenues, and economic vitality for their regions.

Gifts and gratuities

Conflict of interest is usually thought of in terms of abuse of position for direct financial gain. It would be considered a conflict of interest if a scientist used his or her position to unduly influence the decision to buy supplies from a company in which an immediate family member held a direct financial interest. It is wrong to accept an expensive gift as an inducement to select a particular vendor, but the line between inappropriate inducements and acceptable gratuities is not unambiguous. Scientists have considerable influence on procurement decisions, including equipment and services. Vendors use a number of inducements to convince scientists and purchasing agents of the merits of their products or customer services. Exhibitors at national professional meetings hold breakfasts and receptions and give out carrying cases and a variety of mementos to establish product recognition in the minds of scientists. These modest gifts and gratuities have become routine, accepted, and expected. Vendors also give books and videotapes and host formal lunches and dinners. At some point, meals and entertainment cross from modest mementos to serious inducements. At the present time, frequent flyer credits are a widely used inducement about which different employers take different positions. There is no doubt that some scientists select an airline carrier according to accumulation of frequent flyer credits rather than cost or convenience. When this occurs, the scientist has allowed a personal interest to conflict with the interests of the employer (18).

There is no sharp boundary between gifts and compensation. Is the biomedical scientist who is fully reimbursed to attend a conference receiving compensation, a gift, or a gratuity? Scientific leaders are sometimes invited to attend a conference, not to give a formal lecture but to lend prestige and credibility to the program. Bench scientists may be invited to clinical conferences to lend the aura of solid scientific underpinning even though the scientists may have no direct experience with the drug or clinical trial. Local scientists may be invited to a conference to build community goodwill or to fill the audience, or both. There are no universally applicable guidelines to delineate the boundary between professional courtesy and a personal perquisite that has fiscal implications and the potential to influence a decision. Clearly, America's free economy relies heavily upon advertising, promotion, and inducements to influence purchasing choices. Scientists are confronted with the dichotomy that what is proper as an inducement to purchase a home television is usually not proper as an inducement to influence selection of a television monitor at work.

Compensation

Academic scientists are employed by their institutions to teach, to carry out research, and to render service to the institution, surrounding community, and the profession. The relative effort in each activity varies according to the mission of the institution and according to strategies to utilize effectively the talents of the faculty. Faculty members who are actively engaged in research have the opportunity to present their results to colleagues, including those who are employed by for-profit corporations. In general, universities encourage faculty members to present seminars and lectures at other research centers and condone payment of speaker's fees and full reimbursement of travel expenses. Scientists whose research bears upon commercial application of a product may be invited to conferences targeting groups that influence purchasing decisions. A scientist studying the mechanism of action of an antibiotic may be invited to participate in a conference sponsored by the pharmaceutical company distributing the antibiotic, targeted to physicians who will prescribe the drug. The scientist may be paid a generous speaker's fee or honorarium and provided luxury travel and lodging accommodations. There is a broad spectrum of speaker's fees, honoraria, and travel accommodations, some of which have attracted the attention of the Internal Revenue Service as well as the public. Honoraria and speaker's fees above a modest level are increasingly scrutinized by employers, especially academic administrators.

Consultantships are a formal agreement between a scientist and a corporation other than the primary employer, and usually with a for-profit company. Consultants have played critical roles in technology transfer, and academic scientists gain insights into the needs of industry for personnel and basic research. In general, consultantships have been beneficial to all parties: industry, the university, and the individual scientist. Consultantship arrangements are usually reviewed and subject to approval by the employer. There are a number of valid concerns about consultations, however. A scientist-consultant must not transmit to a private business any information, records, or materials generated as a result of research sponsored by benevolent foundations or governmental agencies unless the same information, records, or materials are made readily available to the scientific community in general. This guideline does not preclude appropriate contractual arrangements among the research sponsor, the research institution, and a private firm, particularly in the context of a licensing agreement (12). A consultantship should be based upon the collective knowledge and experience of the scientist and not constitute a means to gain access to privileged or confidential information available to the scientist by virtue of his or her employment or professional activities on advisory boards. A scientist-consultant must assiduously avoid the appearance of a conflict of interest whenever the employer is negotiating a contract with the private organization with which the scientist is a paid consultant. Scientist-consultants have the responsibility to disclose to their employers any agreements to perform consulting services. Moreover, scientist-consultants should not participate as evaluators of grant or contract proposals submitted by companies for whom they serve as consultants.

Multiple pay for one job

As relationships for the conduct of research become more complex, several sources of financial support are used to pay for research, especially that having potential commercial value. A university-based scientist, paid primarily by institutional funds, may conduct research on a project supported by a federal agency such as the NIH. In addition, the scientist may hold a paid leadership position in a venture company which has a contract with the university supporting research in the same laboratory for the same or closely related study. It may not be clear whether or not the scientist is being paid by his or her employer and by a for-profit corporation for technical guidance of the same research. Most employing universities insist on documentation that their employees are not being paid twice for the same job assignment. This usually involves documentation of the management role of the scientist in the leadership of the venture company.

Courseware

Scientists have the opportunity, even the responsibility, to disseminate information. Most often, this dissemination of information takes the form of instructional material: textbooks, computer programs, Web pages, and videotape. The copyright of scholarly scientific works traditionally has been retained by the creator until assigned to a publisher or distributor (see chapter 9). There is no trend to alter the practice of creators' retaining the copyright of technical manuscripts or of works unrelated to job assignments. However, instructional materials, especially those in electronic format, are now being generated as assigned work with considerable investment of resources by the employer (10). Sophisticated software and courseware potentially have substantial commercial value. Conflicts between creators and employers about distribution of revenue from the sale or licensing of electronic scholarly materials are increasing, and many research universities are currently revising their copyright policies to provide for revenue sharing. The issues being addressed include (i) the extent that the current employer may continue to use and share revenue from copyrighted instructional material after the creator leaves the institution or takes another assignment within the institution; (ii) the rights of the creator to use, sell, or license copyrighted instructional material, particularly to a competing organization; (iii) the rights of creators to restrict use of their voice and their personal images in electronic courseware; and (iv) the rights of the employer to assign other employees to modify, edit, and update electronic courseware. During this period of rapidly evolving practices, the employee-creator is advised to develop a memorandum of understanding with the supervisor and the employer's intellectual property officer.

Nepotism

Most state and federal agencies are subject to statutes or have rules that preclude a scientist from hiring or supervising an immediate member of his or her family or of the same household. These statutes are, in part, rooted in strategies to ensure fair access to employment opportunities. One of the most frequent nepotism practices is hiring high school or college-age progeny by an investigator, particularly for part-time and summer work. This practice is clearly contrary to equal access to employment opportunities and career development for underrepresented groups. In addition, selection of immediate members of the family for employment constitutes use of a position of authority for personal gain. Similarly, exercise of an investigator's authority to hire a member of his or her family has multiple implications for equal employment opportunity and personal financial benefit. The boundaries of propriety are not always well delineated. A few organizations prohibit members of the same family from working in the same department, even though neither party has direct authority over se- lection, promotion, or salary. With a growing number of two-career fami- lies, this limits the ability to recruit highly competent professionals to some institutions. A few institutions, on the other hand, have made concerted efforts to recruit two-career families. There are risks, however. The careers of the two individuals may not advance in parallel, and a two-career couple may make personal decisions about their relationship that cause tensions in the workplace. The organizational distance between members of a family or household in the workplace is not well defined. Is a faculty member per- mitted to select a member of the household of a departmental chair, dean, or vice president for a position in the faculty member's laboratory? The def- inition of member of the immediate family or household has occasionally been broadened to encompass individuals with a significant personal rela- tionship but who are not blood relatives or married. Nepotism regulations will undoubtedly remain in a state of flux as the goals of equal access to employment and career opportunities conflict with the career aspirations of two-career families. The American Association of University Professors has called for the discontinuation of policies and practices that proscribe the opportunities for the members of an immediate family from serving as professional colleagues (13).

Scientific conflict of interest

A successful scientist is afforded the opportunity to participate in the decision-making process that influences the allocation of resources. The peer review system, which is considered one of the essential safeguards for the quality of science, can be abused to serve a personal interest. Members of editorial boards have occasionally been accused of delaying publication of the results of a competitor in order to gain priority and recognition that strengthens applications for funding from granting agencies. Members of editorial boards have also been accused of being uncritical of manuscripts presenting results favoring a method or product in which the reviewer has a personal interest. Authors sometimes feel that reviewers have been unduly critical of manuscripts that describe in a favorable light products competing with one in which the reviewer has a personal interest. There is growing concern within the scientific community about the prudence of allowing employees of commercial firms to review manuscripts evaluating methods or products having economic value or potential. Some journals that publish articles related to commercial methods or products are asking both authors and reviewers to disclose their financial interests. Many scientists have felt that these requirements have impugned their integrity and in most instances believe that their financial interests are proportionally so modest that they cannot be considered a "substantial personal interest." Nevertheless, concern about the perception of conflict of interest is growing, especially in biomedical fields, and demands for financial disclosure by scientists are apt to increase.

Most grant review panels and advisory boards have established conflict-of-interest guidelines. The NIH asks individuals evaluating grant or contract proposals and applications to avoid participation in the review of submissions from organizations in which they (i) have a financial interest, (ii) are directors, officers, consultants, or employees, or (iii) are prospective employees or shareholders. The admonition extends to spouses and minor children, and even to circumstances in which there is only a perceived conflict of interest. Members of NIH study sections are not allowed to review applications from their own institution or those of a former student, professional collaborator, close personal friend, or colleague with whom the evaluator has long-standing or personal differences. If the excluded category is too large, the most knowledgeable reviewers are not allowed to participate in the decision-making process. The risk of inept evaluation by less-informed reviewers must be weighed against the adverse effects of a perceived conflict of interest. Unsuccessful applicants for research grants occasionally feel that a competitor on a study section has been unduly critical in order to gain an edge in recognition and future funding. The NIH has developed an appeals process to handle complaints of alleged unfair review of grant applications.

Scientists are increasingly being called upon to serve as experts for executive, legislative, and judicial deliberations. In addition, scientists have sought opportunities to testify before legislative committees that appropriate funds for research and higher education. Legislators, in turn, may sometimes view scientists as lobbyists or trade union representatives, advocating self-interest rather than the public interest. The scientific community itself is divided over the propriety of direct appeals to fund scientific projects outside of the peer review system by congressionally earmarked or "pork barrel" appropriations. Scientists engaged in expert testimony before the courts have encountered an adversarial culture unlike scholarly debate. Sometimes scientists are unwilling expert witnesses who have been subpoenaed to present evidence and research results (9). Scientists who willingly serve as expert witnesses for pay have been accused of conflicts of interest and of giving misleading information. Advisory boards of executive agencies have increasingly insisted on disclosure of past and present financial interests and have excluded persons with financial interests in the product or the company. The threshold for determining a perceived conflict of interest varies and sometimes is set so low that the guideline leads to the exclusion of scientists whose financial interests have been limited to speaker's fees and associated reimbursement of expenses. The assessment of the risk of advice from a less knowledgeable panel against the adverse effects of perceived conflict of interest is often made in the context of the political sensitivity of the issue rather than the needs of the decision-making process (19).

Academic conflict of interest

Academic scientists have special responsibilities to protect academic freedom, to disseminate knowledge, to maintain academic standards, to critique the current state of knowledge, to synthesize existing knowledge, and to apply knowledge to solve basic and applied problems. Faculty members are increasingly called upon to link the educational process to fund-raising and revenue-generating enterprises. Research faculty members are sometimes encouraged to market their expertise by organizing and presenting profitable workshops, particularly for business firms, under the auspices of the university. In other instances, faculty members have independently developed for-profit short courses and used the net earnings as a source of personal income. At some point these entrepreneurial activities, which are not restricted to academic scientists, have the potential to constitute a conflict of interest in which the faculty members utilize the reputation and even the resources of their employer for personal gain. In addition, the time and energy devoted to these activities may lead to a conflict of effort. Corporations and wealthy individuals may want to use their resources to influence the direction of academic programs. An agribusiness corporation may want to endow a chair in human nutrition, and a grateful patient may want to endow a chair in transplantation biology. Universities have developed sophisticated infrastructures to enhance these sources of support. Universities give prizes to alumni and business leaders, not totally without some consideration that the grateful recipients will generously support the university in the future. Gifts that are consistent with the mission of the university are aggressively sought. Agreements that proffer undue personal benefit to the donor, the university, or an employee of the university may constitute a conflict of interest.

Academic degrees have economic value and, not uncommonly, progress toward completion of a degree becomes an issue in a conflict of interest. For example, a faculty advisor might extend the course of study of a student to benefit a corporate sponsor. Companies sometimes use opportunities to obtain advanced degrees as an employee benefit and perquisite to enhance retention. Companies usually place limits on the time that they will pay for educational leaves or release time. The duration of educational leaves is usually inadequate for the average student to complete the degree program in the expected depth. The student and the student's supervisor in the company sponsoring the educational leave may put pressure on the advisor to make exceptions and to waive requirements. These pleas may be linked to hints of benefits to the advisor and institution once the employee graduates and returns to his or her regular or more influential position in the company.

Insider trading

Scientists conducting research sponsored by industry or who are engaged in consulting usually have completed confidentiality agreements. The scientist agrees to avoid discussing proprietary information in the presence of unauthorized parties, including family members and friends. Proprietary information includes, but is not limited to, the company's future plans and ideas, trade secrets, financial information, technical and research data, operating strategies, internal business processes, and technologic improvements that are not generally known to the public. In the course of proprietary research or consultation, a scientist may become aware of information relating to the economic value of a product or potential product. A toxicologist, for example, may be involved in a project in which a serious adverse effect of a marketed drug is discovered, and this result will jeopardize the continued approval of the drug. A chemical engineer who is consulting for a drug manufacturer may learn that the last hurdle to large-scale production and formulation of a new, much-needed drug has been overcome. By virtue of the paid relationship between the scientist and the company, the scientist is an "insider" and is restricted from using confidential information to personal advantage; that is, to sell stock of a company whose drug faces liability suits or loss of market share or to buy stock of a company on the verge of introducing a highly valued new drug (8).

Institutional conflict of interest

Institutions acquire financial interests in the private sector through (i) earnings on intellectual property, (ii) exclusive contracts with industry, and (iii) equity ownership in a for-profit company. In general, the interests of scientist-inventors and their employing institutions are congruent with respect to earnings on intellectual property. When the scientist and the institution share in revenues based upon a predetermined rate, the more successful the product, the better each fares. There are several areas in which the scientist and the institution may have conflicting interests. The scientist may seek a generous consulting arrangement as part of a licensing agreement. The institution may have limitations on this type of consulting arrangement or may seek other concessions from the company seeking a license at the expense of the scientist's self-interest. Similarly, scientists may seek research and development funds for their laboratories as a part of licensing agreements. This entails assigning rights of first refusal to the licensing company, a commitment about which the scientist and the employing institution may have divergent views. In addition, the institution may have restrictions on this type of grant or contract, particularly if it involves assessment of the efficacy of the invention or product. Moreover, the institution and the licensing company may feel that the invention will be developed more rapidly and to a greater extent without the parallel participation of the inventor.

Institutions have been entering into exclusive contracts with industry to give preferential access to research results to a company. The company usually awards the institution a large multiyear umbrella award. Invention disclosures are called to the attention of the sponsoring company, and technology transfer officers of the university may encourage scientists to work in areas of interest to the company. Several conflicts are arising from these blanket agreements between a company and an institution. Any one company, regardless of its size, has a reasonably well-defined scope. Scientists whose inventions lie outside the interest of the company may not receive adequate assistance from their employing institution in patenting and licensing efforts. There is a potential conflict between scientists whose work is supported by other commercial firms and the institution, which is striving to fulfill its contract to the company with an exclusive agreement. There is growing concern, too, that funds from government agencies and from tax-exempt foundations are being used to subsidize preferentially the research and development of for-profit companies, many of which are foreign owned.

Equity interests

Members of the academic scientific community are receiving conflicting admonitions from government, employers, and the public (15). Scientists are urged to accelerate the transfer of basic science knowledge into application and commercialization. The public has expressed concern that science is not sufficiently responsive to public need and that the lag from laboratory discovery to application is too long. National, state, and local governments and business communities have turned to research as the means to maintain economic competitiveness. Scientists quickly learn that most of their research discoveries with potential for commercialization require substantial development before established industry is willing to invest in university-generated intellectual property. Scientists who are convinced of the market potential of their inventions soon find that the patent process and product development are expensive and time-consuming. In addition, most scientists lack experience in writing a business plan and in securing venture capital. Quite often, the scientist will enter into an entrepreneurial corporation as an equity owner. Scientists inevitably feel that they are the most qualified to lead the technical development of the invention. It is at this stage that concerns about conflict of interest arise. In general, public institutions restrict the circumstances under which scientist-entrepreneurs may receive grants or contracts through their universities from a corporation in which they are in management positions or equity owners or both. Private institutions usually have fewer restrictions on faculty entrepreneurship than do public institutions. It is imperative that faculty entrepreneurs disclose possible conflicts of interest to their administration. Failure to do so, or the intentional withholding of information about potential conflicts of interest, constitutes a violation of the rules and procedures of most universities (1).

Universities too are being offered equity interest in entrepreneurial ventures involving faculty members. A research institution that accepts an equity position in a start-up company is likely to offer encouragement to the scientist-entrepreneur at critical times. In addition, the investors are not depleted of cash necessary for successful development and marketing of the product, and the ultimate return to the university has the potential to exceed income from royalties and licensing fees. University administrators, in such circumstances, find themselves called upon to make decisions in which the interests of the venture corporation and those of the university faculty may not be identical. University administrators may become unduly interested in the economic success of the venture company, even at the expense of educational responsibilities of the university. There is also a question of whether or not an institution that holds equity in a commercial venture will allow that financial interest to influence staffing decisions or other allocations of resources. When a position becomes vacant, will the employer preferentially seek candidates who will contribute to the development of the product in which the employer has an interest? When decisions on the allocation of limited resources for the purchase of equipment are made, will research administrators favor those units working on proprietary projects in which the employer has an interest?

Universities are under increasing pressure to take equity in start-up companies based upon the intellectual property of one of its own faculty members. The institution may provide release time for the faculty member, technical assistance for the project, and access to equipment and other research infrastructure in return for substantial ownership in the company. Proponents of equity ownership by institutions emphasize that this is an inexpensive investment with the potential for enormous economic returns. Opponents of equity ownership by institutions argue that institutional resources are diverted to the personal benefit of one or two scientists and the investors in the venture-capital deal. The equity-owning institution has an exceptional interest in the success of the venture and may use its research and public relations resources to promote a venture without adequate safeguards on fiduciary responsibility or critical scientific peer review. It is clear that equity ownership of companies based upon the research of the scientists of an institution is coming under increased public scrutiny, legal challenges from other members of the institution, and restrictive regulations from federal funding agencies (17).

Institutional prerogatives

Universities have a strong sense of self-preservation or self-protection when confronted with issues that are likely to have a major adverse effect on the institution. Universities are reluctant to cancel lucrative contracts when a faculty member is found to have a serious conflict of interest. The reputation of leading research universities is based upon their extramural support and achievements that attract positive public attention, such as patents, prizes received by faculty, and scientific breakthroughs of general interest. Universities are doubly threatened by scientific misconduct: there is the potential loss of grant funds and the loss of prestige. In addition, an investigation of scientific misconduct is expensive. As a result, universities are not eager to invite complaints of scientific misconduct or conflict of interest.

The bureaucrats within the university are reluctant to be drawn into proceedings pertaining to scientific misconduct or conflict of interest. Administrators are insecure about their mastery of the process, are fearful of political repercussions within the institution when a distinguished scientist is the subject of a complaint, and are anxious about criticism from news media that frequently focuses on individuals rather than issues. Colleagues within the university, too, are reluctant to become involved in deliberations about conflict of interest or scientific misconduct because it is perceived as taking sides with the complainant or the alleged perpetrator. Scientists are also aware of the potential financial damage to their institution and the negative effect on the institution's image and feel some need to protect their employer and to attenuate adverse effects of the allegation.

Some argue that universities have failed to take the lead in addressing scientific misconduct and conflict of interest. This idea has been supported by perceptions of news media and some legislators suggesting that universities have been inept and even recalcitrant in assuming responsibility for the behavior of the members of their community. Universities are particularly concerned about the increasing administrative responsibilities assigned to them by state and federal governments, because many of these requirements are perceived as placing university administrators at odds with the attitudes and aspirations of their own scientists. There is little doubt, however, that the public and legislators are increasingly insisting that universities accept responsibility for monitoring the integrity of the science carried out by their employees and trainees, and for the personal interests of employees that may affect the independence of decision making. Judgments on these complex issues are best vested in those who understand the normative standards of the discipline and the particular environment in which the conduct being examined occurs.

Managing Conflicts

It is neither possible nor desirable to avoid all conflicting interests. Successful scientists have many requests for their time, expertise, and attention that compete with their primary missions of creating new knowledge and synthesizing critically, evaluating, and disseminating existing knowledge. Nevertheless, participation in the peer review process, in formulating public policy, and in coordinating activities of their employing organization are important responsibilities of a scientist. Research workers will find it useful to, and have the obligation to, discuss these competing demands with their supervisors and colleagues to determine an appropriate balance between personal scholarship and professional service. In some instances, an employer will decide that selected outside activities are in the best interest of the organization, and it will encourage and reward the scientist for these activities. In other circumstances, the employer may place a higher priority on managing a research program, supervising junior workers, and productivity and will discourage outside activities that detract from these internal goals. Although scientists have considerable latitude in personal interpretation of normative standards for outside professional activity, the supervisor has the responsibility to ensure that allocation of effort is consistent with the guidelines of the organization.

There is a wide range of policies and practices among institutions pertaining to financial return on outside related professional activities. A research worker and the supervisor need to discuss the guidelines for speaker's fees, consulting fees, and other financial incentives. In some organizations, fees for outside professional activities are collected by the unit and redistributed as part of the reward system. More commonly, the research worker is permitted to collect speaker's fees and consulting fees with some sort of disclosure and approval process. Moreover, there is a growing concern that consultants may not be equally critical of products marketed by their benefactor or competitors of their benefactors, and authors and speakers are increasingly asked to identify financial relationships with commercial firms.

It is in the area of commercialization of the intellectual property of a scientist that the rules of the game are evolving most rapidly. Employees of public institutions are subject to conflict-of-interest statutes. Many of these statutes have been amended during the past decade to allow personal interests under stipulated conditions. Virtually all research-intensive institutions have developed policies and procedures for disclosure of potential conflicts of interest and for developing safeguards and processes for managing conflicts of interest. These range from barring the individual with a conflict of interest from participating in certain decisions to establishing an oversight committee that periodically monitors activity for bias in personnel utilization and interpretation of experimental results. The latter approach may be viewed as intrusive and adversarial, but properly implemented it protects the integrity of the relationship between the scientist and commercial sponsor and adds value to the quality of the research program.

Conclusion

Conflict of effort pertains to allocation of time on behalf of the primary employer. Although conflict of effort may arise from the same activity that creates a conflict of interest, more often, a conflict of effort arises from diversion of the commitment of an individual by requests to engage in public service and outside professional activities. At some point, service on advisory boards, governing boards of professional and public organizations, and editorial boards and participation in seminars, symposia, conferences, and workshops will impair the ability of individuals to meet their responsibilities to their employer, subordinates, trainees, and colleagues. Conflict of interest is an umbrella term for a wide range of behaviors and circumstances. Conflict of interest at some level involves use of position or authority for personal gain. Although most attention has been directed toward personal financial gain by individuals, it is also true that universities and other corporations may engage in practices that create a conflict of interest between the organization and individuals, most often its own employees, or with other corporations.

Some financial conflicts of interests are obvious. Others are not necessarily obvious and are defined by statutes. Still others are gauged by normative professional standards that vary with time or across disciplines. Various arbitrary thresholds have been established in statutes, institutional guidelines, and federal regulations that define the level of a financial interest that creates a conflict of interest. Some laws may forbid activities or entering into contracts that create a conflict of interest; for example, an employee of a state agency may not receive more than $10,000 in compensation from an outside contractor doing business with that state agency. Most often, the individual is required to disclose a financial interest that may be perceived as creating a conflict of interest. Increasingly, a symposium speaker receiving a consulting fee from a pharmaceutical company is required to disclose that arrangement to the organizers and audience as a prerequisite to participation in a conference addressing the merits of the company's commercial products.

Scientific conflict of interest involves the use of position to influence decisions on publication of manuscripts, funding of grant applications, and formulation of regulations on the use or commercialization of a product. There is no general agreement at this time on the circumstances that create a scientific conflict of interest. With increased emphasis on commercialization of intellectual property generated by academic scientists, there is growing concern about the effect of financial interest on the direction and interpretation of results. Can a scientist who holds a patent on a polymerase impartially compare the efficacy of that polymerase to a competitor's polymerase when the conclusions will affect royalty income? Will an advisor with substantial funding from the private sector allow trainees the opportunity to explore their own ideas that may not directly relate to the industrial project? Should a scientist employed by a pharmaceutical company be appointed to the editorial board of a journal that publishes articles on the efficacy of therapeutic agents? Should a scientist review the grant application of a collaborator or a competitor? Clearly, the definition of scientific conflict of interest cannot be made so broad as to exclude from the evaluation process most individuals knowledgeable in the field.

Institutional conflict of interest is less well defined than individual conflict of interest. In general, employees assign the rights to commercialize their intellectual property to the employer. The institution has the responsibility for managing the potential conflicting interests of the faculty entrepreneur with respect to supervision of trainees, use of institutional resources, and segregation of projects funded by other sponsors from those funded by the personal venture. An institutional conflict of interest arises when the interests of a university diverge from those of its faculty and staff. Most notable is an exclusive contract between a university and a corporation, giving the corporation preferential access to research results. Universities are increasingly becoming co-owners of companies established to commercialize the results of faculty research. There is growing concern in some sectors that this commitment to economic development is leading universities away from their traditional roles as educational and scholarly sanctuaries.

Case Studies

7.1 Dr. Quick, an internationally renowned biomedical scientist, is an authority in her field. She serves on an NIH study section (10 days per year), a state advisory panel (10 days per year), a federal advisory panel (10 days per year), an international advisory panel (10 days per year), and the board of a private foundation (5 days per year) and gives about one seminar a month at research universities across the country. What conflict-of-interest or conflict-of-effort considerations, if any, apply to this scenario?

7.2 Dr. Pride is a consultant to Pest-Free, Inc., a for-profit firm that manufactures a particular pesticide. He is paid $12,000 per year for his services. As an internationally renowned authority on the toxicology of this pesticide, Dr. Pride is asked to serve on a federal advisory panel that is to address the need to regulate the use and sale of pesticides. What concerns might Pest-Free, Inc., and regulatory agencies have about Dr. Pride's overlapping responsibilities?

7.3 Ms. Lean is a single parent working and studying in the laboratory of Dr. Much at Comprehensive College. She is working on a federally funded project that uses lasers for the noninvasive monitoring of glucose in diabetic subjects. Ms. Lean is having difficulty meeting her financial obligations on her student stipend. Sugarstat Company is a local firm that manufactures an adhesive patch for monitoring glucose in the skin and perspiration. The laser technology, if successful, would compete with the patch technology in the marketplace. Sugarstat Company offers Ms. Lean a part-time position in its quality control laboratory, where she would measure the reliability of the packaged patches. Ms. Lean accepts the part-time position, working 20 hours per week, mostly on Friday evenings and weekends. What conflicts, if any, does Ms. Lean have? Discuss Ms. Lean's responsibility in notifying Dr. Much of her part-time position. Would it be different if Ms. Lean planned to work during the week rather than on weekends?

7.4 Dr. Ami and Dr. Eros were independently recruited to Superior University. Dr. Ami and Dr. Eros, after several years of professional association, develop a romantic relationship that leads to marriage. Subsequently, Dr. Eros becomes the head of Dr. Ami's department. What conflict-of-interest considerations, if any, apply to this scenario? What options are available to this two-career family?

7.5 Dr. Cox at Research University is a renowned clinical pharmacologist. He has a large, productive research program in experimental therapeutics, sponsored by federal grants and industrial contracts. His reputation is such that his participation in a conference attracts news media attention and a large attendance. Big Company engages Dr. Cox for a series of lectures at conferences designed to promote Big Company's new therapeutic agent. Dr. Cox is paid a speaker's fee of $25,000 and all expenses for each presentation. Dr. Cox is free to select his topic and the contents of his presentation. Dr. Cox makes five to six appearances per year at the Big Company's conferences. Comment on potential conflicts associated with the lecture series.

7.6 Ms. Jobs is completing her degree at Research University. She has conducted some successful and exciting research in the laboratory of Dr. Keene. Dr. Keene's project was supported in part by a research contract with Innovations, Inc. Dr. Keene and the members of his laboratory developed new, rapid, accurate assays that can be adapted to kits for direct sale to the public. Innovations, Inc., is considering developing and marketing these kits but has not made a definite decision. Leaper Enterprises offers

Ms. Jobs a position in a new unit of the company to apply her training to develop kits based upon the technology that she learned and helped develop in Dr. Keene's laboratory. Discuss any conflict that Ms. Jobs may have in accepting a position in a company that competes with Dr. Keene's sponsor. How is the situation altered if Ms. Jobs was paid or not paid by funds from Innovations, Inc., while a student?

7.7   Dr. Neet and his colleagues who are team teaching an advanced course on applications of molecular biology to anthropology collate and edit their lecture notes and syllabi into a textbook that they subsequently use as the course textbook. The textbook is well received nationally and is adopted by 20 other institutions. Dr. Neet has also written a monograph on anthropologic mysteries resolved by application of modern technologies. In the course on applications of molecular biology to anthropology, Dr. Neet offers extra credit to students who will buy the monograph and use it as the primary source for a term paper. Compare and contrast the conflicts presented by adopting a multiauthored textbook and a single-authored monograph. Discuss other critical factors, including the offer of an enticement to buy the book authored by the course instructor.

7.8   Dr. Operon, the chair of a molecular genetics department, creates a search committee of three faculty members to screen candidates for an assistant professorship in the department. A national search is conducted, and four qualified candidates are brought to campus for 2-day interviews. After extended deliberation, the search committee recommends Dr. Grace for the position. Dr. Operon offers the position to Dr. Grace, and she accepts. Dr. Hope, one of the other candidates, writes to Dr. Operon and complains that the search committee was not legitimate because two of its members are married. Dr. Hope argues that spouses should never serve together on committees that involve judgment of people (promotion, tenure, faculty searches, etc.). Dr. Hope states he will file a complaint with the Equal Employment Opportunity Office of the university. Is Dr. Hope's argument valid?

7.9   Dr. Campbell is in the final stages of negotiations for an assistant professor's position in chemistry at a small, prestigious college. Dr. Campbell wants very much to land the job, but he is one of three highly qualified finalists for the position. He is troubled about a potential conflict-of-interest issue. Dr. Campbell's sister and brother-in-law are both students at the college. As biology majors they will have to take at least one of the chemistry courses that Dr. Campbell will be teaching should he get the position. Dr. Campbell knows the chemistry department is small (only four faculty), and he reasons that working around the problem of having relatives in one's course might not be conveniently done. Dr. Capis, the departmental chair, is heading the search committee. Dr. Campbell is afraid that if he discloses this potential conflict of interest to Dr. Capis, he will lessen his chances of getting the job. Is this a conflict of interest, and when should he disclose this information to Dr. Capis?

7.10 Mr. Rich, a wealthy businessman concerned about the well-being of his adult daughter, Mrs. Mean, offers to endow a chair in biology, provided that his son-in-law, Dr. Mean, is the first appointee to the chair. Dr. Mean is an adequate but not exceptional teacher and investigator. He is an untenured assistant professor and has several more years to go before being considered for promotion to associate professor with tenure. Mr. Rich points out that the university would have the chair in perpetuity even if Dr. Mean leaves the university, does not earn tenure, or retires. As dean, you must recommend to the university president whether or not to accept Mr. Rich's offer. What is your rationale for your recommendation to the president?

7.11 Mr. Asset, a graduate student of Dr. Bond, has been conducting physicochemical studies on the properties of a new polymer. The research is sponsored by Chemical Industries, Inc., and it is understood by Mr. Asset and Dr. Bond that the results are proprietary, confidential, and cannot be used in Mr. Asset's thesis. Mr. Cash, the technical liaison from Chemical Industries, Inc., meets with Mr. Asset and Dr. Bond and expresses his pleasure with the outcome of the recent studies and observes that the new results are the last data required to market a new generation of fire-resistant electrical insulating material. Mr. Cash further comments that this is the product that Chemical Industries, Inc., needed to regain its market share, and the stock of Chemical Industries, Inc., will soar once investors know of the new product. That evening at dinner with his wife and brother-in-law, an investment banker, Mr. Asset tells them about Mr. Cash's enthusiasm about their recent research results and Mr. Cash's expectations that Chemical Industries' stock will greatly increase in value as soon as the new product is announced. The next day Mr. Cash's brother-in-law advises several of his clients to purchase Chemical Industries' stock. Did Mr. Asset breach his confidentiality agreement by discussing his research results with his wife and brother-in-law? Does Mr. Asset profit by the disclosure of the research results that will increase the value of the stock of Chemical Industries, Inc.? Discuss a scientist's responsibility for maintaining the confidentiality of research results.

7.12 Dr. Wilkins has a modest research program supported by a grant from a local foundation. Dr. Wilkins brings a personal check for

$3,000 into the office of Mr. Cole, the departmental administrator, and says that it is a gift that may be used by the department at the discretion of the chair. When Mr. Cole consults with the chair, Dr. Vaughn, he learns that Dr. Vaughn and Dr. Wilkins have already discussed this arrangement. Dr. Vaughn says she has agreed to let Dr. Wilkins spend this money as it will help him strengthen his research program to the point where he will be able to successfully compete for federal grants. Over the course of the next several months, Dr. Wilkins uses some of the money to purchase a new computer and printer, which he installs in his home. He uses the remainder of the money to attend a meeting in his research field. At the end of the year Dr. Wilkins donates $5,000 to the department. Over the next several months he uses this money to attend two other meetings and to pay for several subscriptions to scientific journals and an electronic database. Comment on any conflict-of-interest considerations of this scenario.

References

1. **Association of American Medical Colleges.** 1990. *Guidelines for Dealing with Faculty Conflicts of Commitment and Conflicts of Interest in Research*, p. 18. Association of American Medical Colleges, Washington, D.C.

2. **Blumenthal, D., E. G. Campbell, N. Causino, and K. S. Louis.** 1996. Participation of life-science faculty in research relationships with industry. *N. Engl. J. Med.* **355:**1734–1739.

3. **Bourke, J., and R. Weissman.** 1990. Academics at risk: the temptations of profit. *Academe* **76(5):**15–21.

4. **Chen, P. S., Jr.** 1992. The National Institutes of Health and its interactions with industry, p. 199–221. *In* R. J. Porter and T. E. Malone (ed.), *Biomedical Research: Collaboration and Conflict of Interest.* The Johns Hopkins Press, Baltimore, Md.

5. **Cooper, T., and M. Novitch.** 1992. The research needs of industry: working with academia and with the federal government, p. 187–198. *In* R. J. Porter and T. E. Malone (ed.), *Biomedical Research: Collaboration and Conflict of Interest.* The Johns Hopkins University Press, Baltimore, Md.

6. **David, E. E., Jr. (Chairman, Panel on Scientific Responsibility and the Conduct of Research).** 1992. p. 67–79. *In Responsible Science: Ensuring the Integrity of the Research Process*, vol. I. National Academy Press, Washington, D.C.

7. **Dustira, A. K.** 1992. The funding of basic and clinical biomedical research, p. 33–56. *In* R. J. Porter and T. E. Malone (ed.), *Biomedical Research: Collaboration and Conflict of Interest.* The Johns Hopkins Press, Baltimore, Md.

8. **Ferguson, J. R.** 1997. Biomedical research and insider trading. *N. Engl. J. Med.* **337:**631–634.

9. **Gillis, A. M.** 1992. The unwilling expert. *BioScience* **42:**160–163.

10. **Gorman, R. A.** 1998. The rights of faculty as creators and users. *Academe* **84(3):**14–18.

11. **Haber, E.** 1996. Industry and the university. *Nat. Biotechnol.* **14:**441–442.

12. **Kreiser, B. R.** 1990. On preventing conflicts of interest in government sponsored research at universities, p. 83–85. *In AAUP Policy Documents and Reports*, 7th ed. American Association of University Professors, Washington, D.C.

13. **Kreiser, B. R.** 1990. Faculty appointment and family relationship, p. 116. *In AAUP Policy Documents and Reports*, 7th ed. American Association of University Professors, Washington, D.C.

14. **Krimsky, S., L. S. Rothenburg, P. Stont, and G. Kyle.** 1996. Financial interests of authors in scientific journals: a pilot study of 14 publications. *Sci. Eng. Ethics* **2:**396–410.

15. **Lomasky, L. E.** 1993. Public money, private gain, profit for all, p. 237–241. *In* R. E. Bulger, E. Heitman, and S. J. Reiser (ed.), *The Ethical Dimensions of the Biological Sciences*. Cambridge University Press, Cambridge, U.K.

16. **Marshall, E.** 1992. When does intellectual passion become conflict of interest? *Science* **257:**620–621.

17. **Morgan, H. M.** 1990. Pickled in brine: the possible costs of speculation. *Academe* **76(5):**22–26.

18. **Porter, R. J.** 1992. Conflict of interest in research: personal gain—the seeds of conflict, p. 135–150. *In* R. J. Porter and T. E. Malone (ed.), *Biomedical Research: Collaboration and Conflict of Interest*. The Johns Hopkins Press, Baltimore, Md.

19. **Roberts, L.** 1992. Science in court: a culture clash. *Science* **257:**732–736.

20. **Scott, M. M.** 1998. Intellectual property rights: a ticking time bomb in academia. *Academe* **84(3):**22–26.

Resources

The U.S. Department of Health and Human Services (DHHS) Objectivity in Research Policy may be found on the NIH Guide Web site

http://www.nih.gov/grants/guide/index.html

using the phrase "objectivity in research" on the site's search engine. The DHHS-NIH requirements are almost identical to those in the corresponding National Science Foundation document.

Universities, research institutes, and other agencies usually have their own guidelines and policies governing conflict of interest. These are typically available on-line and can be searched for using terms such as "conflict of interest" or "objectivity in research." As examples, consult the following:

Virginia Commonwealth University's Policy on Objectivity in Research:

http://views.vcu.edu/views/ospa/news/objres08.htm

University of California, San Francisco Investigators' Handbook:

http://www.library.ucsf.edu/ih/

chapter 8

Collaborative Research

Francis L. Macrina

Overview • The Nature of Collaboration • A Syllabus of
Collaboration Principles • Conclusion • Case Studies
• Author's Note • References • Resources

Overview

> Most of the work still to be done in science and the useful arts is precisely
> that which needs knowledge and cooperation of many scientists ... that is
> why it is necessary for scientists and technologists to meet ... even those in
> branches of knowledge which seem to have least relation and connection with
> one another.

COLLABORATION IN SCIENTIFIC RESEARCH has grown dramatically
in the 20th century. But the above words of the French chemist An-
toine Lavoisier tell us that the importance of collaborative research has
been recognized for over 200 years. Collaborative research can increase the
ability of scientists to make significant advances in their fields in general
and in their own research programs specifically. Because of the specializa-
tion and sophistication of modern research methods, collaborations become
necessary if researchers wish to take their programs in new directions or
if practical benefits are to result. Especially in the biomedical, agricultural,
and natural sciences, collaborative research often mobilizes intellectual and
technical resources in ways that lead to scientific discovery of direct benefit
to society.

Today, advances made in the sciences are rarely the result of the labors of
single investigators. Even the paradigm for the training of new scientists is
a collaboration, with the mentor and the trainee contributing individually
to a working relationship that is expected to produce positive outcomes
for both. To be sure, significant scientific contributions do sometimes em-
anate from the work of single individuals. Geneticist and Nobel laureate
Barbara McClintock was the sole author on over 90% of her more than
70 scholarly publications (6). But historically and currently, collaborative

research has played a dominant role in advancing our knowledge of the world and contributing to the betterment of humankind. Today, the solitary scientist — armed with the tools of a single discipline — seeking to conquer some devastating disease is largely a romantic myth. Whether we are trying to unlock some fundamental secret of life or to turn basic knowledge into a practical application, collaborative relationships usually offer us the best chance of success.

The power of collaborative research

The advantages of collaborative research are obvious. By combining unique expertise, technology, and resources, investigators are able to deal with problems in the sciences that are not amenable to a singular experimental approach. Typically, collaboration allows the investigative team to ask powerful new questions, the answers to which would be otherwise unattainable. Much of what is termed collaborative research is interdisciplinary. Such interdisciplinary approaches often involve testing the same or similar hypotheses by different means, e.g., using both genetic and biochemical approaches. When differing approaches yield data that support the same hypothesis, we have higher confidence in the answers obtained. Fruitful collaborations can create new knowledge that propels an existing field dramatically forward or opens up new fields of endeavor.

In biomedical research, trends to create and enhance collaboration are evident on several fronts. Universities foster interdisciplinary collaboration by forming research institutes or centers that are populated with investigators from different backgrounds. For example, an institute with a focus in structural biology might be composed of investigators with backgrounds and expertise ranging from crystallography to pharmacology to molecular genetics. Such interdisciplinary activities might be organized as a virtual center with collaborators working in their home departmental space. Alternatively, defined space or even a whole building might be dedicated to such a research center or institute. Interdisciplinary training programs at both the pre- and postdoctoral levels also fertilize collaboration. When graduate and postgraduate training is based on interdisciplinary approaches, faculty from various departments and disciplines may be stimulated to explore and pursue collaborations. Such training environments are likely to spawn new researchers with an awareness of the benefits of collaboration and a knowledge of how to implement collaborative arrangements.

Another catalyst of collaboration is the increased emphasis on interdisciplinary research being promoted by funding agencies. Requests for grant applications regularly suggest approaches to problems that are based on collaborative or "integrative" research. And clearly a premium is put on applications to support research training that stress interdisciplinary curriculum and laboratory work. The research centers program started in the 1970s by the National Cancer Institute is predicated on collaborative research, as evidenced in the first sentence of the program's mission statement: "The cancer centers program supports research-oriented institutions across the Nation that are characterized by scientific excellence and their ability to integrate and focus a diversity of research approaches on the cancer problem."

Challenges of collaborative research

The increase of interdisciplinary collaborative research has created some challenges. Research universities generally are organized according to a departmental structure that is based on disciplines. Where traditional departments prevail, collaborations may encounter problems as department heads attempt to deal with issues of space and resource allocation and curricular issues. Collaborations may be seen by some as undermining the integrity of the traditional departmental infrastructure of universities. At the level of peer review for grant funds, collaborative research may also pose challenges. For example, the review infrastructure at the National Institutes of Health (NIH) is organized by discipline, although it reflects more breadth than that seen in university departments. Nonetheless, a grant application could consist of diverse experimental approaches to a complex problem developed by collaborating investigators from disparate disciplines. In such a case, the initial review group (also called a study section), comprising reviewers from related disciplines, might not be able to perform a rigorous scientific review of the entire application. Typically, this problem is solved by inviting ad hoc reviewers to sit with the group and provide the needed expertise to fairly and rigorously evaluate the proposal. The successes of collaborative research indicate that these issues do not present insurmountable barriers. But they must be considered and constructively addressed as scientific research continues to embrace and foster collaboration as a strategy.

Other challenges of collaborative research may accompany special situations. Of note are (i) collaborations between industry and nonprofit institutions (universities and research institutes) and (ii) international collaborations involving investigators with different cultural and professional backgrounds. University researchers increasingly enter into collaborative arrangements with industry. These arrangements may bring with them restrictions on public disclosure and publication of the research. These constraints may be inconsistent with pre- and postdoctoral training philosophies and may have to be carefully weighed in that context. Then there is the issue of sharing research materials with the scientific community. Collegiality and sharing are widely held as normative behaviors in science (10), and these norms may be threatened by collaborative arrangements involving corporate research partners. Again, careful consideration is warranted as the benefits of the research are weighed against conditions imposed by the collaboration.

Challenges associated with collaboration involving international partners can crop up in clinical research (1, 3, 15). Ethical and cultural standards may differ from country to country. A collaboration between scientists in a modern industrialized country and those in a developing country might involve a clinical trial of a new drug or experimental vaccine aimed to control or prevent an infectious disease. The developing country is a desirable location for this research because its population is at high risk for the infection. However, the culture and the ethical standards of this country may influence the seeking of informed consent. For example, a village leader or elder may speak for the members of the community. Because of this, the scientists from the developing country may suggest that informed consent not be sought from each individual out of respect for the cultural traditions of the community. An additional dimension of this problem surfaces if we suppose that in this small village-based society the concept of the germ theory of disease is unknown or is not accepted by its members. Can there be a realistic expectation of informing potential experimental subjects of research concepts and risks under such circumstances? The existence of international guidelines addressing the use of humans in biomedical experimentation, especially the Declaration of Helsinki, should always set the tone for such research (see chapter 5). The position that local traditions should never compromise scientific or ethical standards has been affirmed by some (1). Clearly, there is a need to identify and deal with potential problems linked to ethical and cultural issues that have an impact on international clinical research. Such matters must be carefully discussed by all collaborators before any research begins.

Last but not least in the way of challenges presented by collaborative research are the details of striking and implementing a collaborative agreement. Until recently, little has been written about when to begin, how to proceed, and what details need to be covered. But there is now a growing literature about doing collaborative research, and this subject will be discussed below.

A timely example of collaboration

The theme of collaboration is well exemplified in modern genetic research. The Human Genome Project provides a continuing example of interdisciplinary collaborations as investigators of different backgrounds join forces to link nucleotide sequences with human disease. Basic research on gene structure, location, replication, and repair can be connected to general problems of disease etiology through collaborative efforts. Epidemiologic observations coupled to biochemical and genetic data through collaborative research can produce rapid progress. The resulting molecular understanding of disease opens the door to development of novel diagnostic, therapeutic, or preventive applications.

The discovery of a class of colon cancer genes provides a cogent example of collaboration. Geneticists and molecular biologists studying inherited colon cancer discovered a high incidence of DNA instability in certain patients. Microbial geneticists and biochemists made connections between these observations and molecular events accompanying DNA repair systems in bacteria and yeast. Collaborative studies between all these scientific groups resulted in an explosion of information on the molecular basis for a common form of cancer. Knowledge of the genetic basis of and biochemical pathway for the repair of DNA in single-cell organisms was crucially important. DNA repair gene homologs from bacteria and yeast provided important clues to the etiology and pathogenesis of familial colorectal cancer. Bacterial and yeast genes were used to identify eukaryotic homologs. Chromosomal mapping and nucleotide sequence determination of these homologs then set the stage for their genetic analysis in affected patients. The results demonstrated that mutations in these genes were clearly associated with colorectal cancer. One summary of this story is found in the review by Modrich (13). Other examples of similar research are regularly found in the popular press and in scientific commentary and primary research literature (8).

The Nature of Collaboration

Formalizing collaboration

Some collaborations are established by a simple verbal agreement and a handshake. Others take on a more formal tone, with the participants rendering into writing their roles, responsibilities, and expectations associated with the collaborative research. Today's wide use of electronic communications undoubtedly makes written agreements of collaboration common. The ability to communicate by electronic mail allows an investigator interested in forging a collaborative relationship to easily approach virtually anyone doing science. Perhaps the highest degree of formality is achieved when investigators planning a collaboration decide to seek grant support. In this case, the application to the funding agency will contain a letter from the collaborator describing his or her role in the research. The collaborator's biographical sketch also will be included in the application. There might even be a budget request for the collaborator's salary commensurate with his or her effort, as well as requests for collaborator supplies and travel. If the principal investigator of the proposal and the collaborator are at different institutions, officials from both institutions usually must approve the proposal if it involves budgetary items. In any event, collaborations that are written in grant proposals epitomize formality because their existence is clearly documented in materials that are seen by many people at the institutional level and the funding agency (e.g., program officers, peer reviewers).

Is it collaboration?

Consider the initiation of a collaboration that is clear-cut. Dr. Gladden, a molecular biologist, needs to analyze the peptide fragments produced by the action of a protease he has genetically engineered. He knows that Dr. Harris, an expert physical chemist, will be able to characterize his peptides by mass spectroscopy. When Gladden approaches Harris, we expect that he will propose they set up a collaborative relationship. Assuming Harris is receptive, they will work out the details of their collaborative project. The implications of collaboration are obvious to all parties here. Harris's expertise is critical to moving Gladden's project forward.

Now consider Dr. Frank, who gives a new gene expression system to Dr. Louis. A description of the plasmid and its host strain — extremely useful in protein overexpression — has not been published. Instead, Frank describes the usefulness of the plasmid to Dr. Louis at dinner, rendering a map of the plasmid and its features on a cocktail napkin. Louis welcomes having the strain sent to him and uses it successfully to gain important results that he now intends to publish. Neither Frank nor Louis had mentioned anything about collaboration when they talked at dinner. Louis considers the sending of the strain a professional courtesy, similar to requesting a strain that had been described in print. He believes that simply thanking Dr. Frank in the paper's acknowledgments section is sufficient. Frank, on the other hand, had assumed he was making a critical contribution to Louis's work by allowing him to isolate a protein that was previously impossible to purify in reasonable quantities. Frank demands that he and his postdoctoral associate be coauthors on the planned manuscript. In contrast to the first scenario, a collaborative relationship is not obvious here. Failure to consider collaboration in the beginning now creates problems given the assumptions of the two investigators.

Working out the details of a collaboration after the fact is usually not a smooth process. It is relatively easy to agree on collaboration when the stakes are defined and the outcomes are unknown. But once we are aware of the outcomes, our new vested interests strongly influence our negotiations. Communication among scientists proposing to work together is a necessary first step in deciding whether an arrangement is going to be collaborative. Once this is agreed to, then defining the expectations, activities, and responsibilities of all parties in the collaboration is essential.

A Syllabus of Collaboration Principles

Some institutional guidelines on scientific conduct address the topic of collaborative research (e.g., see references 7 and 9). In some cases, discussions relevant to collaborative research also appear in guidelines for authoring scientific papers (see chapter 4 and its resource list). There also are monographs and publications that discuss collaboration from different perspectives (2, 3, 12, 16), and proposals have been made to rethink authorship credit and accountability (5, 14). Clearly, scientists believe there are behaviors appropriate to successful collaboration. However, there are no prescriptions or rules that will ensure a successful collaborative outcome in every case. The following syllabus draws from several sources in building a foundation of issues and principles of scientific collaboration. The principles of good science and the guidance provided by this syllabus can be used to develop productive and successful collaborations.

Communication, communication, communication

When considering the value of real estate we are often advised that "location, location, and location" are the three most important issues. Just substitute the word "communication" to make the same point about scientific collaborations. Establishing, maintaining, and even terminating communication are critical elements of the collaboration. In establishing communication, the potential partners talk candidly about the possibility of collaboration. Unstated assumptions undermine opening the lines of communication, as illustrated by the example of Drs. Frank and Louis above. If you are considering any interactions with another scientist that involve the exchange of resources — expertise, personnel, data, or materials — it is best to ask up front if a collaborative arrangement is appropriate. Typically, guidelines on collaboration (and authorship) say that if someone provides research materials that are part of published results, the donor of the materials is acknowledged in subsequent publications. Simple provision of materials already described in the public domain usually does not constitute grounds for collaboration. Yet situations involving exchange of materials are not always clear — again recall Drs. Frank and Louis. When in doubt, be open and candid about the interactions you and your colleague may be heading toward.

Once a collaboration is established, sustaining communication is imperative. The free flow of information and interpretations between the partners is critical to collaborative success. Talk about your data. Talk about your ideas. Share everything you can with your collaborators. With effective communication comes the establishment of trust, another key ingredient of successful collaboration. Last, when the work is brought to an end, it may not be appropriate to continue the exchange of information at the level of depth and detail established during the collaboration. For the sake of future priority or proprietary information, one or both sides may wish to change or restrict the lines of communication practiced during the collaboration. Better that this is clearly understood by both parties than to have assumptions or accusations made about whether one side or the other has stopped sharing or contributing to the collaboration. So clarity about when to end collaborative communications is also important.

A final word on communication: be proactive. Don't leave it to the other person. And don't assume that just because you're collaborating with a colleague in your department, proximity will substitute for communication. It won't. When collaborating, you have to work as hard at communicating with someone in the next lab as you do with someone on another continent!

Goals

Once a collaboration is established, both sides need to agree upon goals. Whether there is one goal or several, they will be unique to the collaborative arrangement. In other words, they would not be readily achieved by just one party of the collaboration. In arriving at clear goals, these discussions will give rise to anticipated expectations and outcomes.

Responsibilities

Who will be the leaders of each side of the collaboration? This probably will be the lab chief or principal investigator from each group, but it does not have to be, depending on how the collaboration was initiated. Who will do what in the collaboration? This question is a complex one, because often the initial discussion about the collaboration is between just two people representing participating groups (although there could be more people and even more than two groups). The actual collaboration may well involve several people in each of the participating laboratories. Pre- and postdoctoral students as well as technicians may work on the project. How will the work assignments be allocated at both the intra- and interlaboratory levels? Fleshing this out completely make take several discussions if several people will be involved. The leaders from both sides of the collaboration should also talk about how general decisions will be made. Things like adding new members to the collaborative team, and when and how to terminate the collaboration (owing to either success or failure) should be considered. Guidelines have mentioned the value of having all participants of the collaboration share in the decision-making process, and this is a good idea.

Timing and duration of the project

A time frame for the project should be estimated at the outset of the collaboration. It may also be useful to set up timing checkpoints or milestones: dates or events that prompt scrutiny and evaluation of the project's progress. These can be used to make decisions about continuation, modifications, course changes, or termination of the work. Everybody involved in the collaboration should be informed about these dates and the expected duration of the project.

Accountability

Different layers of accountability may accompany collaborations. First, one or more of the parties to the collaboration may be involved in research that is subject to formal policies, regulations, or laws. Such activities could include working with human subjects, animals, or hazardous substances. All participants in the research need to confirm their compliance with appropriate regulations. For example, a collaboration between a clinical research group and a basic research group at different institutions might involve the sharing of patient data. The basic researchers must be fully aware of and honor all patient confidentiality issues mandated under the approved human use protocol filed by the clinical researchers. Further, the clinical researchers are responsible for informing the basic researchers about any potential biohazards of working with clinical materials of human origin.

Second, collaborations that enjoy extramural support will be subject to grant management regulations mandated by the funding agency as well as the grantee institution. Any regulations regarding the expenditure of funds and reporting requirements to the granting agency have to be met by the responsible parties of the collaboration. Consider federal funds subcontracted from a grant at collaborator A's institution to collaborator B's institution to pay for a component of an investigator's salary. There may be a requirement that such funds be dedicated to this purpose. Rebudgeting this salary money to cover the costs of supplies or travel would be forbidden and would likely have negative repercussions at both collaborating institutions.

Finally, collaborations may have outcomes — planned or otherwise — that have implications for the development of intellectual property. Partners in the collaboration should be aware of the necessary steps involved in protecting research results that might have potential commercial application. By disclosing results publicly and prematurely, one research group might compromise the ability to seek patent protection for something codiscovered by the collaborators. Furthermore, there may be institutional and granting agency requirements relating to the prosecution and ownership of intellectual property, and these should be familiar to all parties of the collaboration.

Last, all collaborators must be aware that the failure of anyone associated with the project to comply with any regulations may carry consequences for all of the scientists involved in the study.

Authorship

Deciding authorship on papers that report the results of collaborative projects should parallel accepted norms (see chapter 4). In short, authors earn a place on the paper's byline by making a significant contribution to the work. Moreover, authorship brings with it acceptance of the responsibility for the work. But some projects may provide challenges when the work involves a specialized technology and related data analysis contributed by one of the collaborators. In such cases it may not be realistic to hold every coauthor responsible for all parts of the paper. A strategy increasingly used to deal with this problem is to specify the contributions of all the coauthors, either in a cover letter to the editor of the journal or in a footnote to the published paper. But in the absence of indications to the contrary, readers of papers reporting collaborative research will assume that all authors are jointly responsible for the published work.

Discussions of authorship should begin with the leaders of the collaborative parties and subsequently involve all other participants in the project. This should be done early in the project, even though these decisions can be difficult to make before the results of the project begin to unfold. Such discussions can be useful if for no other reason than that they emphasize the value placed on credit attribution and establish rational discussion as a means to achieve it. The question of proper credit needs to be addressed at every point in the research process and with every person involved in the effort. In sum, the strategy for assigning credit and responsibility should be established early in a research project, reviewed regularly, and revised as appropriate. Participants in the collaboration need to remain flexible in this regard. Contributions made by various collaborators during the progression of the project may change dramatically. This will change credit attribution, and, in turn, authorship priority.

Discussions similar to those defining authorship are also needed to decide who will be acknowledged in the published paper. This too should involve all members of the collaborating teams.

A final issue related to publication and authorship is the production of the manuscript. Who will take the lead in writing the paper? If parts of the paper are to be written in different laboratories, who will be responsible for melding the parts into an integral manuscript? Who will pay for production costs such as figure preparation, photography, and the like? Who will pay for page and reprint charges? These are all questions better discussed early in the collaboration than when a manuscript is being written.

Conflict of interest

Potential conflicts that might affect the collaboration or the participating investigators should be disclosed. For example, one investigator might be supporting a small part of the collaborative research with a grant from a biotechnology company. Suppose the research results have positive implications for a diagnostic test sold by the company. A collaborative paper is written, submitted, and accepted for publication, but disclosure of the biotechnology company support is not made. This fact becomes known after publication, creating misunderstanding and suspicion that has an impact on everyone involved in the collaboration. Thus, collaborators need to inform one another about all sources of support for joint research projects. Together they must make appropriate decisions about disclosing potential conflicts when presenting collaborative results, preparing papers, writing reports, or submitting new grant applications.

Other potential conflicts of interest can arise as the result of collaborations. Consider Dr. Salley, who chairs the Nicholas Foundation's review panel that recommends funds for postdoctoral fellowships. Salley has just started a collaborative research project with Dr. Strauss. Robert Murphy, a postdoctoral fellow in Strauss's lab, applies for a prestigious Nicholas Foundation fellowship. Because the Salley-Strauss collaboration is new, few outside of their labs know about it. Because Murphy is not involved in the collaboration, Dr. Salley does not consider himself in conflict as a reviewer. He provides a glowing review and Murphy is awarded a fellowship. But later, when the collaboration becomes well known, other members of the review panel suspect Salley of bias in favoring the Murphy application. Perceived or real, this conflict now has negative implications for both sides of the collaborative relationship, including a potentially negative impact on Murphy, a bystander to the collaboration. As above, collaborators need to share information that might create conflicts in peer review or other activities related to the conduct of scientific research.

Data sharing, custody, and ownership

Collaborators must establish ground rules for the sharing of data that emerge from joint research projects. The trust that must accompany a successful collaboration undergirds data-sharing activities. But unexpected situations may arise, and collaborators must be prepared to deal with them. Consider two labs collaborating to clone a transcription factor. Lab A has purified the protein and prepared antibodies; lab B will screen an expression library to identify the clone. Clearly, lab B will receive a portion of the highly specific monoclonal antibody, and the resulting DNA clone will be shared. Will lab B also receive the hybridoma cell line? In a similar vein, consider a case in which lab C has isolated and determined the sequence of a cDNA that appears to encode a new member of a protease family. Lab C collaborates with lab D — experts in that protein family — by sending them in vitro translated protein for characterization. Should lab D also expect access to the cloned cDNA? Cases like these often arise. Sometimes the same answer seems obvious to both parties; frequently it does not. The resolution has obvious bearing on the abilities of the individual labs not only to replicate portions of each other's work, but also to undertake independent work at the conclusion of the collaboration. The advisable course of action is to discuss and settle these issues as soon as they can be foreseen.

It is also necessary that all parties to the collaboration have a clear understanding of data ownership and custody issues. Usually, ownership will be governed by the type and source of funds that have been used to support the research. In the case of NIH funding, the data are owned by the grantee institution (see chapter 9), and this will have implications for collaborative research that is done at different institutions and supported by the individual NIH grants of the collaborating principal investigators. The principal investigators and their respective grantee institutions will be subject to the policies governing ownership, custody, and retention of data imposed by the granting agency. Data books and research data created at one site will thus remain at that site in keeping with the policies governing the grantee institution. But the sharing of materials — both during and after the collaboration — must be worked out by the collaborators. To use the above illustration involving labs C and D, discussion of the sharing of cDNAs is best made while the collaboration is being formulated.

Collaboration between research sectors

Collaborative research can involve partnering of different sectors of the research community. Joint projects involving various combinations of academic, government, industry, and research institute participants should be guided by the above principles. But the operating practices of these different entities can vary significantly, underscoring the need to define and understand the constraints that affect the role of the participants and the overall performance of the research. This is particularly important in the case of any collaboration involving industry. Communication and understanding of requirements that are part of collaborations with industry are critical to the success of joint projects. Special requirements may be imposed on decisions to publish material or on the preparation of invention disclosures and patent applications. Similarly, there may be confidentiality issues that go beyond what a nonindustrial researcher is used to dealing with. There may be implications unique to trainees. Is a project or subproject appropriate as a dissertation or thesis topic? How might this affect the trainee's ability to publish results that might be a requirement for completing his or her degree?

A joint research project involving industry may fall under the aegis of Good Laboratory Practices (GLP). This might be necessary because the industry plans to use the research results to support applications for investigative or marketing permits. Such GLP prescribe procedures for documenting, recording, reviewing, and retaining experimental protocols and results. The nonindustrial collaborator must be made aware of the intentions for use of the data, and, obviously, he or she must implement GLP to complete a successful collaboration.

Collaborations with industry may directly and regularly involve more participants than are usually encountered in other collaborations. For example, lawyers, technology transfer and patent officers, marketing personnel, and sponsored research officials from both sides of the arrangement may be involved in the collaboration.

Finally, there may be restrictions on the sharing of data or research materials both before and after publication. Frequently, industrial research laboratories require the completion of a material transfer agreement (MTA) before sharing research materials. However, such agreements are increasingly used by academic or government laboratories, especially if there is some inherent intellectual property value in research materials. Typically, an MTA will specify the parties of the agreement, designating them as "donor" and "recipient." It will also specify what materials are being transferred to the recipient, possibly describing them in precise qualitative and quantitative detail. Then, depending on the nature of the agreement, various other items will be listed. These may include (i) limitations on use of the material (e.g., the material is only to be used for noncommercial, research purposes); (ii) limitations or restrictions on distribution of the material (usually the recipient is forbidden to transfer, sell, or otherwise make available the material to any third party); (iii) conditions of use (e.g., prohibiting use of the material with human subjects or animals); (iv) conditions of publishing results obtained using the materials (e.g., there may be a requirement to provide any manuscripts to the donor before the submission for publication); (v) conditions for acknowledgment of the donor in any disclosure of research involving the materials; (vi) warranties concerning the material (usually the donor provides no warranty); (vii) a "hold harmless" clause, releasing the donor from any legal liability resulting from the recipient's use of the materials; (viii) conditions for the return of unused material, if appropriate; and (ix) the requirement of any associated fees or financial conditions related to the transfer of the materials to the recipient. Last, the MTA usually must be signed by individuals legally authorized to represent the institution. For example, if the agreement involves a company, the president, chief executive officer, or a designee might sign. If it involves a university, the authorized signator might be a sponsored program or technology transfer official; the principal investigator may sometimes be required to sign the agreement as well.

Miscellanies

Do not assume that previous successful collaborations will ensure the success of future ones with the same colleagues. Positive collaborations sometimes create an environment for working together on subsequent joint projects. But you must forge each new project with previous collaborators using the same care and attention to detail as you did in the past.

Last, a word on collaboration and professional development of scientists is in order. Institutions and review committees find it difficult to allocate appropriate credit for publications generated by faculty in collaborative research projects. Because independent work is the prevailing measure of scientific identity, junior scientists establishing their careers need to recognize the importance of balancing collaborative and independent work.

Conclusion

Locke (11) says that collaboration is a critical component of scientific discovery. In making this assertion, he points out that many Nobel laureates are investigators who have collaborated for prolonged periods. Because Nobel Prizes are given only for major scientific contributions that change paradigms or create new ones, the value of collaboration is compellingly affirmed by Locke's example. Yet some Nobel Prize-winning discoveries have yielded collaborators who felt shortchanged by the attribution of credit (4). Certainly, collaboration is critical to the progress of scientific research. But these observations teach us that the rules of engagement involving collaborative arrangements must always be clearly stated and understood. The misunderstanding that results when we fail to adhere to the normative standards of collaboration creates ill will and, at worst, may inflict professional harm.

Case Studies

8.1 Dr. Otto Max recently was hired as an associate professor in the department of biological chemistry at Hercules University. As part of his recruitment package, the university has purchased a specialized, expensive instrument used to analyze macromolecules. The analytical power of this instrument and Max's expertise have faculty in several departments excited about the application of this technology to their research. Faculty who approach Dr. Max to explore the use of the instrument in their research learn that he is happy to cooperate with them. But he spells out conditions for such collaborative research that have some faculty upset. No one but Dr. Max or his technician may operate the instrument. The original printouts of all data must remain with Dr. Max. Any paper submitted for publication that contains data obtained using the instrument must have Dr. Max's name on the author byline and his technician's name in the acknowledgments. Some faculty complain to Max's departmental chair that these conditions are not collegial and are prohibitive. They argue that if university funds were used to purchase the instrument, its use should benefit all university faculty. As the departmental chair, how do you handle this dispute?

8.2 You have had a radical idea regarding how to get eukaryotic cells to take up DNA fragments much more efficiently than was previously possible. You tell your colleague Mary about your idea and how you plan on testing the hypothesis. Mary is not in your field of expertise, but you spend some time explaining to her the details of your study and the expected outcomes. Mary offers a number of unsolicited suggestions on how to improve the study. Because of her lack of experience, many of her ideas are not practical or are very elementary and part of your study anyway. However, Mary suggests some valuable control experiments involving DNA competition assays, which help you make a compelling case for the novelty and efficiency of your method. Mary talks to you frequently about the project and comes to several of your lab presentations. She comments critically on your work and makes other suggestions, including the idea that you try different cell types to further build your case. She offers to try your method on several cell lines that are routinely maintained in her laboratory. You are reluctant to do this, but you suggest that she give you the cell lines so you can do the experiments. She complies, and the experimental results you obtain with her cells further support your hypothesis. You decide to submit a provisional patent application and then submit your exciting results as a short communication to a prestigious journal. Mary argues strongly that her name should be included as a coinventor on the application and a coauthor on the manuscript. How do you respond? What is the rationale underlying your response?

8.3   A clinical scientist and a basic molecular biologist are collaborating on a series of projects that involve patients and normal control subjects. Each investigator is funded from extramural agencies for work distinct from the collaborative project, and each has separate grants for the collaborative project. The clinical scientist views the patient records and diagnoses as her intellectual property and shares these data only when she is ready to prepare a manuscript. The molecular biologist has prepared and preserved cell lines, probes, and reagents that have been kept in facilities readily available to both collaborators. The molecular biologist believes that there are important results that merit publication. He prepares a manuscript up to the point of inclusion of clinical data. The clinical scientist refuses to provide the clinical data. In the dispute that follows, the clinical scientist asserts ownership of the cell lines, probes, and reagents that were developed from patient samples. The dispute is brought to you to mediate. Discuss the data ownership issues of this collaboration. Who owns the clinical data? Who owns the cell lines, probes, and reagents? Who has access to, and use of, the clinical data and the materials prepared from patient samples?

8.4   The research groups of Drs. Scotland and Whales at Research State University (RSU) have a long history of productive collaboration. Scotland is an enzymologist who works on proteases, and Whales is a physical biochemist. Scotland is principal investigator on a new 5-year NIH grant to RSU to study protease inhibitors. Whales is a coinvestigator on this project. But Scotland has been recruited to the University of Integrity (U of I). Scotland has received unofficial agreement from RSU, the U of I, and the NIH to have the grant transferred to the U of I. In fact, the physical biochemistry group at U of I, headed by Dr. Diffraction, offers superior possibilities for collaborative research with Scotland; their group is much larger than Dr. Whales's group, and they have several pieces of state-of-the-art equipment that neither Whales nor RSU has access to. Some of Scotland's proposed collaborative experiments with Whales can be much better done by now collaborating with Diffraction's group. Whales is upset by the impending events. Over the years he and Diffraction have been in competition with each other. Now he feels he is being unfairly squeezed out of his collaborative opportunities with Scotland. He argues that the study section that rated the scientific merit of the proposal approved his participation in the work, not Diffraction's. Moreover, the grant budget includes 20% of his effort and salary. The situation comes to a head when he confronts Scotland over the plan for transferring the grant. Scotland politely counters that collaborative possibilities with Diffraction will greatly benefit the project. And he reminds Whales that the institutions have the decision-making power in the transfer of the grant, not the investigators. Scotland tells Whales that once the grant is transferred, he will make arrangements to keep him on as a coinvestigator and to pay his salary component for 1 year, via a subcontract. Then he plans to make Whales an unpaid consultant for the rest of the project. Comment on the ethical, legal, and practical implications of this scenario. Are there alternative solutions to this problem?

8.5 Drs. Sterling and Crystal at Research University have been collaborators for a number of years. Each is funded as a principal investigator, with the other as coinvestigator, from federal agencies. Drs. Sterling and Crystal develop strong differences, largely of a personal nature. Dr. Sterling, who is more senior, believes that she owns the data and experimental materials derived from the collaboration. Dr. Sterling takes steps to deprive Dr. Crystal of access to the materials. Dr. Crystal appeals to Dr. Bluff, the research administrator of Research University, to intervene. Dr. Bluff calls Dr. Sterling and Dr. Crystal together and asks them to work it out. Drs. Sterling and Crystal cannot reach agreement, and Dr. Crystal decides to leave Research University. Dr. Sterling charges Dr. Crystal with intent to remove research materials from Research University without authorization. These accusations are brought to the attention of the federal funding agencies. You are asked to conduct an inquiry. How do you proceed?

8.6 Drs. Mulligan and Stevens, both associate professors at State University, are collaborating on research that leads to a grant application. Both investigators prepare the application together. After the grant application is submitted, Mulligan gets an attractive offer to join the staff of a private research foundation in another state. He accepts the position and leaves the university before the disposition of the collaborative grant application is known. Shortly after Mulligan takes his new position, Stevens learns that the application was approved but not funded. Mulligan and Stevens consult by phone over this, and both agree that future collaborations will not be plausible. About a year later, Stevens submits a revised grant application using much of the same language as in the first submission but with some new material based on the comments of the reviewer of the first application. The revised application makes no mention of Mulligan or his contributions. However, about two-thirds of the application consists of the exact same words as the Mulligan-Stevens original proposal. For example, the grant applications' abstracts are identical except for 25 words. Mulligan finds out about the application that Stevens has submitted and formally accuses him of plagiarism. Is the charge justified?

8.7 Along with Drs. Hopkins and Carpender, you have submitted a coauthored paper reporting on the regulation of a gene introduced by transfection into fibroblasts. The paper is returned by the editor of the journal with two very positive reviews, suggesting only minor revisions. While the paper is being revised, one of Hopkins's postdoctoral fellows presents data at a lab meeting demonstrating that the results of the gene regulation experiments are dependent on the concentration of DNA used to transfect the cells. She presents data showing that if the concentration of the gene construct is increased fivefold, the previously reported regulatory effects are completely abolished. In light of these results, Hopkins argues that the paper should be withdrawn and not allowed to go to press. Carpender strongly objects to this. He argues that the results of the paper are reproducible and the interpretations of the results are straightforward. He further argues that the new results may be the basis for a whole new paper and that these data should not even be mentioned in the paper. Carpender argues that the paper should be published with the minor revisions suggested by the reviewers. Do you agree?

8.8 The Biomolecular Technology Study Section of a federal funding agency is reviewing two applications: one by Dr. Bass and one by Dr. Perch. Both investigators have a long-standing reputation for collaboration and coauthorship. In this case, however, neither investigator lists the other as a coinvestigator on the application. During the review process, the study section discovers that the introductory sections of both applications are similar. In fact, several paragraphs in each application are identical. In addition, an inspection of reprints appended with Dr. Perch's application reveals three verbatim paragraphs from one of the papers in Bass's application. Finally, a study section member points out that a major section of experimental methods in each application is remarkably similar. Not only are there clearly identical paragraphs, but identical typographical errors exist in each application's methods section. During a coffee break, informal discussion of some of the study section members reveals that Bass and Perch have had a falling out and no longer talk to each other, much less collaborate. After the break, the study section meets and decides to review each application on its scientific merit and not be concerned with the implications of the investigators' relationship. However, one member of the group objects strongly to this, saying that plagiarism is involved in this situation, even though it cannot be sorted out with the information at hand. He argues that every definition of scientific misconduct he knows of lists falsification, fabrication, and plagiarism as transgressions that constitute misconduct. He accuses the study section of "looking the other way" and neglecting its moral responsibilities. Discuss the issues raised by the study section member.

8.9 Bill Williams has constructed a plasmid that allows carefully regulated expression of genes inserted into it. He has not found the plasmid useful in his own work but has discussed it with his colleague Harry Douglas at a meeting. Harry thinks he can put the plasmid to good use and asks Bill if he can try it. Bill is receptive to this and sends the plasmid to Harry. Harry and his coworkers proceed to use the plasmid to create several novel constructs that provide considerable insight into the function of two previously ill-studied genes. The impact of these studies is great, and Harry and his coworkers write a manuscript for submission to a prestigious journal. In the final stages of writing, Harry calls his group together and asks how they feel about including Bill Williams on the author byline. What would you say?

8.10 Global Pharmaceuticals, Inc., has paid a small DNA sequencing company several million dollars for the exclusive rights to the genomic sequence of a bacterial pathogen. Dr. Amy Samuels is a university scientist whose research is supported by a contract from Global Pharmaceuticals. Her specific aims in this research include the identification of new targets for antimicrobial agents. Global makes the genomic sequence of this bacterium available to Dr. Samuels to assist her in finding new genes. Using the genomic sequence, she identifies several novel genes that encode putative surface proteins. Using gene knockout technology, she determines that one of these genes encodes a virulence factor that is likely to be a very good target for an antimicrobial agent. She writes a major paper reporting her research, and it is submitted to you, the editor of *New Chemotherapies*. You proceed to solicit two ad hoc reviews. One reviewer is very positive and recommends acceptance with minor modifications, but the other reviewer recommends rejection. His decision is based on the fact that Samuels's discovery would not have been possible without access to Global's genomic database. He objects that, besides the company, Samuels is the only person with access to this information. The journal's policy is that all sequence data must be on file in a database freely accessible to the scientific community. You reread Samuels's paper and note that she does not report any DNA sequence data in the paper. She characterizes the gene product and demonstrates that a mutation in the gene renders the organism nonpathogenic. How will you act on this manuscript?

8.11 Bill and Sara meet in an introductory graduate course, and over the span of the upcoming academic year, they fall in love and get married. At the beginning of the second year they select different mentors in the same department and begin their dissertation research. The mentors and their groups frequently collaborate and coauthor publications. They both work extremely hard, but Sara frequently has Bill help her in the lab. On weekends they are commonly seen working together doing experiments that are exclusively part of Sara's research project. Over the course of the next 3 years Sara prepares six senior-authored manuscripts, and all are published in peer-reviewed journals. Bill is not included as an author on any of the papers, but he is acknowledged in five of them. In her last year in the program, Sara wins the prestigious graduate student honors day award and is also selected by the departmental faculty to receive the outstanding graduate student annual award. Recently, Sara has been offered a permanent position in a biotechnology company. Bill is not likely to be finished with his dissertation research anytime soon and has no publications or even abstracts to his name. A small group of graduate students meet with you, the departmental chair, and bitterly complain that Sara has had an unfair advantage during her graduate research career. They claim her publication record is deceptive as it fails to account for all the extra collaborative help she received from her spouse. They claim that both she and her mentor are party to inappropriate practices. They want you to intervene in some way. What do you do?

8.12 A major portion of a student's doctoral dissertation is being prepared as a manuscript for submission to a peer-reviewed journal. The first draft of the manuscript prepared by the student has been heavily edited by the mentor. The mentor has even suggested some additional experiments for the student to do. More important, the mentor suggests that the two recombinant proteins being studied be analyzed by circular dichroism. She suggests that this be done collaboratively with a biophysical chemist in another department. Both the mentor and the student agree that this would be a significant contribution and would add considerable strength to the paper. The collaboration is set up over the course of the next 2 weeks. The mentor then tells the student that she would like him to audit a graduate-level course in biophysical techniques being offered in the next semester. The mentor feels strongly that the student should have reasonable command of circular dichroism (CD) theory if the student's paper is going to contain CD data. The student is within 1 year of completing all of his degree requirements. He strongly objects to the mentor's suggestion, stating that he can gain the necessary working knowledge to defend the collaboratively obtained CD data by reading on his own. Comment on this situation. What are the responsibilities of the student and the mentor in the collaborative arrangement? What, if any, are the responsibilities of the biophysical chemist who will do the CD studies?

Author's Note

This chapter was derived from the ideas and writings of a 1994 American Academy of Microbiology Colloquium on collaborative research. The chapter draws significantly from the colloquium's monograph (12). Parts of the monograph are used here verbatim with the permission of the American Academy of Microbiology. The colloquium was supported by the National Science Foundation and the American Society for Microbiology. I chaired the colloquium steering committee, consisting of Susan Gottesman, Bernard P. Sagik, and Keith R. Yamamoto. The other colloquium participants were David Botstein, Gail Burd, Peter T. Cherbas, David V. Goedell, C. K. Gunsalus, Barbara Iglewski, Caroline Whitbeck, and Patricia Woolf.

References

1. **Angell, M.** 1988. Ethical imperialism: ethics in international collaborative clinical research. *N. Engl. J. Med.* **319:**1081–1083.

2. **Austin, A. E., and R. G. Baldwin.** 1991. *Faculty Collaboration: Enhancing the Quality of Scholarship and Teaching.* ASHE-ERIC Higher Education Report no. 7. The George Washington University, Washington, D.C.

3. **Barry, M.** 1988. Ethical considerations of human investigation in developing countries: the AIDS dilemma. *N. Engl. J. Med.* **319:**1083–1085.

4. **Cohen, J.** 1995. The culture of credit. *Science* **268:**1706–1711.

5. **Council of Biology Editors.** 1999. Authorship Task Force. Is it time to update the tradition of authorship in scientific publications? (Available on-line at http://www.sdsc.edu/CBE/)

6. **Fedoroff, N. V.** 1994. Barbara McClintock (June 16, 1902–September 2, 1992). *Genetics* **136:**1–10.

7. **Flay, B. R.** 1992. *Guidelines/Principles of Collaboration and Authorship*. University of Illinois, Chicago. (Available from Prevention Research Center, University of Illinois at Chicago, School of Public Health, Chicago, Ill. 60607.)

8. **Halim, N. S.** 1999. Multidisciplinary collaboration leads to successful genetic research. *Scientist* **13**:6–7.

9. **Hittelman, K. J., and B. Flynn.** 1995. *Investigators' Handbook*. University of California, San Francisco. (Available on-line at http://www.library.ucsf.edu/ih/)

10. **Korenman, S. G., R. Berk, N. S. Wenger, and V. Lew.** 1998. Evaluation of the research norms of scientists and administrators responsible for academic research integrity. *JAMA* **279**:41–47.

11. **Locke, J. L.** 1999. No talking in the corridors of science. *Am. Scientist* **87**: 8–9.

12. **Macrina, F. L., et al.** 1995. *Dynamic Issues in Scientific Integrity: Collaborative Research*. American Academy of Microbiology, Washington, D.C. (Available on-line at http://www.asmusa.org/acasrc/aca1.htm/)

13. **Modrich, P.** 1994. Mismatch repair, genetic stability, and cancer. *Science* **266**:1959–1960.

14. **Rennie, D., V. Yank, and L. Emanuel.** 1997. When authorship fails: a proposal to make contributors accountable. *JAMA* **278**:579–585.

15. **Robison, V. A.** 1998. Some ethical issues in international collaborative research in developing countries. *Int. Dent. J.* **48**:552–556.

16. **Schwartz, J. P.** 1997. Silence is not golden: making collaborations work. *The NIH Catalyst* **5**:3.

Resources

An extensive Working Group Report on sharing research tools (submitted to the director of the NIH on June 4, 1998) may be found on-line at

http://www.nih.gov/news/researchtools/index.htm/

You can find information about material transfer agreements (MTAs) at many university Web sites. Often you'll be able to download or print the forms used for MTAs. Use the term "material transfer agreement" to search the site of your choice. The Massachusetts Institute of Technology has a good selection of information and sample forms. Search the MIT site at

http://web.mit.edu/search.html/

The Uniform Biological Material Transfer Agreement published in the *Federal Register* on March 8, 1995, may be found at

http://intramural.nimh.nih.gov/techtran/ubmta.htm/

The material transfer agreement adapted for use by the U.S. Department of Health and Human Services may be found at

http://intramural.nimh.nih.gov/techtran/ubmta.htm/

A brochure, "Materials Transfer in Academia," may be found on-line at the Web site of the Council on Governmental Relations:

http://www.cogr.edu/

chapter 9

Ownership of Data and Intellectual Property

Thomas D. Mays

Introduction • Review of Ownership of Research Data • Trade Secrets • Trademarks • Copyrights • Patents • Patent Law in the Age of Biotechnology • Seeking a Patent • Conclusion • Case Studies • Author's Note • Suggested Readings • Resources • Glossary

Introduction

INTELLECTUAL PROPERTY IS A UNIQUE CREATION of the human mind. It neither has tangible form nor exists apart from the context of the applicable governmental jurisdiction. An observation of a natural phenomenon may not constitute intellectual property. However, commercial utilization or graphic or electronic representation of such a phenomenon would represent intellectual property. In fact, intellectual property only exists as an exercise of a legal right of ownership conferred under statute or common law. Intellectual property is usually categorized by associating it with the laws covering its use and protection. Such classification yields four types of intellectual property: patents, copyrights, trademarks, and trade secrets. The protection of intellectual property was guaranteed in 1787 by the United States Constitution, which provides that:

> The Congress shall have Power ... To promote the Progress of Science and useful Arts, by securing for limited Times to Authors and Inventors the exclusive Right to their respective Writings and Discoveries ... (U. S. Constitution, Article 1, Section 8)

In 1980, a U.S. Supreme Court ruling had an important impact on biotechnological intellectual property. Specifically, the Court ruled in *Diamond v. Chakrabarty* (447 U.S. 303) that nonhuman life forms could be patented if there was an evidence of human intervention in their creation (see Appendix V).

Every scientist who pursues a course of research using the analytical methodology of observation along with hypothesis formulation and testing follows a long tradition of experimental study. It has been the hallmark of civilization that written records communicate observations, personal impressions, and experimental designs to others geographically and temporally distant to the immediate observer. Through such records, subsequent researchers are able to build upon the work of others. This reflects the central characteristic of scientific discovery; it is a process that builds knowledge incrementally and then pieces that knowledge together in ways that lead to major discoveries. Such discoveries contribute to our understanding of the world, and they often can be applied to practical situations, leading to advancements that improve the quality of life. This serial advancement in scientific and technological fields has acted as an engine of change that has helped transform societies from agrarian villages to robust industrial centers. While this engine of progress may be fueled by curiosity and personal interest, without a means of engagement, much like the operation of a clutch in an automobile, the progress of science and the useful arts would stall or would have little forward movement. The creators of the U.S. Constitution, in true "serial advancement" fashion, borrowed from and improved upon the experiences of Europe dating back to the 13th century. Specifically, they authorized the protection of ownership of intellectual property by authors and inventors.

The two decades following the U.S. Supreme Court's decision in the *Chakrabarty* case have witnessed an explosion in the commercialization of biotechnology. The certainty of intellectual property ownership in its products has been cited as of utmost importance in preserving competitiveness in the biotechnology industry. Biotechnology is viewed as one of the most research-intensive industries in the world. The U.S. biotechnology industry alone is reported to have spent $9 billion in research and development in 1997. The Biotechnology Industry Organization reports that the top five biotechnology companies spent an average of $99,800 per employee on research and development, which compares with an average of $30,600 per employee for the top pharmaceutical companies.

The potential for biotechnological application makes a basic understanding of intellectual property important to scientists in the biomedical disciplines. Of course, other scientific disciplines and areas of research — including software development, electronics, and materials science — have been similarly stimulated by rapid commercial growth and investment. Such growth and development depend in large part on the protection of new technologies as intellectual property. While this chapter will highlight those aspects of intellectual property that relate to the biomedical sciences, this in no way is intended to suggest that intellectual property and data ownership are limited to the biomedical sciences. Many of these principles can be easily applied to new organic chemical processes, novel superconducting ceramics, devices for the high-speed transmission of data, and other research and development areas.

In this chapter the principles of intellectual property will be discussed, distinguishing between the ethical obligations and the legal rights of ownership in the results of scientific research. We will begin with a discussion of the ownership of research data as a basis for building upon the concepts of intellectual property. Through the use of the case study method, the reader is encouraged to consider critically the responsibilities of the scientific researcher under the principles relating to intellectual property rights.

Review of Ownership of Research Data

Ownership of research data

Dictionaries typically define data as facts or information that serve as the basis for decision making, discussion and reasoning, or calculation. In the biomedical sciences, intellectual property is almost always grounded in one or more data sets. Thus, we will consider the basic tenets of data ownership before discussing the various categories of intellectual property. The analysis of ownership of research data begins with the question: Who collected the data? However, equally important is the question: Under whose intellectual direction and guidance were the data collected? If the answers to both questions are the same, that person(s) is the tentative owner. The third question that must be asked is whether or not there was a valid obligation to assign the rights in the data to another. This follows the old common law doctrine that workers are entitled to the benefits of their work product, unless they are obligated to give that work product to another, whether in exchange for money, under terms of employment, or under the terms of some rule or law (e.g., the "work for hire" doctrine; see below).

When the National Institutes of Health (NIH) of the U.S. Department of Health and Human Services awards a research grant to a university, any and all data collected as part of that funded project are usually owned by the university (commonly called the grantee institution). For example, the data books of the principal investigator, predoctoral and postdoctoral trainees, and other staff members working on the project are the property of the grantee institution. Trainees should not be allowed to take their original data books with them when they complete their training programs and leave for new positions. However, the removal of copies of original data or data books may be permitted on a variety of grounds, including duplicative safekeeping and availability of information for manuscript and report writing. Removal of duplicate copies of data should be subject to the approval of the principal investigator. If an investigator were to leave his or her institution during the tenure of an NIH research grant, original data generated as a result of the funded research would still remain the property of the grantee institution. Grants can be transferred from one institution to another when such relocation occurs, but this transfer must meet with the approval of the original grantee institution as well as the NIH. If a principal investigator does not elect to initiate the transfer of the grant from his or her present institution to the new location, then the original grantee institution must petition the NIH to appoint a new principal investigator who would thereafter serve in that capacity.

The scientific community and the public can gain access to original research data obtained as part of federally funded research grants or contracts under the Freedom of Information Act (FOIA). This law allows one to request nonclassified information that is available at any agency of the federal government. Before passage of the Omnibus Appropriations Bill for fiscal year 1999 (Public Law 105-227), a key consideration regarding the data was whether they were in the possession of a federal agency, such as the NIH, a component of the U.S. Department of Health and Human Services. Thus, data records prior to 1999 that were not in the possession of the funding agency were not subject to an FOIA request. However, under current law (with final implementing regulations yet to be adopted as of the time this chapter was written), even records only in the possession of the grantee institution (i.e., the laboratory of the principal investigator) must now be produced, if not otherwise exempt, in response to a request under the FOIA. This situation would be the case for any data that had not been reported to the NIH. Many university and other research groups receiving federal funding are urging the U.S. Congress to eliminate this new requirement under the FOIA. Examples of such reported data routinely found in the possession of a federal granting agency would be those contained in a final report or a progress report that accompanied a new, competing, or continuation grant application. An FOIA request may be denied if the information is classified under a specified exemption (e.g., trade secrets, commercial or financial information, or intrinsically valuable data used to support a patent application or to support a request to the Food and Drug Administration for approval of a new drug).

The current rule regarding retention of research data provides that the data be retained for 3 years from the date of the last expenditure report filed with the granting agency. However, the rights to data access of the granting agency exist for as long as the grantee is in possession of these records. For example, if one should retain data books from an NIH project that ended 17 years previously, the NIH would still have access rights to them throughout that period. Finally, the NIH has the right at any time to inspect any records of the grantee that are pertinent to the award "to make audit, examination, excerpts, and transcripts." Such regulations for data retention may vary from agency to agency (e.g., public funding agency, private foundation). Principal investigators should always be aware of the pertinent rules and regulations that are applied by their funding sources.

Ownership is, in reality, an exercise of a property right (i.e., who is able to exert control over the data, at what times, and under what conditions). As in the exercise of any property right, the ownership is dependent on the context of the property. The context of the property in turn depends on how one protects the data, and this is defined by intellectual property law.

Legal forms of protection of research data

The United States and many other countries recognize four specific forms of intellectual property for which legal protection is available to the owner. These include: (i) trade secrets, (ii) trademarks, (iii) copyrights, and (iv) patents. The current body of laws providing for ownership or the exercise of property right over these forms of intellectual property have developed over the past 200 years. Under the federal system of government in the United States, the states exercise primary jurisdiction over enforcement of trade secrets and, to an extent, share jurisdiction with the federal government over trademarks and copyrights. It should be noted that the Copyright Act of 1976 provided that federal law would exclusively govern the protection and enforcement of almost all copyrights. Patent law has been the exclusive purview of the federal government since the passage of the Patent Act of 1790. While the original colonies granted patents (and some granted copyrights), federal law quickly replaced that of the various states.

However, the legal right to exercise control over research data is a different consideration from when and how to ethically exercise such rights. Since scientific research is based upon the sharing of research data and materials following publication, researchers may find that the failure to share published information and materials may run counter to the publication policy of the journal in which they publish and the agreement between themselves and the journal publisher.

Trade Secrets

Legally defined, a trade secret means information, including a formula, pattern, compilation, program, device, method, technique, or process that:

i) derives independent economic value, actual or potential, from not being generally known, and not being readily ascertainable by proper means by other persons who can obtain economic value from its disclosure or use and

ii) is the subject of efforts that are reasonable under the circumstances to maintain its secrecy.

In other words, a trade secret is information that is not publicly known, but that confers an economic value upon its owner *and* its owner takes reasonable steps to maintain the information as secret. The protection of trade secrets is governed by individual state laws, not federal laws. Traditional legal protection of trade secrets is founded upon principles of contract law and civil misappropriation but does not cover unauthorized use per se. However, legal action can be taken against someone who fails to keep the secret as obligated under contract or a fiduciary relationship or against someone who obtains the secret illegally. A Federal Trade Secrets Act provides criminal penalties for a federal employee who discloses without permission information that concerns or relates to trade secrets provided to the U.S. government.

For information to qualify as a trade secret, the courts, in actions brought for infringement, have based their decisions on such issues as (i) the information was not readily available by independent research; (ii) the information must have been used in business operations; and (iii) the information provided a competitive advantage. Other issues used by the courts in determining the status of a trade secret have included the cost of developing or acquiring the trade secret, who within the business knows the trade secret, and what the business has done to ensure that the information remains secret. However, independent research and "reverse engineering" approaches have been determined to be legitimate means to obtain trade secret information.

The Economic Espionage Act of 1996 (Public Law 104-294) is a new federal criminal statute enacted by the U.S. Congress, which provides for monetary penalties, incarceration, and forfeiture of property for the theft or misappropriation of trade secrets. While this legislation was primarily intended to prevent foreign governments and businesses from illegally obtaining trade secrets of U.S.-based commerce, its definition of trade secret casts a wide net.

Unlike other forms of intellectual property, there is no expiration date for a trade secret. It is in force as long as the information remains secret. This imposes a significant burden upon the owner to take reasonable precautions to ensure that trade secrets do not become publicly known. For example, the recipe for the Coca Cola brand soft drink has been maintained for many years as a trade secret. However, the moment the company fails to maintain the information as a secret and the information becomes public, the owners will lose the protection of the trade secret. Trade secrets may be assigned or licensed to other parties in the same manner that any other form of intellectual property may be sold or leased. Such arrangements require that the recipient be legally bound to keep the information secret.

Sophisticated and powerful chemical, physical, and biological analytic procedures make the use of some trade secrets impractical, especially in the biomedical and biotechnological industries. Today it would be difficult, if not impossible, to maintain a genetic cell line, sequence, or other biological composition as a trade secret. Unlike purely chemical compositions, many biological materials have the unique ability to replicate faithfully in vivo (e.g., cell line propagation) or in vitro (PCR amplification of DNA sequences), thus lending themselves to analysis in ways that can yield secret information. In short, trade secret protection for most biotechnological intellectual property is impractical because of the resolving power of modern analytic technology.

Trademarks

Trademarks embody pictures, sounds, writings, devices, or objects that allow the owner to identify and distinguish some idea, concept, service, or product from those of a competitor. Trademarks protect an idea that conveys the good will or reputation of a product or service of the owner. Consumers often rely upon trademarks to know what they can expect if they buy the product or service. This affords a degree of predictability in commerce that is important to business. A related mark is the service mark, which serves the same purpose as a trademark but denotes a service rather than a product. Trademarks may be registered at both the state and federal levels. Alternatively, a trademark can be used without any type of legal registration; however, enforcement against an infringer of the mark may then be limited.

Federal trademarks are issued by the U.S. Patent and Trademark Office (PTO) for a fee upon the filing of an application by applicant and search conducted by the PTO. Trademark registration lasts for 10 years but can be renewed indefinitely for 10-year periods (with fees and the filing of an application). A mark registered with the PTO should be identified by the letter R in a circle (i.e., ®) displayed next to the mark. Foreign trademark protection must be sought separately in the foreign jurisdiction in which protection is desired. The unauthorized use in commerce of a mark (trademark or service mark) owned by a first party may constitute infringement by a second party if the latter's use creates a likelihood of confusion as to the source of the goods bearing the mark. The courts have considered various defenses to an action against an infringer, including (i) whether or not there was a likelihood of confusion, (ii) whether the mark was valid, (iii) whether the use was authorized, and (iv) whether the mark was merely a descriptive term.

Copyrights

A copyright protects the *expression or presentation* of an idea, but it does not protect the idea itself. Work to be copyrighted must be fixed in some type of tangible medium. This includes material that must be accessed in some way with the assistance of a machine (e.g., audiotapes, videotapes, and computer diskettes). Anyone can use your ideas even if they're protected by copyright.

A copyright comes into existence the instant the author's words or actions are rendered into some tangible form. Although formal action beyond this is not needed, it is recommended that appropriate forms be filed with the U.S. Copyright Office. In addition, a small fee and deposit of the work with that office are necessary. Copyrighted works produced after 1977 by individual authors are protected for the life of the author plus an additional 50 years. Copyrighted works created on a "work for hire" basis (employee's creation, but assigned as work by employer; see below) are protected for 75 years from the date of publication or 100 years from the date of creation (whichever comes first). Copyrights on material copyrighted before January 1, 1978, were in force for 28 years after initial registration; these were renewable for an additional 47 years.

What may be copyrighted falls into two categories: original works and derivative works. Original works include all forms of tangible expression created independently by the author and not copied from any previous work. An original manuscript prepared on your research findings that contained text, figures, and tables is a good example of an original work. Derivative works would include those works created by the author while relying upon other works but does not include the mere copying of those works relied upon. As an example, consider a review article that contains numerous previously published tables and figures from the literature, along with the derivative author's original text interpreting, explaining, or discussing the published literature. Copyright permission would have to be sought and granted to use the figures and tables, but as expressed in your manuscript they would be covered by the copyright protecting your review article. Similarly, your review might discuss the research findings of several papers of others by paraphrasing their writings. This is not a copyright infringement. Moreover, your new written expression of their ideas enjoys its own copyright protection.

It is important to distinguish the requirement of *originality* for copyright purposes over the requirement of *novelty* for patent purposes (discussed below). Work that comprises material that is entirely in the public domain cannot be copyrighted (e.g., common mathematical tables, calendars). The U.S. Constitution provides that only an author is entitled to secure copyright protection. The courts have reasoned that authorship conveys a requirement of originality. The copyright statute similarly provides protection only for *original works of authorship*. While originality may appear to be the same as novelty, "originality means only that the work owes its origin to the author, i.e., is independently created, and not copied from other works." This requirement is in contrast to the prerequisite of novelty for the patenting of an invention. All inventions to be patentable must be novel; that is, the invention must not have been known or used by others in this country nor have been patented or described in a printed publication in this or a foreign country. The requirement for originality is not as rigorous or high a standard as novelty. The copyright originality requirement is not as difficult to satisfy as the patent requirement of novelty. Because originality is easier to meet, the validity of a copyright based upon the work's originality is easier to defend than the validity of the patent grant based upon the invention's novelty. Conversely, the proof of copyright infringement is more arduous and requires evidentiary showing of not only substantial similarity but also the act of copying.

Consider the following example, which invokes the principles of originality and novelty. Laboratory technician Smith creates a computer software program that calculates the half-life of radioisotopes and monitors the inventory of those isotopes in storage using data from a scintillation counter. Ms. Smith's intellectual property could be patented *and* copyrighted. The copyright protection would cover the actual written program (not the idea). The patent would protect the concept of calculating radioisotope half-life and inventory by using the scintillation counter data. The originality of the software would be easily established, since the concept originated from Smith. The novelty of Smith's invention may not be so easily satisfied if another had published a similar (but not the same) invention that used the same elements or components of Smith's invention. If a copyright and patent were each granted to Smith, the validity of the copyright would be difficult to challenge unless the challenger provided evidence of Smith's having copied the work of another. However, the challenge to the validity of the patent may not be as difficult if a challenger were to provide the written description of another's invention that used the same elements or components as Smith used and claimed in her patent.

The owner of a copyright has exclusive rights over reproduction, distribution, sale (or other transfer), and, if appropriate, public performance of the work. The copyright owner also may authorize others to do the same. Copyright is explicitly indicated by the symbol © along with the year of publication or creation. The word "copyright" can be substituted for or used in addition to the © symbol. The author's name should appear along with this indication, if not obvious elsewhere on the work. Indicating copyright in this manner is recommended (but not required) even for unpublished work. Language indicating restrictions is frequently included. Examples of such restrictive language include the following.

- *Copyright © 1991 by Jane Smith. All rights, including the right of presentation or reproduction in whole or in part in any form, are reserved.* This would have special meaning for a work of drama, for example. Even one scene from the play could not be performed publicly without permission from the author.

- *Copyright © 1992 by Jane Smith. All rights reserved. No part of this publication may be reproduced, stored in a retrieval system, or transmitted in any form or by any means, electronic, photocopying, recording or otherwise, without the prior written permission of the publisher and authors.* This language speaks to the prohibiting of electronic scanning (or retyping) of material into an electronic format that could be accessed by computer.

Coauthors own the copyright on their part of the work. If partitioning of this sort cannot be plausibly done, then the authors are equal co-owners of the copyright. They must let each other use the work, but it cannot be licensed to another party without the permission of all the others. Of course, as with any property right, the true owner(s) may assign his or her rights to another. However, assignment may not be required, if the work constitutes a "work for hire." A work for hire is that work prepared by an employee within the scope of his or her employment. Where the employer is the hiring party and the employee has created a specifically assigned work within the scope of employment, the employer will own the copyright. Alternatively, work may be prepared on a special order, commission, or contractual basis, and such work is also considered work for hire. In this case certain requirements must be met. Specifically, a written agreement must exist that provides that the copyright will vest in the hiring party. Furthermore, the work must fall into one of nine categories. These include works or writings prepared as (i) a contribution to a collective work, (ii) an audiovisual work (e.g., a motion picture), (iii) a translation, (iv) a supplemental work (i.e, something written to accompany a primary work, such as a book foreword), (v) a compilation, (vi) a textbook intended for instructional use, (vii) a test, (viii) answer material for a test, and (ix) an atlas.

If an employee is assigned to write an instruction manual for a company instrument, then the copyright belongs to the employer. If, however, the employee writes such a manual without being asked or specifically assigned, then the employee owns the copyright. One academic institutional intellectual properties policy affirms this in the following way: "Assigned duty is narrower than 'scope of employment', and is a task or undertaking resulting from a specific request or direction. The general obligation to engage in research and scholarship which may result in publication is not an assigned duty. A specific direction to prepare a particular article, laboratory manual, computer program, etc., is an assigned duty" (Intellectual Properties Policy, Virginia Commonwealth University, Richmond, 1988). Thus, in the context of this language, faculty who prepare original articles on their research findings hold the copyright to such material. When accepted for publication, the author(s) usually assigns the copyright to the publisher of the journal in which it will appear. The NIH and funding agencies in general encourage the publication of research results. The NIH specifically provides that appropriate material created under a grant may be copyrighted by the grantee. In practice, this usually means the principal investigator (and any coauthors) hold the copyright. However, as with ascertaining any legal right, competent legal counsel should be sought in order to understand the effect of all applicable laws and regulations.

Current copyright law provides that *fair use* of copyrighted material will not constitute an act of infringement. An individual may copy from a protected work as long as the value of the work is not diminished and such activity is nonprofit in nature. Fair use activities must be related to (i) criticism, (ii) news reporting, (iii) teaching, or (iv) research or scholarship. Other considerations of fair use include the nature of the work, the quantity and substance of the material being copied as compared with the copyrighted work as a whole, and the possible effect of such use on the potential market for the copyrighted material. Photocopying an article from a scholarly journal for your personal (nonprofit) use is generally recognized as a fair use practice. On the other hand, preparing a compendium of photocopied chapters from several textbooks for use in a graduate course and distributing such documents at a fee to cover the copying costs would likely represent copyright infringement. Such use could be reasoned to diminish value (i.e., students would not buy the books). Thus, the market for the books would be negatively affected. Similar arguments can be made for the photocopying and use of articles from serial publications. Indeed, court rulings have been clear in finding copyright infringement where one who does not hold the copyright and, without permission of the copyright holder, distributes photocopied compendia of those works as well as third-party copying and distribution of serial publication articles. The interpretation of fair use under the above-mentioned criteria holds that the copying and use must be of a personal (nonprofit) nature; that is, articles are copied by the individual who intends to use them under one of the categories related to fair use.

Computer software applications usually are covered by copyright law. Inspection of program diskettes or accompanying literature will reveal program copyright information. Usually, commercially available software is marketed under a so-called *end user's* agreement. Such agreements between you and the software seller provide that you observe copyright law as it pertains to the computer program. Their language usually indicates that the software is being issued to you under a limited, nonexclusive license. This always means you cannot electronically copy the program and provide it to other individuals for their use under any circumstances. Transfer of the software or documentation in whole or in part to another party is often explicitly prohibited. In some cases such agreements specify the conditions for personal use of the software. For example, you might be able to install the program on no more than one or two of your personal computers. Wording published on software packages often states that you agree to the terms of the software license when you break the seal on the software package or open the envelope that holds the electronic or optical media. Thus far, courts have been readily inclined to enforce these so-called "shrink-wrap licenses" where the software licensee breaks the seal on the software package and installs the software on his or her computer.

Some software is marketed under agreements called site licenses. This commonly applies to educational and business institutions and involves the authorization of multiple users for a software program. In this case the license is made to the institution, and the individual agrees to honor the copyright that protects the software. Site licensed software can be used only at the institution that holds the license. So-called *copy protected* software makes the unauthorized use of software difficult, if not impossible. Copy protection may be part of the software system itself or may involve a hardware device that is sold with the program. Such protection prevents copying or use of the software on machines other than the one on which initial installation took place. Copy protection is used by some manufacturers for specialized or costly programs. However, the majority of contemporary software packages do not come with significant copy protection, although upon installation many current software products require that the installer enter a specific serial number that is provided under the license. Without the serial number, additional copies of the software may not be subsequently installed. Thus, users of such software are entrusted with ensuring the appropriate legal operation of purchased programs. Transgressions of computer software copyrights are morally and legally wrong.

Patents

The term "patent" is derived from the Latin *patens* meaning "to be open." This term refers to the royal grants of the British monarchy which were "letters open" or *litterae patentis*. The early British patents granted during the 14th through 16th centuries were in fact royal grants of monopoly in a specific field or for a specific product. A corrupt practice of selling royal grants for tribute brought such patents into disrepute.

The modern patent is a grant by a national sovereign government to an applicant for a specific and limited period of time during which the grantee has a legal right to exclude others from making, using, or selling his or her claimed invention in exchange for the grantee's providing a full disclosure as to how the invention may be made, may be used, or functions. This is the classic example of the *quid pro quo* (this for that), a contractual exchange between parties. One party is the sovereign, acting on behalf of society, who provides this period of exclusivity to the second party, the patentee, in exchange for the patentee's providing a full disclosure of novel, nonobvious, and useful inventions. This exchange is viewed as one of the most powerful forces for advancing the technological basis of a nation's economy. All developed nations have national patent statutes and are signatories to international patent treaties.

A patent is governed by explicit law. The United States patent law can be traced to legislation presented before the first session of the First Congress. The U.S. patent statutes are the product of several major revisions and recent amendments. Current patent statutes are codified at Title 35 United States Code (Supp. 1999) from the Patent Act of 1952. Under U.S. law, a patent grants an individual (coinventor) or group of individuals (coinventors) the legal right (personal property right) for a defined period of time to exclude all others from making, using, or selling the invention as claimed. The United States amended its patent statutes in 1994 in order to bring the term of a U.S. patent into conformity with those of other nations. For those utility patents or plant patents filed on or after June 8, 1995, the term begins on the date the patent issues and continues for 20 years from the filing date of the earliest filed application (e.g., the term of a patent issuing on January 11, 1996, from an application filed July 11, 1995, expires on July 11, 2015; note that this is an enforceable term of $19\frac{1}{2}$ years). For those utility and plant patents filed before June 8, 1995, the term is the longer period of 17 years from the date of issue or 20 years from the filing date of the earliest filed application (e.g., a patent issuing on January 11, 1996, from an application filed on May 11, 1992, would have an enforceable term of 17 years, which is the greater of 17 years from date of issue or 20 years from earliest effective filing date). If a patent claims a composition of matter or process for using a composition of matter that has been subjected to a regulatory review by the Food and Drug Administration, the term of the patent may be extended up to 5 years beyond the original 17 or 20 years. Design patents have a term of 14 years.

In return for this property right, the inventor provides full and complete instructions regarding the claimed invention: how to make or use it, its useful purposes, and, to an extent, how it functions. So a patent is a reward for disclosing something of social value to the public. The law states that:

> Whoever invents or discovers any new and useful process, machine, manufacture, or composition of matter, or any new and useful improvement thereof, may obtain a patent therefor, subject to the conditions and requirements of this title.

This law is specific to individual countries, but there is much interest in "harmonizing" patent statutes globally. Patent protection is guaranteed only in the country where the patent has been issued. A U.S. patent on a specific invention does not preclude others from making, using, or selling the invention in Japan, for example. However, a U.S. patent that claims a *process for making* a composition or product may be enforced and preclude the import of the composition or product even if the acts that would otherwise infringe the patent if performed in the United States were performed in another country.

Contrary to common thinking, under the patent statute, a patent does not give someone the right to make, sell, or practice the invention. It simply permits the inventor to *exclude others* from making, selling, or using the invention. However, common law provides a right to the inventor to practice his or her invention. This right may be dominated by patents held by others. For example, a patent claiming the use of a recombinant plasmid for the overexpression of a gene could dominate a patent claiming the use of that vector for the isolation of large quantities of a novel enzyme. In such a situation, the parties involved would need to cross-license with one another to practice their own invention or risk an infringement action. Because a patent is considered personal property, it can be sold or transferred (assigned) to another or it may be rented (licensed) in whole or in part for the full or partial term of the patent.

For a subject matter or invention to be patentable it must be useful, new or novel, nonobvious, and reduced to practice. Reduction to practice must entail either the actual reduction to practice by the creation of a working model (which is operable) or the constructive reduction to practice by the filing of a patent application. A patent may issue only from an application that provides a comprehensive description enabling one "skilled in the art" to practice the claimed invention. Inventorship of patentable subject matter requires both the conception and the act of reduction to practice. The inventor(s) of an invention who applies for and receives a patent is recognized as the patentee or patent owner; his or her rights under a patent are considered personal property rights and are assignable.

In the absence of a written agreement to the contrary, the patentee owns the patented invention. The employer may obligate assignment of invention rights if the employee is hired to specifically perform research and invent. Under the "shop right" state laws, the employer may own a personal, nontransferable, royalty-free nonexclusive license to the patent *if the employee used the employer's time, materials, or facilities* in the course of inventing. The scope of the shop right is determined from the nature of the employer's business, character of invention, circumstances of its creation, and law of the specific state of jurisdiction.

The point in time to file a patent should be as soon after the invention is actually reduced to practice or as soon as the inventor is able to provide the full and complete disclosure that is required to achieve the constructive reduction to practice. In the United States, the applicant is permitted to file an application *within* 1 year of the first disclosure (publication of a scientific paper or, in many cases, presentation before a public meeting). However, publication or public disclosure will most likely result in the loss of foreign patent rights, unless either a provisional or regular utility patent application is filed prior to the public disclosure. A filing in the U.S. PTO can protect the foreign patent rights if a subsequent foreign patent application(s) is filed within 1 year of the U.S. filing.

Research sponsored under a federal funding agreement (grant, cooperative agreement, or most contracts) that gives rise to an invention can become the property of the funded nonprofit organization or small business ("contractor"), if the contractor elects to take title to the subject invention and notifies the funding federal agency. When the contractor elects title, it (i) is required to periodically report to the federal agency on the utilization of inventions, (ii) is required to place a notice in the patent specification (description) identifying the federal support, and (iii) must provide a share of the royalties of any licensed subject invention to the inventor and utilize its royalties for scientific research or education. In the event that the contractor declines to elect title to the subject invention, the federal agency determines whether it wishes to retain title. If the inventor requests, the federal agency may waive title to the inventor, who may commercialize the invention.

The United States and a few other countries operate under a "first to invent" policy. Patent prosecution and litigation are based on who can demonstrate that they were the first to invent. Most countries operate under a "first to file" system where patents are awarded and litigated based on who files the application first. There is international interest in harmonizing the national patent laws to provide the same standard throughout the world. However, in the United States there is a long tradition of granting the patent to the individual who invents first. The United States recently amended its patent statutes in order to further implement the General Agreement on Tariffs and Trade. Effective June 8, 1995, an applicant may file a provisional patent application, which is not examined by the PTO and does not require the same degree of formality as the regular utility patent application. The provisional patent application may merely consist of a copy of a scientific manuscript prior to its publication. However, crafting a provisional application with an eye toward filing a regular patent application provides a good foundation for continuing to seek protection. For example, such a provisional application might contain claims drawn to the invention or subject matter. Generally speaking, the inventor is best served by a provisional patent application that is as complete as possible so as to provide an enabling or full disclosure of the invention that is subsequently disclosed and claimed in the regular utility patent application. If a regular utility patent application and any foreign patent applications are filed within 1 year of the date that the provisional patent application is filed, patent rights in the United States and internationally may be generally preserved.

Historically, the public disclosure of an invention or subject matter to be patented allowed an application for a U.S. patent to be filed within 1 year of the disclosure date. However, this disclosure immediately precluded the possibility of seeking patent protection outside the United States. By filing a provisional patent in the United States, the inventor gains a year of protection, even if his or her invention is disclosed during that time. In effect, the provisional patent provides a 1-year grace period during which the rights to file for foreign patents are preserved, despite a subsequent public disclosure. It is important, however, that the provisional patent application be filed before any public disclosure occurs (e.g., manuscript publications, oral presentations). In summary, the provisional patent provides a convenient and inexpensive way to maintain protection of inventions in terms of foreign patent rights.

A compromise appears to be under consideration that would permit "prior user rights" for companies or individuals who have developed and used patentable products or technology but were not the "first to file." However, this proviso would not help university researchers, since most inventions are not commercially practiced in that environment. Instead, universities usually license their patents to companies, and the proposed prior use rights could not be transferred to licensees. Current patent law amendments in this area are being argued by some as having severe negative effects on universities.

The types of subject matter that can be patented include processes, machines, products, or composition of matter. Patents can also be obtained for modifications or improvements to any of the above. Any new and distinctive variety of plant that is asexually reproduced (excepting plants of the tuber-propagated family or plants propagated by seed) is considered patentable subject matter under a plant patent. Sexually reproduced plants and tuber- or seed-propagated plants can be registered by the U.S. Department of Agriculture under the Plant Variety Protection Act. However, one recent court held that sexually reproducible plants are also patentable subject matter under 35 U.S.C. § 101 (see *Pioneer Hi-Bred International Inc. v. J.E.M. AG Supply Inc.*, N.D. Iowa, No. C-98-4016-DEO, August 19, 1998). This case is currently on appeal to the U.S. Court of Appeals for the Federal Circuit. It remains to be seen whether this decision will be overturned or sustained. A recent amendment to the patent statutes did include *parts* of asexually reproduced plants as well as the whole plant as claimed under the plant patent statute.

Finally, design patents provide protection for any new, original, and nonobvious design for a product (e.g., a new automobile body).

Patent Law in the Age of Biotechnology

Recent developments in U.S. patent jurisprudence may have a significant effect upon the development of new technologies. A number of controversies have erupted over the patenting of life forms or their components. For example, a recent report described the filing of a patent application for a method for making creatures that are part human and part animal by combining embryos of both and implanting these hybrid or chimeric embryos into surrogate mothers. While the report noted that the inventor did not intend to make such creatures, his goal was principally to provoke public debate and possibly initiate a case that could reach the U.S. Supreme Court, concerning the morality of patenting life forms and engineering human beings. The PTO released a "media advisory" entitled "Facts on Patenting Life Forms Having a Relationship to Human." This statement by the PTO outlined the agency's responsibilities to issue patents that meet the statutory requirements, including the utility requirement. The PTO further noted that inventions directed to human–nonhuman chimera may, under certain circumstances, not be patentable because they may fail to meet the public policy and morality aspects of the utility requirement. Such a strong view of public policy on morality grounds under the utility requirement is not universally embraced among members of the patent bar.

This is understandably a highly charged political issue; however, the PTO's position is that it can distinguish a legitimate medical research animal from a monster. Research scientists and patent attorneys may not be so sure. Numerous patents are issued that cover transgenic animals, cell lines, and other compositions that contain human genes. It is by no means clear what constitutes the threshold amount of human genetic material required to trigger such a holding of lack of utility on moral grounds. The PTO's position is based upon an 1817 court decision which states that an invention is patentable unless the invention cannot be used for any honest and moral purpose. In this connection other courts have held that the law requires that the invention not be frivolous or injurious on either practical or ethical grounds. Current law provides a minimum threshold of the utility requirement and gives little weight to any consideration of the morality of the use of the invention.

Another aspect of patentable biotechnology research relates to gene therapy of the human germ line. Both human and nonhuman animals are made of somatic and germ line cells. The germ line cells — egg cells and sperm cells — have reproductive capability, while somatic cells do not. The combining of the germ line cells during fertilization results in the genetic composition of the embryo. So the genetic sequences of the germ line cells are inheritable, being passed from parent to offspring (see chapter 10 for additional background material). Genetic therapy directed to the germ line may in some instances be more technically effective in replacing or repairing mutations that cause disease. However, modifications in the germ line may affect generations, while the somatic modifications affect only the individual. Although patents have issued with claims regarding gene therapy of somatic cells, this is not true for any gene therapy involving germ line cells. Since patent applications are confidential unless the applicant makes the application public, those applications directed to human germ line gene therapy would not become public until they issue. It remains unclear how the PTO will address claims drawn to gene therapy of the human germ line. It may be more difficult for the PTO to deny such inventions on utility (morality) grounds, particularly in view of the potential medical benefits to patients suffering from inheritable genetic diseases. On the other hand, without the incentive to invest in the costly and time-consuming research to create new medical treatments provided by secure patent protection, development of vectors and other compositions useful in human germ line gene therapy may be discouraged. Critics of genetic therapy could view this inhibition of development as a way to protect the natural evolution of human genetics.

Along these same lines, there is concern over the patenting of expressed sequence tags and single-nucleotide polymorphisms, which are partial genetic sequences. Many critics of the patenting of genetic sequence view patent protection of large numbers of partial genetic sequences as interfering with scientific research by impeding the free exchange of materials and information, although many patent applicants also make their genetic sequence databases accessible. Others have expressed concerns that the commercialization of human genetic sequences raises ethical issues. Patent law is ill equipped to address such policy issues. Statutes providing property rights in intellectual property are a mechanism to achieve social goals, such as promoting technological and commercial development as well as international economic competitiveness. Whether those goals should be restricted or left open to competitive enterprise continues to be debated.

On a more technical level, the patenting of genetic sequences, like the patenting of any other composition of matter, requires that the invention be a useful, novel, and nonobvious composition. Further, the applicant must provide an adequate written description of the invention and provide an enabling disclosure of how to make and use the invention. In 1991, the NIH filed a patent application for 351 genetic fragments sequenced from brain tissues. The PTO rejected the application in 1993, and the NIH chose not to appeal the decision. The courts have clearly stated that an applicant's general disclosure of a genetic sequence that fails to provide an adequate written description of the invention will not support the patenting of specific genetic sequences. In *Regents of the University of California v. Eli Lilly* [119 F.3d 1559, 43 USPQ2d 1398 (Fed. Cir. 1997)], the court found that claims to a *human* DNA-encoding insulin were not adequately described by the disclosure teaching a *rat* DNA-encoding insulin. Therefore, an applicant's written disclosure of a partial genetic sequence may not be sufficient to support claims drawn to the complete gene sequence. In 1997, the PTO issued its first patent that claims expressed sequence tags encoding portions of novel protein kinases. The issue of the utility of expressed sequence tags is still open to question in the minds of some and is likely to be the subject of continuing debate.

Beyond the legal requirements for the patenting of cell lines, genetic constructs, and transgenic animals and plants lie the cultural issues that raise the question of whether such materials should be patented, even if patentable. International debate has been stimulated by the patenting of human cell lines isolated from clinical samples of indigenous peoples, the patenting of plants used in religious rituals and considered sacred by Amazonian people, the patenting of new varieties of plants that have been considered cultural assets, such as basmati rice of India, and the construction of transgenic animals and plants used for medical research and agriculture. As the world evolves a more integrated economy, many of these intercultural views raise religious, economic, and sociological issues that require ethical as well as legal analyses.

In the United States and many developed countries, the biomedical research community, government leaders, and others have considered whether the commercialization of biotechnology may be hampering the sharing of research tools. Some note that proprietary genetic constructs are not accessible to the research community, while complex commercial license arrangements may be needed for the distribution of gene chips or cDNA library arrays. Others consider that increasing competition for research funding and an increasingly competitive global economy may exert undue pressure upon universities and other nonprofit organizations to seek patent protection and commercialize research inventions. Still, the patent system appears to remain a grand experiment that provides incentive to the inventor through the grant of a limited period of exclusivity, which in turn has stimulated the development of exciting new technologies and greatly advanced the quality of life for millions throughout the world. The patenting of biotechnology inventions remains a challenge to scientists and nonscientists alike, but one principle remains clear: new inventions will always arise. This will inevitably result in the continued evolution of patent laws, which must take into consideration new societal needs and concerns by changing in some instances from traditional precepts to more responsive policies.

A final relevant anecdote of traditional patent lore has held that patents may not be obtained for methods of conducting business. Further, the patenting of computer software has been fraught with requirements that the software to be patentable must involve the transformation or representation of a physical object. Recently, the court held that a general-purpose computer programmed to implement a business-oriented process qualifies as patentable subject matter [*State Street Bank & Trust Co. v. Signature Financial Group, Inc.*, 149 F.3d 1368, 47 USPQ2d 1596 (Fed. Cir. 1998)]. This court ruling has major implications for the protection of computer software in general and for software used in research laboratories specifically. As mentioned above, most software programs are now protected via copyright, which in effect is licensed to the end user. The use of patents to protect software becomes legally plausible under this court ruling. But the extent to which patent protection will affect the software industry and the use of software in the future remains to be seen.

Seeking a Patent

To obtain a patent in the United States, one files a patent application with the PTO in Washington, D.C. (the office complex is physically located in Crystal City, Virginia). Prosecution of a patent application generally takes from one to several years. In some fields of technology, particularly biotechnology, it may take from 3 to 5 or more years before the patent is granted. Patent applications may be prepared and prosecuted before the PTO by registered patent attorneys or registered patent agents. While the inventor is always entitled to prepare and prosecute on his or her own behalf, no one else may represent the inventor before the PTO unless he or she is admitted to practice before the PTO. The law states that: "Whoever, not being recognized to practice before the United States Patent and Trademark Office, holds himself out or permits himself to be held out as so recognized, or as being qualified to prepare, or prosecute applications for patent, shall be fined not more than $1,000 for each offense" [35 U.S.C.§33 (1984 and Supp 1998)]. The requirement for patent attorneys or agents to be registered by the PTO is to ensure that only qualified practitioners represent inventors. Patent prosecution procedures are highly regulated with myriad rules, regulations, and deadlines. The failure to meet a deadline may cause the applicant to lose his or her right to obtain a patent. Generally, in the field of biotechnology, an uncomplicated patent application (e.g., utility patent) prepared by a law firm may cost from $10,000 to $20,000. In contrast, a provisional patent application, similarly prepared, may cost much less. However, if a provisional application is poorly prepared and not fully enabling for the invention as claimed in the later filed regular utility application, the provisional application may be a waste of time and money and result in the loss of patent rights. Submission of a patent application is no guarantee that a patent ultimately will be issued.

The usual first step in the preparation of filing a patent application is for the inventor to file an Invention Disclosure with the inventor's employer or patent attorney. This is key to securing protection of intellectual property in a patent. Invention disclosure forms vary from institution to institution. The scope of information required by such documents is exemplified by the information required on the invention disclosure used at Virginia Commonwealth University (Office of Sponsored Programs,

Virginia Commonwealth University, Richmond, VA 23298). Such required information includes the following.

The invention

1. Title of invention.
2. Attach a concise description of the invention, which should be sufficiently detailed to enable one skilled in the art to understand and reproduce the invention, and should include construction, principles involved, details of operation and alternative methods of construction or operation. Attach drawings, photos, manuscripts, sketches that help describe the invention. Is it a new process, composition of matter, a device, or one or more new products? Is it an improvement to, or a new use for, an existing product or process?
3. What is novel or unusual about this invention? How does it differ from present technology? What are its advantages?
4. What is the closest technology currently available, upon which this invention improves?
5. What disadvantages does this invention have? How can they be overcome?
6. What uses do you foresee for the invention, both now and in the future?
7. Has any commercial interest been shown in the invention? Please give company and individuals' names, and addresses if available.
8. What other companies, or industry groups, might be interested in this invention, and why?
9. Please comment on any preferences or ideas you have for a good way to commercialize this invention.
10. What additional work is needed to bring the invention to a licensable state? Please estimate time and cost.

Timeliness and sponsorship

11. Has the invention been described in a "publication" (journal articles, abstracts, news stories, talks)? Please provide details including dates and copies of written material.
12. Do you plan to publish within the next six months? Please provide approximate date, and any abstract, manuscript, etc.
13. Dates of record, demonstrable from lab notebooks, correspondence, etc.:
    Earliest conception_____
    First disclosure_____ to whom:_____
    First reduction to practice_____
14. Please list all sources of support contributing to this invention.

Besides the above, information must be provided concerning the inventor(s) (name, address, etc.), including the percentage of the contribution of each inventor to the invention.

It is essential that the inventor maintain a properly kept laboratory notebook. In addition to being crucial to preparing an invention disclosure or patent application, the research laboratory notebook is frequently used in responding to challenges either during the prosecution of a patent or in postpatent litigation. Maynard (7) has proposed some questions to consider relating to whether or not to file a patent application. Many of these questions can be simply answered by examination of the laboratory notebook. A few of these questions include:

- What is the nature of the invention?
- When was the invention made and where is the experiment recorded?
- Has the invention been disclosed publicly?
- What is known about the prior work in the field?
- How complete are the data?

Kanare (6) lists additional important points of laboratory notebook writing relating to invention disclosures and patent applications. The conception of an invention that follows from work should be clearly stated. This should be done in a way that documents your own work and compares it to prior work and knowledge in the field. Your laboratory record keeping needs to document that you have worked diligently to reduce your invention to practice. To do this, you must demonstrate that you have worked on your invention continuously. In other words, at no point did you set it aside or abandon it. Having your work witnessed by someone who understands it, but is not an inventor him- or herself, provides important evidence in both the filing of a patent and in postpatent litigation.

Conclusion

Intellectual property law has always been relevant to scientific research. The ability to protect intellectual property by patenting has been a driving force in the application and commercialization of basic research. In today's global economy, no existing or new area of technology can truly prosper and have its maximum impact without the benefit of intellectual property law. This is especially true for the biomedical and biotechnological sectors. We continue to reap the benefits of the biological and digital information revolutions of the last quarter of the 20th century. The commercialization of numerous discoveries in both these areas can be traced to many small companies whose competitive position was made possible by the powerful use of intellectual property protection.

Case Studies

9.1 As is sometimes the case, the student and advisor are not seeing eye to eye. The relationship grows increasingly acrimonious, but the student completes her dissertation and successfully defends it. The student finds a good postdoctoral position and is glad to be moving on. Bitter over the way she believes she has been treated by her advisor, she informs him that she is not going to publish any more of her dissertation work. Further, she informs her advisor that she has copyrighted her dissertation and that he is not allowed to publish any of the work in it either. The advisor is frantic; he received a federal grant to do the work. He needs to show that the proposed work has been carried out by publishing the data in peer-reviewed journals. Yet he is afraid that if he uses any of the data from the copyrighted dissertation he will be sued! What copyright issues and data ownership issues are relevant to this scenario? Imagine you are the faculty member's chair. He comes to you for advice on how to handle this situation. How do you respond?

9.2 A faculty member directs a core laboratory facility at his institution that supplies researchers with custom-made tissue culture media. This valuable service saves investigators time and money. Media can be picked up on 4 hours' notice, and the institution underwrites the facility by paying for the salary of the core lab technician. Thus, the cost of the media is significantly below what can be purchased from commercial suppliers. The rest of the expenses for running the lab (supplies, etc.) are paid for from funds generated from the selling of the media. The faculty director of the core lab realizes that a clinical facility in his university's medical center is discarding 1-liter glass bottles that contained tissue culture reagents purchased for use by the clinical lab. He reasons that he can save additional money by washing, sterilizing, and using the discarded bottles to package the tissue culture media sold by his core laboratory. The washing process removes the paper label affixed to the bottle but does not remove a painted logo used by the company as a trademark. He creates his own labels and affixes the labels on the bottles. The company trademark logo is still visible, however, after the label is in place. Is this a trademark infringement in your view? Would it make any difference if the labels were affixed to the bottles in such a way as to cover the original trademark?

9.3 A recently arrived faculty member is setting up his laboratory at an academic institution. He has just assembled his personal computer and has purchased six different software application programs with grant funds. He is preparing to install these programs on the machine when a colleague drops in on him. She suggests that he save time by simply letting her use a portable tape backup system to install the various application software packages on his machine. She says she owns all of the same software and it will take a couple of hours for him to make all of the necessary installation and adjustment settings. She can "dump" all of the same software onto his machine in about 20 minutes. She argues that since he has purchased the identical software for his use, installation of her software on his machine will not be a breach of any copyright or user's agreement. She indicates that while she is at it she will install several other software programs that the faculty member does not own so that he can try them out. She says that the conditions of this "trial" will be that if the faculty member thinks that he will be using the software, he must go out and purchase a copy for himself. Comment on the legal and ethical implications of this scenario.

9.4   A postdoctoral fellow and his mentor have coauthored a paper describing their research results. This paper appears as a preliminary report in a copyrighted monograph. One of the figures in the paper is a computer-generated graph that describes data on a series of bacterial growth curves. The postdoctoral fellow and mentor are now preparing a major paper for submission to a peer-reviewed journal. They both agree that the growth curve data in the monograph article are crucial to the story they are telling in the present manuscript. Accordingly, they decide that this same figure must be included in their present writing. Because they are aware of potential copyright violations, they generate the exact same figure using different typeface fonts and different line thicknesses for the ordinate and the abscissa. They have decided that since this is not the exact same figure that appeared in their monograph article, the use of it will not constitute a copyright infringement. They also plan to indicate in their manuscript that this figure has been "adapted from" the one initially published in the monograph article. Comment on what these authors are doing. Do you view it as copyright infringement? If so, are there conditions of modification of tables or figures that would sufficiently change them in a way that avoids copyright infringement?

9.5   Dr. Clancy has been invited by Dr. Cook to write a chapter on protein structure. Cook is editing an introductory biochemistry text to be published by The Dawson Publishing Company. Clancy is paid a one-time honorarium of $600 for his chapter. He signs a property transfer agreement assigning the copyright for his manuscript to the publishing company. The book does exceptionally well in its first edition, and Cook signs a contract with Dawson Publishing to edit a second edition. Because Cook was not happy with Clancy's original chapter, he invites Dr. Pearson to write the protein structure chapter for the second edition. Pearson writes the chapter using three illustrations taken from Clancy's chapter. He also includes several of the end-of-chapter problems written by Clancy. Most of the text of the second edition chapter was written by Pearson, but there are several instances where parts of paragraphs are verbatim copies of those from Clancy's original chapter. Clancy had been unaware that a second edition was being written. He has just received a complimentary publisher's copy and is incensed. He tells you he plans to file scientific misconduct charges against Pearson for plagiarism. How do you advise him?

9.6 Dr. Harold Hefner subscribes to a popular scientific journal that is published weekly and is also available on the Internet. Hefner receives both the printed and on-line versions of the journal. To access articles on-line, he must log on to the journal's home page with his user name and password. Hefner's research group is composed of several pre- and postdoctoral trainees. He makes his user name and password available to each of his lab trainees, claiming that this is no different from circulating his printed journals using a routing list. He encourages his trainees to print copies of relevant articles appearing in the on-line journal. He cautions them that they should make copies only for their personal use in order to be consistent with the fair use doctrine of copyright law. Some of Hefner's trainees regularly peruse the on-line journal and print papers for use in their research. Others in his group refuse to use the on-line journal, arguing that such a practice is different from using the printed journal to make a photocopy for their personal use. From class discussion and your reading, do you think that Hefner's policy is legal? Is it ethical?

9.7 A newly characterized infectious agent is under intense study, and a number of its genes have been cloned and sequenced and the results have been published in the peer-reviewed literature. The organism, an obligate intracellular bacterium, is fastidious, and identifying it in clinical material is difficult. Dr. Greg and his group are studying the properties of fresh clinical isolates of this bacterium. Jeff Charles, one of Greg's postdoctoral fellows, discovers a newly issued patent in which the inventors (who are competitors of Greg) report the use of specifically designed oligonucleotides to amplify a combination of three different genes and, in doing so, provide definitive identification of the bacterium. The gene sequences from which the oligonucleotides were designed are all published and available in public domain databases. But the nucleotide sequences of the oligonucleotides derived from the genes are disclosed only in the patent. Jeff purchases a copy of the complete patent from a clearinghouse. He uses the information to synthesize the necessary oligonucleotides and finds that they work as predicted. Jeff proceeds to use this technology to sort out dozens of clinical isolates that were previously in question. When he presents his results at a departmental seminar, a heated discussion ensues about what he has done. Several people argue that he cannot legally use the information in the patent for his research, because the patent claims these oligonucleotides and their use in a method to identify the bacterium. They contend that a patent gives you the right to prevent others from practicing your invention. Others in the audience say that because Jeff is using these oligonucleotides for research and not to make a profit, what he has done is permissible. Comment on the legal and ethical aspects of this scenario.

9.8 Jim Stocking is serving a 4-year term as a member of an NIH study section. His service is a matter of public record, and his name appears on a roster distributed with all written critiques to grant applicants. In preparing his own grant application, Stocking reproduces a table and a figure taken from the background and significance sections of two applications he has reviewed. He indicates the origin of both items in his own grant and attributes them to their authors. Is this legal? Is it ethical? As the scientific review administrator of the study section, you learn what Dr. Stocking has done. What, if anything, will you do?

9.9 Dr. Art Murray, a new faculty member in the chemistry department, is assigned directorship of the laboratory safety course. This course is required of all graduate students in several departments in the School of Natural Sciences, including chemistry, biology, physiology, cell biology, and genetics. The course has no syllabus, and over the next 2 years Dr. Murray writes a complete syllabus containing useful reference material, well-documented procedures, and problem sets. He publishes a Web site that contains all the syllabus material in a useful format. During his fourth year as an assistant professor, his chair, Dr. Janet Bell, tells him that his faculty contract will not be renewed. She explains that the department is losing a position because of budgetary cutbacks, and Murray's position must be vacated in order to balance the budget. Murray is very upset but lands a new job at another university. Murray removes the course syllabus from the university computer and uses it in a comparable course at his new institution. The next year Dr. Bell decides to teach the laboratory safety course and intends to use the electronic syllabus written by Murray. She is surprised to find it missing from the university's computer. She finds that Murray has taken all the files for the syllabus Web site. He claims he holds the copyright and that Bell's university can license the site from him for a fee of $1,000 per year. Dr. Bell is angered by this and reminds him that she assigned him the course directorship; thus, she considers the Web site as being done on a work for hire basis. She concludes that her institution holds the copyright on the laboratory safety course Web site. Comment on the legal aspects of this scenario. Regardless of legal interpretation, do you consider Murray's actions to be ethical?

9.10 Ron Roman, a postdoctoral trainee whose work is funded by a research grant on which his mentor is listed as principal investigator, develops a powerful computer algorithm using a commercially available spreadsheet program purchased with the mentor's grant funds. The particular analysis routine that Ron has developed works completely within the spreadsheet application software. It is a sophisticated routine that has required many hours of design and testing. Moreover, Ron has made it available to all members of the mentor's lab and, based on their comments over several months, has introduced many refinements and improvements to the routine. In short, the system can take raw data from enzyme assays and, together with physiological measurements made in animals, statistically analyze data sets and present the results in multiple graphic formats. The application software used for this project was purchased under an academic institutional site license. The software package is copyrighted by the manufacturer. Ron is considering protecting his algorithm as intellectual property before he distributes it to anyone outside of the lab. Can he copyright the algorithm? Can he patent the algorithm? Can he do both? Will this serve any useful purpose? What advice would you give him?

9.11 Dr. Apple, a researcher working under a National Science Foundation grant, is studying the replication of bacteriophage in *E. coli*. Dr. Apple attends a lecture of a world-renowned scientist, Dr. Ball, who discusses her studies on the replication of a particularly useful bacteriophage that infects *E. coli*. Dr. Apple requests a sample of Dr. Ball's bacteriophage. Dr. Ball declines to provide a sample, even after several persistent and strongly worded telephone calls from Dr. Apple. Dr. Apple, obsessed with securing Dr. Ball's bacteriophage, has a plan. Dr. Apple writes a letter to Dr. Ball and again requests the material. At the conclusion of the letter, Dr. Apple pleads, "If you insist on denying me this virus, at least give me the courtesy of a written response to this letter." Dr. Ball quickly responds with a one-page, one-sentence response: "Forget it!" After receiving Dr. Ball's letter, Dr. Apple (knowing Dr. Ball's propensity for performing all her working tasks at the lab bench) places the letter in a blender, making a slurry using sterile buffer, and spreads the slurry on lawns of bacteriophage recipient strains of *E. coli*. Soon, Dr. Apple isolates the long-sought strain of the bacteriophage. Were Dr. Apple's actions appropriate? Should Dr. Apple's actions give rise to an investigation of possible scientific misconduct? If the bacteriophage were used in a commercial pharmaceutical process and if

Dr. Ball were employed by the pharmaceutical company, would Dr. Apple have illegally obtained a trade secret from Dr. Ball?

9.12 Susan Barnes, a cell biologist working in a pharmacology department of a university, has isolated a novel soil microorganism with powerful apoptosis-inducing activity against eukaryotic cells. She tells Jesse Packard, a colleague of hers at a biotechnology company, about her discovery. In turn, Jesse tells the vice president for research at the company, who then invites Susan to give a seminar there. After her seminar the vice president asks Susan to prepare a five-page proposal and says that the company should be able to provide a grant to support some of Susan's work. The anticancer implications of this agent have commercial importance to the company. Susan writes a proposal that says she aims to purify the activity and test it against various cell lines. The grant application is submitted, and an appropriate agreement about intellectual property is executed. The company will have first right of refusal to license the compound from Susan's university, pending her results. The grant is paid as a one-time $75,000 award. The grant provides that Susan should share research materials with the company on a nonexclusive basis. About 1 month into the project, Jesse asks Susan to send him a culture of the microorganism, and she honors this request. A team of scientists at the company have come up with some predictions about enzymes that are likely to be involved in the synthesis of this apoptosis-inducing agent. Over the course of the next several months, they clone the corresponding genes and determine that the pathway for synthesis of the compound is composed of the products of 19 linked genes. They determine the nucleotide sequence of this 35-kb operon. Who owns patent rights for this important biosynthetic operon? Based upon your reading, do you think that the company and its scientists acted legally? Did they act ethically?

Author's Note

This chapter does not purport, nor is it intended, to provide legal advice. The reader is advised in all instances to seek advice from competent legal counsel to ascertain his or her legal rights regarding intellectual property.

Some of the cases in this chapter have solutions that impinge on intellectual property law. Discussants are cautioned against assuming that their proposed solutions to these cases — based on reading and class dialogue — may be legally definitive. Typically, such cases that require legal solutions would depend on the analysis of *all* facts and consideration of current law. This is usually not possible in the scientific integrity classroom. The cases present limited fact patterns designed to provoke discussion based on the general outline of intellectual property law discussed in this chapter.

The references section contains a few publications cited in the text but generally should be considered a reading list to assist the student in seeking additional information on the topics discussed here. A glossary has been included to provide the reader with a convenient source of commonly used legal terms.

Suggested Readings

1. **Doll, J. J.** 1998. The patenting of DNA. *Science* **280:**689–690.
2. **Eisenberg, R. S.** 1997. Structure and function in gene patenting. *Nat. Genet.* **15:**125–130.
3. **Fishman, S.** 1998. *The Copyright Handbook: How to Protect and Use Written Works.* Nolo Press, Berkeley, Calif.
4. **Foltz, R., and T. Penn.** 1999. *Handbook for Protecting Scientific Ideas and Inventions.* Penn Institute, Cleveland, Ohio.
5. **Grisson, F., and D. Pressman.** 1996. *The Inventor's Notebook.* Nolo Press, Berkeley, Calif.
6. **Kanare, H.** 1985. *Writing the Laboratory Notebook.* American Chemical Society, Washington, D.C.
7. **Maynard, J. T.** 1978. *Understanding Chemical Patents.* American Chemical Society, Washington, D.C.
8. **Pressman, D.** 1998. *Patent It Yourself.* 7th ed. Nolo Press, Berkeley, Calif.

Resources

The Internet is a rich source of information on intellectual property law. But the information must be viewed and used cautiously. As disclosed on most home pages, the contents of such Web sites are not meant to provide legal advice. Using a Web search engine, the phrase "patent primer" will usually turn up a variety of such pages. One particularly comprehensive site may be found at

http://www.patents.com

Information on patents and other forms of intellectual property can be found at the Web site of the U.S. Patent and Trademark Office:

http://www.uspto.gov

Another database of patents is maintained commercially and can be searched by a variety of techniques:

http://www.patents.ibm/com

This site enables you to read patent abstracts and allows you to purchase patent specifications at a nominal cost. Patent specifications can be obtained by FTP download or as files on CD-ROM.

The Web site of the U.S. Copyright Office

http://lcweb.loc.gov/copyright/

contains much general information about copyrights as well as a search engine for finding copyright registrations.

Glossary

Civil misappropriation Taking and using of the property of another without permission for the sole purpose of capitalizing unfairly on the good will and reputation of the property owner.

Common law Generally refers to principles of law developed through litigation in the courts, rather than statutes enacted through the legislative process.

Contract law Subset body of law developed as common law and statute that relates to agreements between parties, including rights and obligations of parties.

Copyright A property right over intangible intellectual property concerning original works of authorship fixed in any tangible medium of expression.

Derivative work Work that is compiled by the author from preexisting works; a copyright to a derivative work extends only to that material contributed by the author and not to the preexisting work.

Fair use Statutory protected form of noncommercial use of work under copyright that includes use of work for purposes of criticism, comment, news reporting, teaching, scholarship, and research.

Freedom of Information Act Statute requiring U.S. government agencies to provide upon request documents in the possession of the agency and those whose research is supported under a federal funding agreement and all research data produced therefrom, not otherwise exempted from release under statute [5 U.S.C. §551 *et seq.* (1977 and Supp. 1998)].

Grantee Institution, organization, individual or other person designated in the grant; the legal entity to whom a grant is awarded. In the context of federal funding, the party receiving a grant of financial assistance, as provided under 45 C.F.R. Part 74, for grants from the U.S. Department of Health and Human Services.

Patent — Design Design patents provide a 14-year period of protection for the ornamental features of an article of manufacture.

Patent — Plant Plant patents provide the same term as discussed below for utility patents. Plant patents provide protection for those plants (and parts thereof) that the inventor discovers and is able to reproduce asexually, other than tubers (e.g., potatoes).

Patent—Utility For those patent applications filed on or after June 8, 1995, the term begins on the date the patent issues and continues from 20 years from the filing date of the earliest filed application (e.g., the term of a patent issuing on January 11, 1996, from an application filed July 11, 1995, expires on July 11, 2015; note this is an enforceable term of $19\frac{1}{2}$ years). For those patent applications filed prior to June 8, 1995, the term is the longer period of 17 years from the date of issue or 20 years from the filing date of the earliest filed application (e.g., a patent issuing on January 11, 1996, from an application filed on May 11, 1992, would have an enforceable term of 17 years, which is the greater of 17 years from date of issue or 20 years from earliest effective filing date). Utility patents provide protection for those inventions that are useful, novel, and nonobvious and that constitute a process, machine, manufacture, or composition of matter, or any new improvement thereof; this includes the invention claimed as a drug or claimed as a use of a drug.

Principal investigator A single individual designated by the grantee in the grant application and approved by the Secretary of the U.S. Department of Health and Human Services, who is responsible for the scientific and technical direction of the project.

Provisional patent application An informal patent application filed with the PTO that is less expensive to prepare than a regular utility application. The provisional patent application is not considered by the PTO but remains on file for 1 year. Once filed, this document precludes a subsequent public disclosure of the application's subject matter from destroying the patentable novelty of the invention. Disclosure without provisional patent application protection might otherwise result in forfeiture of patent rights. A regular patent application must be filed by the end of the 1-year period of the provisional patent application, or the opportunity to patent the invention will be lost.

Statute An act of the legislature declaring, commanding, or prohibiting something; a law.

Trademark A distinctive mark that indicates the source of a particular product or service.

Trade secret A formula, pattern, device, or compilation of information that is used in one's business and that gives one opportunity to obtain advantage over competitors who do not know or use it.

chapter 10

Genetic Technology and Scientific Integrity

Cindy L. Munro

Introduction • Genetic Screening and Diagnosis • The Human Genome Project • Manipulating Genes • Conclusion • Case Studies • References • Resources

Introduction

THE KNOWLEDGE AND TECHNOLOGICAL ADVANCES that have emanated from biomedical research during the past 25 years have been remarkable in a variety of ways. For example, our technological ability to isolate, analyze, replicate, change, and generally manipulate genetic information has jumped several quantum levels since the 1970s. Although recombinant DNA technology began as a research technology, it has quickly been applied to the clinical setting. DNA-based reagents are rapidly emerging as tools with unprecedented power in diagnosing and predicting susceptibility to human diseases. Such diagnostic technologies can be applied at stages that range from conception through adulthood.

The Human Genome Project, which promises to provide genetic mapping and DNA sequence information on the estimated 100,000 human genes, will have immeasurable effects in advancing both DNA diagnostics and therapeutics. It is likely to have applications as yet unimagined. Serious possibilities for abuse of the technology and information exist. In recognition of the magnitude of issues related to the science, the Human Genome Project includes a subcommittee to consider ethical, social, and legal implications. Our view of ourselves and our relationship to other species may be profoundly influenced by knowledge of the human and other genomes.

The isolation and manipulation of genes have launched experimental somatic cell gene therapy and led clinicians to begin to debate the merits and dangers of germ line gene therapy. In addition, genetic manipulation could be used to alter or enhance phenotypes not generally associated with

211

diseases. Interesting questions arise regarding the appropriate uses of genetic manipulation in humans.

Controversy has surrounded the issue of property rights related to genetic information. Some forms of genetic information are patentable. The impact of patenting sequences on the development of genetic biotechnology is an area of debate.

Genetic Screening and Diagnosis

Detection of gross changes in the morphology or number of chromosomes has been used in postnatal diagnosis of genetic disease since the observation in 1957 that children who had Down syndrome also had three copies of chromosome 21. Prenatal karyotype analysis for chromosomal abnormalities was first reported in 1967, and amniocentesis was clinically available in the early 1970s. Prenatal diagnosis of chromosomal diseases early enough in pregnancy to permit termination quickly became an option available to parents concerned about bearing children free from chromosomal abnormalities.

The development of new techniques in molecular biology has fueled a revolution in genetic testing. Many diseases result from alterations in the DNA that are too small to be seen in a karyotype analysis. Changes in a single base of the DNA may result in formation of an abnormal product in the cell and systemic disease. Examples of diseases that can result from single base changes are sickle cell hemoglobinopathy and cystic fibrosis.

Before the advent of technology that enabled direct testing of fetuses for genetic problems, carrier testing was used to provide prospective parents with information about the likelihood of genetic disease in their offspring. This methodology gave prospective parents information they could use in decisions regarding whether or not they would choose to have children, but it did not provide information specific to a particular pregnancy. Information about the genetic health of a fetus can be obtained via a sample of DNA from cells of the chorionic villus at 10 weeks' gestation, or from cells in amniotic fluid after 16 weeks' gestation. Results obtained can inform decisions regarding termination of the pregnancy. It is now possible to analyze the DNA of a single cell and to detect changes in DNA as small as a single chemical base. Researchers are exploring the feasibility of isolating fetal cells for amplification and DNA analysis from maternal peripheral blood very early in pregnancy; this could provide testing for single gene disorders much earlier in pregnancy without the risks associated with invasive testing such as chorionic villus sampling or amniocentesis.

Advances in DNA amplification, DNA testing, and in vitro fertilization technology permit assessment of the genetic health of human blastomeres cultivated in vitro before they are selected for implantation. In this process, ova and sperm are harvested from the parents or donors, and conception occurs in vitro. After culturing for 3 days, the fertilized cells have divided to the blastomere stage; each is composed of eight genetically identical totipotent cells. One cell can be removed from each group of eight cells for analysis without disruption of the growth and development of the embryo. DNA sequences of interest can be amplified from the removed cell by PCR, and the presence of particular sequences associated with disease can be determined. By implanting only blastomeres that are free of the disease sequence, parents can avoid initiating pregnancies that would result in a child with a particular genetic disease and avoid issues of pregnancy termination. Blastomere analysis before implantation has resulted in the birth of infants free of cystic fibrosis and Marfan syndrome. In most cases, parents who are consumers of blastomere analysis technology do not have fertility problems that would prevent natural conception; rather, they choose in vitro fertilization for the express purpose of genetic testing of products of in vitro fertilization before initiation of a pregnancy.

The current emphasis on primary health care and prevention of disease is congruent with an emphasis on presymptomatic disease testing. In cases where genetic predispositions to disease are known and modifiable risk factors or effective therapies exist, diagnosis of disease before the advent of symptoms can prevent the occurrence of symptoms. This strategy is illustrated by the well-established newborn screening programs for phenylketonuria (PKU). PKU, a deficiency of phenylalanine hydroxylase inherited in an autosomal recessive pattern, is entirely genetic in etiology, and pathology is entirely preventable. If children with genetic susceptibility to PKU are provided with a diet low in phenylalanine throughout their early development, they grow and develop normally. If, however, phenylalanine is not limited, severe mental retardation and shortened life expectancy result. Since mental retardation in untreated children is evident by the age of 1 year and is not reversible, it is clearly in the best interests of a child to be diagnosed before development of symptoms. Postnatal screening programs, mandated by many state governments, were often based on detection of the presence of phenylalanine by-products in the urine or phenylalanine levels in the blood. Since the gene for phenylalanine hydroxylase has been cloned, it is now possible to test directly and prenatally for the defect. Similar strategies are becoming available for some adult-onset diseases with a genetic component.

Initially, genetic testing of adults was a vehicle for informed decision making regarding reproduction. Current applications of genetic testing relate not only to prediction of the health of potential offspring, but to disease susceptibility and prognosis in the individual. Hereditary hemochromatosis is one example. Iron uptake is enhanced in this autosomal recessive disease. Over time, excess iron accumulates in internal organs, and untreated persons ultimately succumb to cardiac or hepatic failure. In presymptomatic stages (before iron overload), the disease is preventable through reduction of dietary iron and phlebotomy. Presymptomatic diagnosis is facilitated by genetic testing that recently became clinically available for hemochromatosis mutations. As additional genes are implicated in adult-onset diseases, it may become possible to tailor preventive activities to individuals based on genetic profile. For example, dietary or activity modifications to reduce risks of heart disease would be particularly important for individuals who have genes that would predispose them to heart disease.

The use of DNA-based screening for diseases that have a genetic component can pose particularly difficult dilemmas. In most cases, the ability to identify particular genes precedes a thorough understanding of the implications of the presence of a defective gene and effective treatment. It might be argued that providing individuals with knowledge of the potential for disease promotes autonomy; however, the person's welfare may or may not be enhanced by knowing that one has a predisposition toward a disease for which there is currently no preventive therapy and no cure. Would it benefit or harm the person to know, years in advance of the development of symptoms, that the future is likely to hold Huntington's chorea, breast cancer, or colon cancer?

Although it may be possible to predict genetic or chromosomal disease, the variability of individuals in disease course and severity complicates the issue of decision making. Clinicians have been able to accurately predict trisomy 21 prenatally since amniocentesis became widely available in the 1970s; it is still not possible to predict for parents from a karyotype whether their fetus with Down syndrome would grow to be a mildly retarded adult capable of functioning independently or a severely retarded individual who would require extensive and expensive care.

Further complications are posed by the uncertainties of future options. It is not possible to predict what therapies for management or cure of disease may be developed, or when these therapies will be available to patients. For example, many innovative therapies are currently available for cystic fibrosis, and both life span and quality of life have improved considerably for patients in the last decade. However, these advances were not predictable to clinicians providing prenatal genetic counseling a decade ago.

Quality of life for individuals who develop a particular disease is not predictable by genetic tests. Not only do course and severity of illness vary on an individual basis, but individuals at the same level of severity may have very different perceptions of how burdensome the disease is and the degree to which it affects the quality of their lives. This individual variation in requirements for and perception of quality of life may not be recognized by others and may lead to assumptions about what is necessary for a meaningful life. Arguing in favor of the benefits of prenatal diagnosis and selective termination of pregnancy, Jackson (12) states, "It is intended to relieve suffering and improve health. It is obviously admitted that it would be better to have curative approaches to genetic diseases, or successful treatment approaches if cure is not possible. However, all practicing physicians recognize the incredible burden of chronic diseases and genetic disorders, especially those beginning early in life. The fact that caring parents would wish not to burden their offspring with such extraordinary difficulties simply represents the good attitude of one human being toward another." DeRogatis (6), a nurse who speaks about her own disability, offers a different viewpoint. She says, "Our culture does not reflect the ways in which people with disabilities experience and value our bodies and our lives . . . I understand that it may be difficult for able-bodied people — particularly those in the health professions — to believe that disability may be experienced as different, not less."

Concerns exist about the confidentiality and use of genetic information and results of genetic tests. Unlike many other specimens, genetic information can be stored for long periods in the form of a frozen blood sample. This sample provides a source of material that can be analyzed for factors other than that for which it was originally intended and at a time removed from the collection and consent process. It is vital that informed consent be elicited from patients in health care settings and subjects in research settings; those from whom specimens are collected should give express permission for analysis and should be made aware of confidentiality safeguards in storage and future use of the material.

Valuable data could be obtained by collection and analysis of DNA from large populations. A project is currently under way to collect genetic data from the population of Iceland, which is stable and homogeneous. The Icelandic project, financed by a for-profit company, includes extensive procedures for the protection of privacy of individual subjects. The Human Genome Project identified examination of human sequence variation as a new goal for the 1998–2003 project period. DNA samples will be anonymous but will remain linked to phenotypic and geographic information; sex and ethnicity of sample donors will not be part of the database. This will enable examination of genetic similarities in specific subsets of the American population. Such information could provide guidance in detection, prevention, or treatment of diseases, but there is potential for misuse of the information as well. The Human Genome Project acknowledges that ethical, legal, and social concerns raised by this goal require further consideration.

The use of stored genetic information as an individual identifier will increase in the future. The Department of Defense plans to store a blood sample from each active-duty service member to serve as a source of "genetic dogtags," permitting identification of remains. The Department of Defense has provided assurances that this genetic information will not be used for any other purpose and will be securely stored. The FBI and some states have demonstrated interest in maintaining samples of DNA as part of penal records for individuals convicted of crimes.

The appropriateness of use of genetic information by employers for pre-employment screening or job assignment is debatable. It has been argued by some that screening of individuals for susceptibility to injury in a particular workplace (for example, genetic susceptibility to disease related to chemicals in use at the job site) promotes the health of individual workers and reduces job-related morbidity, thus reducing employers' health care costs. Such information about susceptibility might be used in a variety of ways: to counsel employees regarding risks, to institute more careful safeguards against exposure, to guide frequency of health monitoring, to assign jobs, or to influence hiring decisions. Public health benefits may also accrue from an ability to identify those who might be more likely to develop a workplace disability that would endanger coworkers or the public. However, genetic information is often erroneously applied and may be discriminatory. Employers may confuse those who carry one allele of a recessive disorder (carriers) with those who have the disease. Treating those who have a genetic predisposition as if they were already ill (or inescapably destined to become ill) can lead to discriminatory practices. One large company was found to be discriminating against persons who had sickle cell trait although the trait had not been shown to be associated with higher workplace disability (14).

Discrimination has occurred in health, disability, and life insurance coverage as well. In one reported instance, a health maintenance organization attempted to limit postnatal coverage of a fetus that the parents elected not to abort following a positive genetic test for cystic fibrosis (10). Billings et al. (3) described 41 separate incidents of discrimination against individuals that occurred "solely because of real or perceived differences from the 'normal' genome." Concerns persist that participation as a research subject in predictive genetic studies may adversely affect access to insurance.

Recent federal legislation has been enacted that will reduce the likelihood of genetic discrimination in health insurance. The Health Insurance Portability and Accountability Act of 1996 prohibits denial of coverage or assignment of higher premiums based on genetic information. Additionally, the act prohibits the use of genetic test results in defining preexisting conditions in the absence of a corroborating medical diagnosis.

The Human Genome Project

The Human Genome Project intends to provide genetic maps, physical maps, and nucleotide sequence data from the human genome and the genomes of several model organisms. The project has stayed remarkably on time and in budget; interim goals for the first and second 5-year project periods were met on schedule. Since the initiation of the project in October 1988, approximately 5% of the budget has been directed to consideration of the social, legal, and ethical implications of the project. The information generated by the project will be a valuable tool to researchers in localizing and isolating DNA sequences associated with diseases or other traits. Public access databases have already been developed that permit electronic access to primary data, maps, marker information, and reference information. As more specific DNA sequences are associated with particular diseases or other phenotypes, the ability to detect genetic predisposition to disease (and attendant problems addressed above) will explode. We will also gain insight into the genetic component of human traits that are not generally associated with disease.

Advancements in sequencing technology derived from the Human Genome Project have been applied to microbial, plant, and animal genomic sequencing projects as well. The Institute for Genomic Research published the first complete sequence of a microorganism in 1995, and sequences for a number of other organisms are now available. Complete sequences are also available for three of the model organisms (*Saccharomyces cerevisiae*, *Caenorhabditis elegans*, and *Escherichia coli*) targeted by the Human Genome Project. In 1998, the National Science Foundation began a plant genome initiative focused on economically important plants. The first complete plant sequence, that of *Arabidopsis thaliana*, is an outcome of that initiative. Although the commercial value of *A. thaliana* is limited, this species is an important model system in plant biology.

Advancing knowledge of the genome may fundamentally alter our view of humanity. Investigations regarding genetic influences on behaviors are currently stirring controversies about the extent of choice and responsibility in behavior. Identification of a genetic component to behavior does not predetermine how we will interpret or apply such information. For example, Hamer and coworkers found evidence of linkage between a region on the X chromosome and male homosexuality (9, 11); how does this demonstration of a genetic component to homosexuality affect or inform our views of homosexual behavior? Provided with the same data (that male homosexuals differ from male heterosexuals in a particular genetic region), one might conclude either that homosexual behavior is a normal variation in human sexual expression or that it is an abnormal "disease" allele. Increased information about genetics may lead to either increased acceptance or renewed rejection of particular groups of individuals.

Comparisons of the genomes of humans and other organisms may affect our associations with the larger world. Murray (15) suggests that in light of examination of our genetic relatedness to other species, "we may reevaluate not only our molecular but also our moral relationship with nonhuman forms of life."

International consensus regarding the ethical problems posed by the Human Genome Project has begun to develop. Although there may be some agreement regarding broad ethical issues related to human genome research, the interpretation and evaluation of specific challenges continue to generate lively debates.

Manipulating Genes

The notion that we may be able to directly manipulate human genetic material and effect changes in the function of genes has been enlivened by the Human Genome Project and advances in molecular biology. Technology that permits direct intervention at the molecular level, coupled with increased knowledge about the location, structure, and function of genes, provides possibilities for changing sequences to alter the function of particular genes. Such interventions are currently being explored in somatic cell gene therapy and could also be applied to germ line cells. Current efforts focus on prevention or treatment of diseases, but the same techniques could be applied to alter genes not commonly associated with diseases.

Somatic cell therapy

Somatic cell gene therapy is currently in clinical trials. Inherited disorders such as cystic fibrosis are obvious targets for gene therapy. However, somatic cell manipulation has also been proposed as therapy in cancers (5, 20), cardiovascular diseases (4), and infectious diseases such as human immunodeficiency virus (HIV) infection (21). Interestingly, the majority of current clinical trials are not focused on monogenetic inherited disorders. A 1998 National Institutes of Health update (16) indicated that 62% of clinical trials were targeted toward cancer, while 13% involved monogenetic inherited diseases (half of which centered on cystic fibrosis).

Different methods for gene delivery are being tested on the basis of target cell typology. For example, in clinical trials of the treatment of cystic fibrosis, a functional copy of the *CTFR* gene can be delivered to respiratory epithelial cells by inhalation of an adenoviral vector carrying the gene. Adenoviral vectors do not require active target cell division, and this makes them appropriate vectors for terminally differentiated cells of the respiratory tract. In this case, the therapeutic effect is lost when the treated cells die; this gene therapy provides treatment but not cure. Retroviral vectors are able to integrate into a chromosome of a target cell, and future generations of that cell will inherit the introduced gene. If the targeted cell is a pluripotent stem cell, it could provide populations of cells with the introduced gene over a long period of time. In this case, it might be appropriate to speak of the gene therapy as a cure. In somatic cell therapy, alterations introduced affect only the individual recipient of therapy; they are not inherited by offspring of the treated person.

Some somatic cell therapies have been targeted to treating genetic diseases. Gene therapy for single-gene recessive disorders requires only gene addition for treatment, providing the ability to make adequate amounts of a functional gene product results in amelioration of the disease. Strategies for dominant disorders and multigene disorders will be more difficult. In dominant disorders, disease results not from the absence of a functional gene product but from the presence of an abnormal product. Treatment would be contingent upon ceasing production from the affected gene(s) and might also involve providing a normal copy of the gene if none is present.

A variety of approaches have been suggested to harness gene therapy to treat HIV infection (21). These approaches can be broadly classified as either intracellular immunization or immunological gene therapy. Intracellular immunization involves transfer of genes that inhibit viral replication; introduction of these genes into hematopoietic cells could reduce the spread of HIV within an individual. Immunological gene therapies are those that are designed to enhance the natural defenses of the infected individual.

Genetic manipulation of somatic cells for the purpose of treating or preventing disease poses fewer ethical dilemmas than does manipulation aimed at germ cells or alteration of nondisease genes. Although in vitro laboratory studies and animal experiments precede all clinical trials, and initial human trials have been encouraging in many gene therapy protocols, it is important to remember that all current gene therapies are experimental. It is important for both clinician researchers and subjects to understand the experimental nature of the work. Apprising subjects of possible risks as well as potential benefits is an essential component of obtaining informed consent. The novel nature of these therapies makes it difficult to anticipate risks. Somatic cell gene therapies may have both immediate and delayed unanticipated complications. Several safety issues are of potential concern in somatic cell therapy. Depending upon the method of gene delivery, family members and those caring for the gene therapy patient may inadvertently be inoculated. Present methods do not permit targeting of genes inserted by retroviral vectors to a particular chromosomal location. Insertion of the introduced gene could theoretically result in a harmful mutation at the insertion site, resulting in loss of a critical cell function or loss of growth control. Regeneration of infectious particles from viral vectors is thought to be unlikely but is a potential problem. There are concerns that an immune response may be generated against target cells following gene therapy. However, the risk to date remains hypothetical; no pathologic events or malignancies related to a gene therapy retroviral vector have been reported in animal or human subjects.

Germ line therapy

Manipulation of DNA in germ cells raises additional issues. Manipulations that affect ova, spermatozoa, or totipotent cells such as blastomeres are heritable. Changes made to the DNA in these cells have the potential to affect not only the treated individual but also his or her offspring. Germ line manipulation has been accomplished in mammals, and the use of genetically manipulated animals in research is widespread. In the production of transgenic mice, for example, the genetic material of an embryonic stem cell is the target. The cell is then cloned and used to establish a lineage of mice carrying the added (or altered) genetic material.

The benefits and risks of germ line gene therapy in humans have been widely debated. Many of the concerns raised in the consideration of somatic cell therapy apply to germ cell therapy as well. Even when restricted to prevention of and therapy for severe genetic diseases, germ cell therapy poses additional questions. The opportunity to do good for many potential individuals in a single intervention is great; it would be much more effective to prevent disease in all of the future branches of a patient's family tree than to treat each descendant individually for the disease. This potential to maximize beneficence is attractive to health care providers. However, accompanying the potential for good is a potential for harm. Any untoward effects of germ cell manipulation may also be propagated to the patient's descendants. Concerns have also been expressed regarding the balance of individual autonomy (the argument that decisions about genetic manipulation are best made by the individual involved or parental surrogates) and society (the argument that since germ cell manipulation potentially affects more than the individual and may have far-reaching future effects, society has a legitimate interest in availability and use of the technology). Experts at an international symposium on human germ line engineering in 1998 predicted that germ line gene therapy will be a reality within 20 years (23).

Enhancements

The preceding sections addressed manipulation of genetic material of either somatic cells or germ cells in efforts to prevent, treat, or cure disease. Many have argued that genetic manipulations should be reserved for the treatment or prevention of serious disease (1).

Techniques being developed to permit modification of phenotypes associated with disease could be used to modify other characteristics. Such modifications depend upon an exquisite knowledge of the location and operation of genes, knowledge that is not currently available for most traits. The Human Genome Project will provide valuable information to spur the research of those interested in traits not currently associated with disease. Speculation about the ramifications of manipulation of the human genome to alter or enhance particular nondisease traits is an active area of discourse.

It is likely that some healthy individuals will seek genetic manipulation. For example, some athletes have sought improvement in their oxygen-carrying capacity by a variety of means. High altitude training has been used to elevate hemoglobin levels. Blood doping, via removal of blood one month before a critical sports event and autotransfusion just before the event, has been used to increase the number of circulating erythrocytes during competition. (This practice is prohibited by most organizations governing athletic competitions.) Biosynthetic erythropoietin, developed for the treatment of chronic anemias, is of current interest to some athletes; it has been classified as a performance-enhancing drug by the International Olympic Committee. Those athletes who are willing to use pharmacologic and invasive methods in an effort to improve performance might view genetic manipulation as an additional tool to maximize oxygen capacity.

In the example of erythropoietin, techniques currently being developed for use in treatment of hematologic disease may be appropriated by healthy individuals for the purpose of enhancement. In other instances, individuals may seek cosmetic alterations. Somatic cell gene therapy has sometimes been viewed as an alternative method for producing therapeutic results that we would otherwise seek from surgery or pharmacologic agents (17). This view of somatic cell gene therapy as equivalent to other therapeutic modalities complicates the issue of cosmetic enhancement. Is genetic enhancement akin to cosmetic surgery? Both pharmacologic agents and surgical methods are used to alter the appearance of individuals who are within the range of normal appearance and function before intervention. We do not limit the autonomy of individuals to undergo cosmetic surgery, except in special circumstances where such surgery might negatively affect physical or mental health. Indeed, many of those who choose current methods for enhancement of appearance experience positive benefits such as an improvement in quality of life and enhanced self-esteem. If techniques can be developed that are relatively safe and effective, similar benefits might accrue to individuals who achieve cosmetic results via somatic cell gene manipulation.

The manipulation of germ line material outside of its use in disease prevention and treatment has generated much controversy. Issues of personal and parental autonomy and of consent are more problematic when changes may affect future generations, particularly when changes are not initiated in response to potential or actual disease. Decisions may have broader societal effects, and the specter of the development of a genetic class system has negative implications. Peters (18) states, "The growing power to control the human genetic makeup could foster the emergence of the image of the 'perfect child' or a 'super strain' of humanity; and the impact of the social value of perfection will begin to oppress all those who fall short." Not all agree that use of germ line manipulation to affect nondisease traits will necessarily have a negative effect on individuals or society. It is possible

that different parents would prefer and select different traits, without a societal trend for some selections to be labeled as preferable by the society as
a whole. Caplan said in regard to germ line manipulation, "The question
of whether I should be able to pick blue eyes or brown or tall people or
short—I don't think there's anything wrong with that fundamentally" (2).

Cloning

Cloning is a special case of genetic manipulation in which a replica, or genetic copy, is made. Two techniques (blastomere separation and somatic
cell nuclear transfer) are currently available in mammals; blastomere separation has already been demonstrated with human cells. Human cloning
is not a new topic for bioethical debate; the U.S. House of Representatives
held hearings on the topic in 1978. However, it continues to be a difficult problem for those concerned with scientific integrity, as scientific and
technological abilities outpace a consensus regarding appropriate use of the
technologies.

In blastomere separation, a fertilized ovum is developed in vitro to an
early multicellular (up to 32-cell) stage. Each of the blastomeres is totipotent
at this stage, and careful division of the cell mass yields multiple cell masses,
each capable of developing into a genetically identical organism. For example, a 16-cell embryo can be divided to yield two 8-cell masses (resulting
in identical twins) or four 4-cell masses (resulting in identical quadruplets).
Blastomere separation was first demonstrated with mouse embryos in 1970
and in cattle embryos in 1980. In the context of infertility research, nonviable human embryos were duplicated using this technique in 1993; news
reports generated a great deal of public debate regarding the ethics of the
technology (19).

Adult somatic cells have recently been used to produce cloned animals.
This technique, somatic cell nuclear transfer, involves transferring genetic
material from an adult somatic cell into an enucleated unfertilized ovum
and subsequent development in a surrogate mother. The first successful
production of an animal through this process (the sheep named Dolly) was
reported in 1997. The report generated a great deal of discussion, as the
public considered the potential for application of the technique to humans.
President Clinton issued a ban on federal funding for human cloning and
appointed a National Bioethics Advisory Commission to report on issues
related to potential cloning of humans via somatic cell nuclear transfer. The
commission concluded that "at this time it is morally unacceptable for anyone in the public or private sector, whether in a research or clinical setting,
to attempt to create a child using somatic cell nuclear transfer cloning" (22)
and urged both federal legislative prohibition of human research in this area
and continued public discussion of the issue. Concerns were voiced in the
scientific community that proposed legislative moratoriums were too broad

and would inhibit research in related areas such as regeneration of diseased or damaged human tissues. A proposed bill failed, but public debate concerning human cloning was further inflamed by the announcement of one researcher that he intended to apply somatic cell nuclear transfer cloning to the problem of human infertility. Debate regarding the ethics of human cloning continues. It is imperative that researchers remain sensitive to public concerns regarding human cloning.

Finally, some interesting issues have been raised in regard to ownership of genetic information. Kevles and Hood (13) express a common sentiment when they state, "if anything is literally a common birthright of human beings, it is the human genome." However, genetic material in some forms can be owned; it is patentable, and the limits of patent protection are being tested (7, 8). Intellectual property issues related to genetics and genetic engineering are discussed in chapter 9.

Conclusion

Genome and biomedical research has exploded in the recent past, but even more revolutionary developments are on the horizon. Somatic cell gene therapy for a variety of diseases is currently under way. Integration of somatic cell therapies into regular medical practice may not be far distant, and germ line gene therapy is an active area of discussion. Whether gene therapies will be limited to serious diseases or used over the same spectrum of care as current medical and surgical interventions remains an issue. As the Human Genome Project progresses, it is certain to have wide-reaching effects on research and knowledge about our genetic selves. Coupled with the ethical and moral decisions posed by genetic technology, legal and economic layers of complexity are added by the potential impact of the patent system. Many issues have been raised in this discussion. Some of the questions that will be most crucial to the impact of the technology cannot be envisioned at our current level of understanding. Formulating the questions and articulating the issues is a beginning step in preparing to meet the challenges to health care and scientific integrity posed by the expanding area of genetic biotechnology in these contexts. Both researchers and clinicians will need to involve themselves in public discourse. In answering the newest questions, it will be necessary to look beyond the scientific significance of research and address its broader societal implications.

Case Studies

10.1 The prospects of genetic screening raise interesting and sometimes controversial issues when applied to the family setting. Disclosing information may violate a sibling's right to privacy, but withholding it may cause harm, too. How should information discovered by genetic technology in one family member be treated if it could affect other family members?

10.2 Consider a diagnostic DNA probe that could provide information on the nature of some types of neoplastic disease. Specifically, the probe in question can determine whether the specific cancer is aggressive or relatively slow growing. Proponents of the clinical use of such a probe argue that it promises to provide important prognostic information and potential guidance toward the effective use of anticancer therapies. Opponents argue that such information is likely to make its way into databases that can adversely affect a patient's ability to secure various types of insurance and in general threatens these individuals' right to privacy. Discuss the scientific and ethical implications of this scenario.

10.3 A recently enacted state law requires the state's Division of Forensic Science and Investigation to analyze, classify, and store the results of DNA identification characteristics ("DNA fingerprints") from all convicted felons. Thus, blood samples must be taken from felons in order to implement this plan. The results must be maintained electronically in a DNA data bank that is available for criminal investigation. Only 10% of the available blood samples have been processed and DNA data entered when the state police develop a suspect in a series of rape cases. The suspect had been previously convicted and sampled, but his blood has not yet been analyzed so his DNA pattern is not yet in the data bank. The investigators do not have sufficient probable cause to get a blood sample from the suspect, so they ask that the division's blood sample be processed now and analyzed together with the evidence from the crime scene. Do you think this is legal? Is it ethical? How should the results be reported in the event of a match? A nonmatch?

10.4 Recent results from the Human Genome Project have been coupled with epidemiological data to identify a locus that can be used to predict predisposition to a specific genetic disease. The gene probe that has emerged for this disease is particularly useful in diagnosing carriers of this defect who may pass the gene on to their offspring. Dr. Dell, a researcher at a large urban medical center, decides to evaluate the incidence of this defect using the new probe. She learns from the director of her medical center's blood bank that she can purchase expired whole blood from several blood service facilities in smaller towns in the region. She is able to obtain a large number of samples from one blood bank that has a catchment area consisting of several small rural communities. She extracts DNA from blood obtained from approximately 150 donors. To her surprise the incidence of carriage of the suspect locus is approximately 18% as opposed to the 1 or 2% seen in studies involving general populations. Comment on the human use experimentation implications of Dell's work. Does she have any moral obligations to report this work? If so, to whom?

10.5 An apparently asymptomatic 39-year-old man whose father died of Huntington disease (HD) at age 50 comes to the Medical Genetics Clinic for presymptomatic DNA testing. He is paying out-of-pocket to avoid alerting his health insurance carrier and insists that absolute confidentiality of test results be maintained. Regardless of the outcome, he intends to tell no one and go on with his life exactly as before. You reassure him that this is standard practice in HD testing and counseling and that the results will be given to no one else without his express written consent. One week later the diagnostic molecular genetics laboratory informs you that his test is positive for the HD trinucleotide expansion mutation, with alleles of 18 and 48 repeats. When you retrieve his clinic chart to schedule a return visit for results reporting and counseling, you happen to notice an entry on the intake form that had been overlooked before: Occupation: Pilot; Employer: TransCoastal Airlines. Does this realization alter your disposition of the test results with regard to confidentiality? What do you intend to do under these circumstances?

10.6 A 26-year-old woman who has just had surgery for removal of a parathyroid adenoma is found to carry a mutation in exon 11 of the RET proto-oncogene. You inform her that this test result is diagnostic of multiple endocrine neoplasia, type 2. Since the inheritance pattern of this disorder is autosomal dominant, each of her six siblings as well as their children are at risk and should be screened so that a life-saving prophylactic thyroidectomy can be performed on those who test positive. However, the woman informs you that she is on very bad terms with the rest of her family and refuses to contact them with this genetic information. Should you contact the relatives and urge them to be tested?

10.7 A pregnant woman whose mother and grandmother died of breast cancer undergoes DNA testing and is found to carry a mutation in the BRCA1 gene. She now requests amniocentesis so that the fetus can be tested for sex determination and for the BRCA1 mutation, with intent to terminate the pregnancy if the fetus is found to be female and a mutation carrier. How would you proceed with this case and counsel the woman?

10.8 You are sitting on a blue-ribbon panel that will provide recommendations to Congress regarding the writing of legislation on genetic privacy. The issue of human tissue in paraffin blocks is being discussed as a valuable database. Several of the panel members have argued that the research use of such materials is permissible under federal law. They argue that such law permits research usage of clinical materials as long as the specimens are provided to researchers without any identifying marks (e.g., patients' names, codes). Others on the panel say that it is impossible for embedded tissues to be anonymous, because DNA fingerprinting could be used to definitively associate any tissue specimen with its donor. They argue that even though such identification would be costly and tedious, it is formally possible so it eliminates anonymity. What is your position with regard to this dilemma? Also consider the suggestion that the "middle ground" solution is to obtain informed consent from the patients from whom the tissues originated. Can this be plausibly and cost-effectively done in your view?

10.9 In the course of your work as director of a blood bank, you have been collecting and storing many hundreds of frozen blood samples from donors in order to perform some population-based immunologic studies. Now a geneticist colleague asks if he can have access to some of the samples to examine the prevalence of mutations in several genes. Under what, if any, conditions would you release the samples for such purposes? Are there any sorts of genetic studies you would not allow? If a clinically significant mutation were to be found in one of the samples, under what conditions would you feel obligated to recontact the donor and report the finding?

10.10 You have been named to a task force of a national biological society of which you are a member. The principal duty of this task force is to develop a position with respect to human genetic diagnostics. This position will be voted on by the entire society. The basic premise being addressed concerns the increasing availability of gene probes that provide the direct or indirect means for the diagnosis of genetic diseases or genetic states that predispose to disease. It is obvious that in many cases such diagnoses involve diseases that cannot be cured, prevented, or even treated. The first meeting of the task force opens with the chair proposing that the committee embrace the following concept. The members of the society will be asked to vote on a resolution stating that no gene probe should be used in a human clinical diagnostic situation involving a disease for which there is not a significant therapeutic intervention that either improves the patient's quality of life or prolongs it. What would be your arguments either for or against this proposal, and what is the rationale underlying your arguments?

10.11 Near the end of World War I a virulent influenza virus swept the globe. The infection it caused was called the Spanish flu, and between 20 and 40 million people died from this disease. The graves of several fishermen, documented to have succumbed to the Spanish flu, have been identified in a Norwegian cemetery located in the hard-frozen, permafrost region. The bodies of these fisherman are expected to be cryogenically preserved. You are sitting on a review panel considering funding a proposal to investigate this Spanish flu virus. A team of scientists has proposed to exhume these bodies, extract lung tissues, and study the virus. They will use a combination of PCR and recombinant DNA to clone viral genes, and they will also attempt to cultivate the virus in animals and tissue culture. This will be done under conditions of strict biological containment. The rationale for this work is that by studying this virus, the scientists will be able to learn things that may prevent such an epidemic from happening again. Comment on the medical and ethical implications of this proposal. Will you support it?

10.12 The editor of an American biomedical journal arranges a teleconference call with you and seven other associate editors on the editorial board. She seeks guidance on the handling of a submitted manuscript. The paper in question was submitted by an interdisciplinary group working in a foreign research institute. The paper reports on recombinant DNA experiments that modify a virulent bacterial pathogen capable of causing fatal infections in humans. The only available preventive measure for this disease is an attenuated whole-cell vaccine. In their paper, the authors demonstrate that the animals infected with this genetically engineered strain died more rapidly than those infected with the wild-type strain. More important, immunization of the animals with the current whole-cell vaccine fails to protect them against lethal infection with the genetically engineered strain. The authors argue that this work will open new doors for the understanding of the disease process and ultimately will lead to the development of more effective vaccines. The pathogen in question is believed to be stockpiled as a biological weapon by certain countries. The editor is considering rejecting the paper on ethical grounds. What advice will you give her?

References

1. **Anderson, W. F.** 1998. Human gene therapy. *Nature* **392**:25–30.

2. Arthur Caplan discusses issues facing the growing field of bioethics. (October 17, 1984.) *Scientist* **8**:12, 25.

3. **Billings, P. R., M. A. Kohn, M. de Cuevas, J. Beckwith, J. S. Alper, and M. R. Natowicz.** 1992. Discrimination as a consequence of genetic testing. *Am. J. Hum. Genet.* **50**:476–482.

4. **Clowes, A. W.** 1997. Vascular gene therapy in the 21st century. *Thromb. Haemostas.* **78**:605–610.

5. **Dachs, G. U., G. J. Dougherty, I. J. Stratford, and D. J. Chaplin.** 1997. Targeting gene therapy to cancer: a review. *Oncol. Res.* **9**:313–325.

6. **DeRogatis, H.** 1993. A different reflection. *Nurs. Outlook* **41**:235–237.

7. **Doll, J. J.** 1998. The patenting of DNA. *Science* **280**:689–690.

8. **Eisenberg, R. S.** 1997. Structure and function in gene patenting. *Nat. Genet.* **15**:125–130.

9. **Hamer, D. H., S. Hu, V. L. Magnuson, N. Hu, and A. M. Pattatucci.** 1993. A linkage between DNA markers on the X chromosome and male sexual orientation. *Science* **261**:321–327.

10. **Holtzman, N. A.** 1992. The diffusion of new genetic tests for predicting disease. *FASEB J.* **6**:2806–2812.

11. **Hu, S., A. M. Pattatucci, C. Patterson, L. Li, D. W. Fulker, S. S. Cherny, L. Kruglyak, and D. H. Hamer.** 1995. Linkage between sexual orientation and chromosome Xq28 in males but not in females. *Nat. Genet.* **11**:248–256.

12. **Jackson, L. G.** 1990. Commentary. Prenatal diagnosis: the magnitude of dysgenic effects is small, the human benefits, great. *Birth* **17**:80.

13. **Kevles, D. J., and L. Hood.** 1992. *The Code of Codes: Scientific and Social Issues in the Human Genome Project.* Harvard Press, Cambridge, Mass.

14. **Murray, T. H.** 1991. Ethical issues in human genome research. *FASEB J.* **5**:55–60.

15. **Murray, T. H.** 1993. Ethics, genetic prediction, and heart disease. *Am. J. Cardiol.* **72**:80D–84D.

16. **National Institutes of Health.** *Human Gene Therapy Protocols.* October 14, 1998.

17. **Nolan, K.** 1991. Commentary. How do we think about the ethics of human germ-line genetic therapy? *J. Med. Philos.* **16**:613–619.

18. **Peters, T.** 1994. Intellectual property and human dignity. *In* M. S. Frankel and A. Teich (ed.), *The Genetic Frontier; Ethics, Law, and Policy.* Association for the Advancement of Science, Washington, D.C.

19. **Robertson, J. A.** 1994. The question of human cloning. *Hastings Center Rep.* **24**:6–14.

20. **Roth, J. A., and R. J. Cristiano.** 1997. Gene therapy for cancer: what have we done and where are we going? *J. Natl. Cancer Inst.* **89**:21–39.

21. **Savarino, A., G. P. Pescarmona, E. Turco, and P. Gupta.** 1998. The biochemistry of gene therapy for AIDS. *Clin. Chem. Lab. Med.* **36**:205–210.

22. **Shapiro, H. T.** 1997. Ethical and policy issues of human cloning. *Science* **277**:195–196.

23. **Wadman, M.** 1998. Germline gene therapy 'must be spared excessive regulation.' *Nature* **392**:317.

Resources

National Human Genome Research Initiative. Information about the Human Genome Project, with links to the committee that examines ethical, legal, and social implications of the project.

http://www.nhgri.nih.gov/index.html

National Institutes of Health, Office of Recombinant DNA Activities. Includes information regarding gene therapy clinical trials.

http://www.nih.gov./od/orda

U.S. Department of Energy Human Genome Program. Follow links to an excellent primer on molecular genetics, information about the history of the Human Genome Project, and updates on the status of microbial genome sequencing.

http://www.er.doe.gov/production/ober/hug_top.html

GeneClinics. Provides extensive information, organized by disease, about genetic testing and the diagnosis, management, and counseling of individuals and families with inherited disorders. The site also lists clinical and research laboratories performing testing for heritable disorders.

http://www.geneclinics.org

Scientific Record Keeping

Francis L. Macrina

Introduction

P ROPER RECORD KEEPING IS CRUCIAL TO SCIENTIFIC RESEARCH. But the accepted practices of record keeping and policies on custody and retention of data are usually learned passively by most scientists. Informal surveys often reveal that trainees receive little instruction in the principles of scientific record keeping. When mentors do not communicate their expectations on the subject, trainees learn the practice of record keeping by trial and error and by having mentors correct their mistakes. Moreover, most of the granting agencies that support graduate training in the biomedical sciences fail to provide any guidance on record-keeping practices.

Discussions of scientific record keeping run the risk of implying some uniform prescription for the process — a rigid method for the one correct way to do things. However, like the resolution of the case studies in this book, there are multiple right ways to keep scientific records. So, although this chapter will have much to say about keeping a laboratory data book, its message is not an exact prescription or set of immutable rules. On the other hand, there are important principles that create a foundation for good record keeping (Table 11.1).

The nature of the research, the form and amount of data generated, and the preferences and experiences of individual scientists influence the record-keeping process. Thus, there are many styles and permutations of record keeping that are appropriate and effective. Equally important, there are practices that are improper or even scientifically irresponsible.

Howard Kanare's *Writing the Laboratory Notebook* (7) is a definitive work on this subject. Its technical and detailed style is informative and useful.

Table 11.1 Data book zen

Useful data books explain:
- Why you did it
- How you did it
- Where materials are
- What happened (and what did not)
- Your interpretation
- What's next

Good data books:
- Are legible
- Are well organized
- Allow repetition of your experiments
- Are the ultimate record of your scientific contributions

Kathy Barker's *At the Bench: A Laboratory Navigator* (1) is another useful book containing discussions of laboratory organization skills and record keeping. These books are good resources for both trainee and principal investigator.

Why Do We Keep Records?

Kanare (7) defines and describes the laboratory data book as "a bound collection of serially numbered pages used to record the progress of scientific investigations. . . . It contains a written record of the researcher's mental and physical activities from experiment and observation, to the ultimate understanding of physical phenomena." Such records provide the platform for analysis and interpretation of results obtained in the field or the laboratory. They are the basis for scholarly writings, including reports, grant and patent applications, journal articles, and theses and dissertations. Laboratory data books are the definitive source of facts and details. Good record keeping fosters the scientific norms of accuracy, replication, and reliability. Corroboration and verification of scientific results using primary data contained in a laboratory data book may involve individuals other than the primary data book keeper. A scientist or scientist-trainee may take over a project, and it will be necessary for him or her to understand precisely the laboratory data book contents in order to continue the work. Thus, a specific data book may become a key research tool for someone else in the laboratory group, or even someone outside the laboratory or the institution. This makes clarity and completeness of the laboratory data book essential to its usefulness.

Proper data book keeping also has legal implications and responsibilities. The National Institutes of Health (NIH) has the legal right to audit and examine records that are relevant to any research grant award. It follows that recipients of research grants have an obligation to keep appropriate records of experimental activities even though the granting agencies rarely provide guidance on how to do this. Providing primary research data is often a component of the approval process for new drugs or medical applications (e.g., data submitted to the U.S. Food and Drug Administration). Record-keeping requirements for this type of research are usually explicit. Failure to conform to such specifications can compromise the validity of the data and the utility of the research. Finally, scientific record keeping is a critical element in proprietary issues. As one seeks the protection of intellectual property by applying for a patent (see chapter 9), it may become necessary to disclose data book contents to the patent examiner. Such disclosure might be related to requests for additional supporting data, dates of experiments or discoveries, verification that the records have been properly witnessed, or proof of reduction to practice. Properly kept data books continue to be important after a patent is issued. Patents can be legally challenged once they are issued. Litigation involving such challenges may require that original data books be inspected as part of the legal proceedings. Patents in whole or in part can be nullified as the result of such legal activities.

Defining Data

What do we mean by data? Simply stated, data are any form of factual information used for reasoning. Data take many forms. Scientific data are not limited to the contents of data books. Much of what we would call data contained in data books is commonly classified as being intangible. That is, it contains handscript or affixed typescript that records and reports measurements, observations, calculations, interpretations, and conclusions. The term "tangible data," on the other hand, is used to describe materials such as cells, tissues or tissue sections, biological specimens, gels, photographs and micrographs, and other physical manifestations of research.

Data are said to have authenticity and integrity. Authentic data represent the true results of work and observations. When data deviate from this standard because of carelessness, self-deception, or deliberate misrepresentation, they lose their authenticity. Integrity of data is dependent on results being collected using well-chosen scientific methods carried out in the proper manner.

During the course of experimentation some kinds of data evolve into different forms. Let's say you set out to do an electrophoretic analysis of some proteins. Your experiment results in a polyacrylamide gel in which a mixture of several proteins has been electrophoretically separated in a single lane. One lane of the gel contains reference proteins of known molecular weight and concentration. You visualize the protein components by staining with Coomassie blue dye. Then you desiccate the gel and seal it in a clear plastic envelope. You photograph the gel, and the resulting print and negative are placed in plastic sleeves and taped into your data book; the desiccated gel is also taped to a data book page. Next, you calculate the apparent molecular weights of the proteins by comparing their migration relative to the standards. You do this by making measurements on both the gel and the photograph. In both of these cases, the data become transformed into handwriting in the data book. Then you enter your measurements into a computer, which generates a numerical data set that is fixed as a printed copy; it is also maintained as an electronic file. You use a computer algorithm to determine the apparent molecular weights, and you compare the results obtained by the different methods. Can you ascribe value to the various forms of the data which have come from this work? Is the gel itself the most important piece of data? Or, could the gel be discarded once it is recorded photographically? This scenario can be made more complex. For example, you scan the photographic negative using a digital scanner, resulting in its image being captured in an electronic file, which can then be printed. You use these electronic data to quantitate the proteins by comparing them with the concentrations of the proteins present in a control lane on the gel. You also use these data to make measurements electronically, enabling the program to compute the molecular sizes of the proteins.

All the forms of the data being considered—desiccated gel, photographic, electronic, and written or printed formats—are legitimate. Electronic technologies are changing how data are acquired, handled, and stored. The questions of identifying legitimate data strongly affect data analysis. Some forms of data may be better used for measurements and calculations than others. In the example given, it can be argued that measurements made from an optically or electronically generated image are more uniform from experiment to experiment than are those taken directly from the gel. This example also raises issues about data storage. Is it better to emphasize the long-term storage of desiccated gels or to rely exclusively on a photographic or electronically derived image?

Terms like "raw data," "original data," and "primary data" are often used by scientists, but their definitions are elusive and their use can be confusing. The changing face of data collection, now strongly affected by electronic technology, requires careful consideration of what constitutes legitimate and valid data. Thus far, definitions of scientific data have been of limited scope and usefulness. Yet, the definition of data is central to scientific integrity. Scientists need to recognize the importance of multiple data forms and to strive to clarify and define their importance. When doing sponsored research, scientists should be aware of and comply with all agency and institutional requirements concerning data custody and storage, removal and duplication, and disposal.

Data Ownership

Let's revisit the topic of data ownership, which was extensively discussed in chapter 9. It's safe to say that the details and implications of data ownership are not foremost in the minds of most researchers when they are writing grant applications or doing experiments. However, many funding agencies that sponsor research are clear on the issue of data ownership. As the primary and largest funding agency for biomedical research in the United States, the NIH under the aegis of the U.S. Department of Health and Human Services (DHHS) provides guidance on data ownership related to work supported by its research grants. As a matter of both policy and practice, the DHHS recognizes the grantee institution as the owner of the data generated by the NIH-funded research (8). NIH grants are made to institutions, not to individuals. The individual who submits the grant on behalf of the institution is called the principal investigator. In practice, the principal investigator is the steward of the federal funds and of all aspects of the research that is sponsored by that support. The principal investigator assumes the primary responsibilities for data collection, recording, storage, retention, and disposal. Grantee institutions (e.g., universities) usually operate so as to give maximum latitude and discretion to principal investigators. However, the discharge of these duties does not impinge upon, nor should it cloud, the issue of data ownership. For example, if the principal investigator resigns his or her position to take another one at a different university, the grant award, the equipment purchased from the grant funds, and all of the data are required to remain at the institution that initially received the award. However, permission is usually sought to transfer the grant award, some or all of the equipment, and the data to the principal investigator's new institution. The process to do this is formal and requires mutual consent of the involved parties: the granting agency, the current grantee institution, and the proposed grantee institution. If for some reason an agreement is not reached, the initial grantee institution can keep the award, assuming it identifies a new principal investigator who is acceptable to the granting agency. The principal investigator as an individual never legally has ownership of the data. The transfer of data ownership, when it occurs, is between grantee institutions.

In summary, neither the principal investigator nor any member of the laboratory research team owns the data generated under an NIH research grant. Informing trainees and staff about practical issues of record keeping is the responsibility of the principal investigator.

Data Storage and Retention

The NIH requires that data obtained under the aegis of an NIH grant be retained for 3 years beyond the date of the final financial expenditure report (8). Requirements for the amount of time research data must be retained may vary for various public and private funding agencies. Because of this it would be impractical, if not impossible, for a major research university to organize, implement, and maintain a uniform data storage system for all of its research projects. Such logistical problems at most universities and research institutions place the responsibility for the storage of data squarely on the principal investigator. Therefore, it is essential that investigators have a clear understanding of their granting agency's policies governing data ownership issues and data retention.

Tools of the Trade

Keeping original results and observations for significant periods of time requires the selection of appropriate materials for recording and storing data. An entire chapter of the Kanare (7) book is devoted to "The Hardware of Notekeeping."

Paper

Kanare's discussions on the quality of data book paper are thorough, technical, and may be summarized as follows. Make sure your data are recorded on acid-free paper as the best insurance for permanence. Selection of data books composed of paper that is considered permanent can be aided by consulting data book suppliers or manufacturers. Often, paper composition is printed on the bound data book cover. The longevity of laboratory data books is facilitated by proper storage. Strong light sources, especially sunlight, high humidity, extremes in temperature, and excessive dust can have unwanted and undesirable effects on stored laboratory records.

Ink and pen type

Kanare's recommendations on instruments for writing in data books are simple. Never use pencil. Do not use pens with aqueous-based inks. Graphite smudges over time, and even a little water can obliterate the inks in many popular pens (e.g., felt-tip, fountain, rollerball). Kanare's testing of various inks and pens led him to conclude that a ballpoint pen with black ink is best for scientific notekeeping. Colored inks are not desirable because their decomposition promoted by light is significant when compared with black ink. However, the usefulness of varying the color of inks when drawing diagrams, for example, may be essential in some types of work. Inventories of pens for laboratory use should be sufficient for short-term (a few months') use. Long-term storage of ballpoint pens is undesirable because of ink component partitioning within the ink cartridge, which can result in problems of ink flow.

Bound versus loose-leaf data books

Most, if not all, industrial research laboratories mandate the use of bound (sewn or glued binding) data books with serially numbered pages. Variations on this theme include bound data books with duplicate numbered carbon pages, which may be detached and stored separately as backup. Any other type of binding — plastic comb, wire spiral, or ring binder — is considered unacceptable because pages can be intentionally inserted, removed, or accidentally ripped out or lost. This could damage the integrity of the records, compromising, for example, the ability to gain patent protection. Outside of industry, however, the use of loose-leaf notebooks is commonly seen along with bound data books. Although the typical three-ring loose-leaf binder offers advantages of being able to logically organize ongoing and completed experiments, the above-mentioned drawbacks are important.

Bound, page-numbered data books have features that argue compellingly for their use. Their integral construction is consistent with preservation of data authenticity because intentional page deletion or insertion becomes immediately obvious. Quality control of paper composition is easier when compared with the vast array of papers available for loose-leaf books. Data books of uniform size and shape also are more amenable to efficient and organized storage. Numbered volumes, with serially numbered pages, may be readily indexed, making the task of locating stored data relatively easy. In sum, bound data books provide organization and ease of use that makes sense for the responsible custody of scientific data. As a practical matter, the use of a bound data book with chronological ordering of experimental protocols and results, supplemented with a looseleaf notebook containing all original data and electronic printouts, with each page dated, serves the purpose of most academic research laboratories.

Laboratory Record-Keeping Policies

Principal investigators and laboratory leaders are well advised to develop policies for record keeping. No guidance on scientific record keeping amounts to a tacit approval of slipshod practices that threaten the authenticity and integrity of scientific data. Ideas for developing such data-keeping policies and practices can be obtained from a variety of sources, including books (5, 6) and some published lab manuals (3) and by contacting various industrial research laboratories or research institutes. The latter frequently have printed documents that cover record keeping and data book maintenance. Academic institutions may have such guidelines or policies, but the challenge of covering widely divergent research areas makes the development of uniform policies difficult. The University of California, San Francisco, has developed such guidelines, which can be found on its Web site (6).

Policy documents need not be complex or lengthy. They may reflect the experiences, training, and personal preferences of the principal investigator or group of principal investigators who write them. Group efforts are useful in writing guidelines. The experience and wisdom of several investigators will give a valuable perspective to your guidelines. Once in place, such documents should be regularly reviewed and modified as necessary. A clear statement about data ownership and retention should be part of such documents.

Suggestions for Record Keeping

Drawing from the references of the types cited in the previous section and from personal experiences, the following is an overview useful for thinking about and developing laboratory record-keeping policies.

Data books

The case for using permanently bound laboratory data books with consecutively numbered pages has been made previously, but the discretion of the principal investigator should prevail in selecting specific data book types and mandating their use. Hereafter in these discussions, use of bound data books will be assumed. Some investigators like to control the distribution of data books. For example, data books are given out as needed by the principal investigator or the lab manager. At the time of distribution, a record is made of the date, data book user, and project; at this point the data book can be coded with a designation (e.g., a volume number), which will allow for its tracking while in use or storage. This strategy has merit in laboratories where there are multiple trainees and staff working on a variety of projects, funded from different sources. Data book users should clearly understand the lab policy for data book storage, retention in the lab, and any requirements for duplicating data book pages and other forms of data.

Organization

The first several pages of an individual's data book should be reserved for a table of contents. The first entry before beginning the table of contents should consist of the name of the data book user and other relevant information; especially for work with potential proprietary implications, the location (room, building, institution) of the laboratory in which the experiments are being performed is recommended. Financial sponsorship should be identified by stating the title of the grant proposal, its agency identification number, dates of support, and the name of the principal investigator. Experiments listed in the table of contents should have concise but descriptive titles. The numbering of experiments chronologically facilitates cross-referencing experiments. A glossary of abbreviations, symbols, or common designations may be included after the table of contents or, alternatively, can be listed at the end of the data book. Leave enough space for this information in order to be able to make additions to the glossary throughout the project.

The maintenance of a master data book log may be desirable. Such a central record (essentially a standard data book or perhaps even a computer-based word-processing or database algorithm) contains a listing of all experiments performed by the research team. Individuals are responsible for maintaining the log by entering experiment titles, dates, investigator names, and location of relevant data. A second type of laboratory-based reference resource is the methodology notebook. These notebooks are a compendium of all standard laboratory methodology. Compilation of such books works best when it involves all laboratory members. Experimental methods should be described in sufficient detail to be useful even to the novice investigator. A printed copy of the complete book (in this case, loose-leaf or comparable binders are acceptable) can be kept in a central location, or duplicated copies can be distributed to lab members. Alternatively, copies of the methods notebook can be distributed in electronic format for the use of lab members. If a laboratory methods notebook is to be kept, it is critically important that the master copy, controlled by the principal investigator, be updated regularly — perhaps on a yearly basis. Again, this can be done as a group effort, benefiting from improvements and refinements made by individuals using the techniques. Updated copies of new methodology notebooks should be distributed to replace old versions. The previous version of the master methodology log should be stored in an unaltered state. This allows for methods that have been updated or discontinued to be saved; referring back to methods, even discontinued ones, is sometimes necessary. Archiving such methods should be accomplished so that the date of revision or replacement of the method is obvious. Even if a central methodology book is maintained in the laboratory, it is a good idea that the data books of individual investigators describe regularly used procedures. These can be transcribed into the data book. Alternatively, typed copies can be prepared on high-quality paper and attached to the pages of the data book using archival-quality tape or glue. Obviously, any specialized techniques or methods used in research projects (which might not be appropriate for a central methodology book) should be recorded in the individual's data book.

Finally, consider a methods book kept separately by each member of the laboratory. In other words, investigators compile their own methods books and modify them as needed, leaving the original copy with the rest of their data books when they leave the lab. This might be practical in a laboratory where strikingly different methods are used in various projects. All of the above considerations apply to the maintenance of such a methods book. Decisions relating to whether to use centralized or decentralized record keeping should be made by the laboratory leader. Modern biomedical research frequently involves methodologies and interdisciplinary research that requires the centralized organization of methods commonly used by the group. Such organization and maintenance facilitates the teaching of novice trainees and staff, ensures quality control, and helps in the troubleshooting of technical problems.

Tangible data and the data book

Tangible forms of data such as photographs, negatives, autoradiograms, and printouts should be included in the data book when this is physically possible. The use of archival-quality glues and tapes is suggested for affixing such materials into data books. Materials that cannot be glued or taped directly into the book should be inserted into plastic sleeves, which are then fixed in the data book. Printed material, especially that produced by photocopying or laser printing, should not come in contact with plastic material of any type. Over time the ink will transfer its image to the plastic and this will obscure, if not ruin, the printed data. Information that is collected on tape, printouts, thermofax paper, or any paper stock of low quality should be photocopied onto high-quality paper before being glued or taped into the data book.

Certain materials that contain or represent data cannot be practically included in the laboratory data book. These include, for example, oversize photographic or autoradiographic material, magnetic media, embedded specimens or tissues, or some data obtained by light or electron microscopy. For proper storage of such materials, one should consider such factors as humidity, temperature, light, security, and ease of accessibility. For example, oversize X-ray films contained in protective sleeves that are appropriately coded can be stored in metal cabinets of some type. Pressed board boxes also are useful for storage. Such containers come in varied sizes and shapes, but only those composed of acid-free materials should be used. Ordinary cardboard boxes, even those commercially sold for storage purposes, are inferior and can release damaging acids over time. When using remote site storage, it is important that a description of the data storage system, the storage location, and the coding scheme be described in your laboratory data book. As a rule, an individual who inspects the data book should be able to locate all forms of data relevant to the experiments presented simply by reading its pages. For example, if centrally stored electron microscope grids or tissue sections cannot be located from reading the data book, then repeating certain experiments or observations may not be possible.

For maximum longevity, prolonged storage of data books and related materials such as photographs, negatives, or oversized documents should ideally occur under conditions of controlled temperature ($68 \pm 3°F$) and relative humidity ($<50\%$). Basements, attics, and poorly ventilated storage rooms are notoriously bad places for long-term storage of data and data books.

Format

Investigators should plan how experiments will be recorded in the data book. Some argue that writing should be concise. Although this is a reasonable guiding principle in data book writing, it should never compromise capturing any part of the experiment. For example, if an observation requires an explanation that is complex and must be described at extraordinary length, then this should be done without reservation. The same is true for interpretations and for thoughts on plans for additional work. Presentation and detail must be complete and comprehensible. All entries in the laboratory data book should be made legibly.

Purpose.   Each experiment should begin with a brief but instructive statement of the purpose of the experiment. This is done no matter how routine the experiment. Whether the experiment is to test some elegant hypothesis or simply to isolate cellular DNA, its purpose should be recorded. No experiment is too trivial to warrant a written purpose. An investigator might want to know how many independently isolated preparations of a plasmid DNA were used in performing genetic mapping studies. His or her job will be easier if each preparative run can be traced to a clearly recorded experiment that begins with a statement of purpose.

Materials and methods.   A description of any methods not found in the laboratory central methods manual should be included in the data book. The appropriate literature from which methods are derived should be cited. Assuming a central methods book exists as described previously, methods used may be cited by referring back to the central laboratory source book. Specific reference to the exact book (likely designated by date, e.g., "1994 version") should be made so that the precise method may be located in the future. If there are deviations from referenced procedure, such changes must be precisely indicated. To eliminate any confusion, it may be necessary to write the modified method in the data book.

The materials and methods section of the experiment should also document materials being used. The grade, sources, and lot numbers of specialized chemicals, reagents, and enzymes should all be recorded. If there is any question about the name recognition of the supplier (e.g., the supplier of a rare chemical or unusual enzyme), the name, address, and phone number of the supplier should be included. In the case of biological materials such as cell lines, bacterial strains, or animals, specific information on properties (e.g., genotypes and phenotypes) and source should be recorded. If working laboratory designations have been used for convenience, a full explanation of the material's original designation should be included.

Each repeat of an experiment should be written up separately in the data book. In the case of materials and methods, it is acceptable to record this section with appropriate detail and completeness the first time the experiment is performed. Assuming no changes in methodology are implemented in future runs, it is acceptable to refer back to the materials and methods section recorded in the first experiment of the series. If changes are made, reference to the original methods can be made and the modifications noted. When recording changes made to established or previously tried protocols, it is a good idea to present the rationale for the change.

If an experiment requires the use of specialized equipment, relevant information should be recorded in the data book. For example, if several electron microscopes are available for the work, which one (type, location) was used in the experiment? If calibration of a piece of hardware is required, information on the calibration process should be recorded.

Observations and results. Data should be recorded directly into the data book as soon as they become available. Original data recorded in handscript are always entered directly into the data book. Data should never be written on loose sheets of paper and then transcribed later into the data book. This practice risks the incorporation of errors during transcription and threatens the authenticity of the data. Direct recording of data requires organization at two levels. First, any writing that will facilitate data entry should be planned and carried out in advance of doing the experiment. For example, a matrix drawn and labeled to receive written data from instrument readings greatly assists data collection. The second organizational consideration involves the physical availability of the data book to the investigator while the experiment is being performed. The data book should always be conveniently accessible to the investigator. This may mean arranging bench work space ahead of time so as to accommodate the physical tools of the experiment, including the data book itself.

In addition to recorded data, the observations and results section should contain all renderings of the data, including calculations and organized presentations such as tables and graphs created using the data. Calculations should be explained. Tables and graphs should be clearly labeled. Photographic materials should be affixed to the page using archival-quality glue or tape. Any related materials not included in the data book should be catalogued and their storage location identified. For example, photographs attached to the data book may have their corresponding negatives stored in an appropriate file (see below). Negatives should be contained in glassine envelopes and stored at room temperature away from sources of high humidity, excessive light, and temperature swings (i.e., avoid proximity to windows, water baths, incubators, ovens, or autoclaves).

Discussion. Each experiment should be discussed following the recording of observations and calculations. It may be necessary to enter discussion comments at various places in the experimental write-up. In other words, the discussion for a single experiment need not be organized to appear at the conclusion of the write-up. It is appropriate to include comments that capture impressions and present interpretations at various places in the written experiment. This is convenient and ensures the most effective use of space in the data book. The standard formal presentation usually required by scientific journals, with its clear separation of the actual results and their discussion, is not usually applied to data book keeping.

The last entry in the completed write-up of the experiment should state the conclusions of the work. This should be done even if it repeats comments previously written into the data book. Conclusions logically belong at the end of the experiment. Just as we look to the beginning of the write-up of an experiment to find a purpose, conclusions are logically sought at the end of the write-up. A conclusion should be written no matter how trivial or routine it is thought to be. Future reference to the data book is aided by written experiments that have a clear beginning and a clear end.

There is no clear agreement about the style of the discussion section. For example, making comments that editorialize on the results has been debated. Some investigators urge refraining from this on the grounds that it may create confusion and mislead others at a later time. Moreover, editorializing is generally inconsistent with the overall recommendation of recording notes in as concise a fashion as possible. Others argue that the data book should record all of the mental and physical activities of the investigator. Accordingly, if something is important enough to record, a note of it should be made. Interestingly, some industrial research data book policies admonish investigators never to make comments that could be subject to misinterpretation by others. Specifically, investigators are cautioned against using phrases like "the experiment failed" or describing a yield of some biological material as "no good." This is argued on the theoretical grounds that the interpretation of a single experiment is usually not enough on which to base a far-reaching conclusion. Repetition and confirmation are always necessary and hence subjective statements about individual experiments are considered ill advised and are vulnerable to incorrect interpretation. On practical grounds, such statements are potentially damaging to a planned or existing intellectual property position (e.g., a patent application).

Sample data book pages.   Pages 244 and 245 contain photographs of laboratory data book pages. They illustrate the style of a single person but exemplify a number of features important to good record keeping. Each numbered page in the bound data book is dated. The experiment is titled (title and page are also recorded in the book's table of contents) and begins with a statement of the objective. The opening remarks contain a literature citation for reference. DNA oligonucleotide primers being used in this experiment are not described at the sequence level, but the data book page on

7-14-99   PCR analysis of transformants

Want to see if I can perform PCR on cultures or colonies
without performing DNA preps. In I & I 66(10):5620-29 PCR was
performed on cell pellets of S. pneumoniae after heating them 5 min
at 96°C in 100μl H₂O. However S. pneumoniae lyses more readily
than S. mutans. Will try boiling cells from a plate for 5 min.
Will use transformants 1 & 2 streaked out on TSA on 7-12 and still
at 37°C anaerob. If that doesn't work, can try growing the
cells on BHIT rather than TSA and/or growing them in
liquid BHIT.
    Also want to try out a new primer & new conditions
for PCR.
DNA samples:
V403 DNA 7-30-97, 5ng/μl (used many times for PCR)
"My V2404 DNA from yellow-top tube" 5-1-97, 5ng/μl
V2613 DNA 6-10-98 prep diluted today to 5ng/μl
V2613 transformant 1 - colony resd in 100μl H₂O, boiled 5 min, cent. 2 min
   "   "   2   "   "   "   "   "   "   "

Primers:
-334 (new sequencing primer; see p. 120
-833 new primer, but like Salmut1; see p.120
1260R used several times for PCR along with Salmut1; see p. 70-71, 114-115
Diluted -334 to 25pmol/μl. Already had 25pmol/μl stocks of the
other 2 primers. Want to try -334 in place of -833 (or Salmut1)
because it will reduce the size of the products about 500 bp, which
should make the reaction more robust.
    Expect: full-size fimA (V403)   deleted fimA (V2613)
-334 + 1260R   1599 bp   870 bp
-833 + 1260R   2098 bp   1369 bp

Set up rxns:
                         ×5   Used same program as on
BRL Supermix (4/99) 22.5μl   113μl   p. 115. Set volume to 25μl.
    Primer 1   .5μl   2.5μl
    Primer 2   .5μl   2.5μl
    (DNA)   2μl   23μl → ea tube (containing 2μl DNA)

Tubes 1-5 primers = -334 + 1260R Template V2404 → V403 → V2613 → #1 → #2
  "   6-10   "   = -833 + 1260R   "   = "   "   "   "   "
(Tubes 5 & 10 received ~ 23μl supermix + primers rather than the full 23μl)
   See next page for gel

128

1-14-99 *Gel of PCR analysis of transformant colonies

1.0%, TBE, EtBr, 70V 2:35 - 4:35

Conclusions

- I included V2404 because my rim A sequence comes from there, so it's possible that some of my primers won't work with V403 but will with V2404.
- It looks like -833 probably works better with V2404 than V403 (though it appears more specific in V403).
- Even though -344 produces a smaller product, -833 appears to be a better primer when used with 1260R.
- All of the primary products (the largest in each case) are the expected sizes.
- There appears to be a faint band the size of the 2613 band in the -833 + 1260R amplification. However, I need better results

Modifications to hopefully produce better results with PCR ing cells

- streaked out / patched V2613 & all 5 mutants onto BHIT. Used single colonies & toothpicks. (I'm not including em because I don't need pVA838 to be maintained.)
- Also started o/n cultures in (freshly prepared) BHIT broth ~ 5 ml ea. Inoculated w V2613 & transformants 1-3 (from the plates that were streaked out 2 days ago).
37°C anaerob. 4:30 o/n

which this information may be found is noted. A digitized, computer-labeled image of an ethidium bromide-stained agarose gel has been printed and taped to the page. A series of conclusions are listed, and modifications for a future experiment are proposed.

Good data book keeping

For single projects (e.g., a dissertation research project), data books should be used consecutively; do not start multiple data books. Once appropriate pages are reserved for a table of contents and abbreviation list (if necessary), make entries in the data book in a continuous and chronological fashion. Do not skip pages. Date each experiment, and date the entry of all recorded data and your comments. Many suggest writing each page in such a way that little or no margin space is left available for after-the-fact note taking. If an alternative explanation of the data becomes apparent, begin a new entry at the next available point in the data book. Then cross-reference the new entry with the original experiment (page and experiment number). Unused portions of any data book pages should be marked through with pen.

Mistakes in the data book should be marked through with a single line and a full explanation of the error provided. For mistakes that can be corrected instantly, this practice presents no problem with available page space. For mistakes discovered at a later date, there may not be enough space to provide an explanation. Thus, an investigator marks a line through the error and writes: "see page XX for explanation." Never obliterate mistakes with ink or cover them with any type of correcting fluid. First, their legibility may be important to you in the future, as such incorrect entries may provide needed information. Second, to the casual observer, practices that appear to remove data from the data book may suggest that such actions were taken for improper reasons.

Witnessing data and interactions with other people

Witnessing of data is a required procedure in the industrial research laboratory. The need to protect inventions and potentially patentable ideas necessitates this practice. Witnessing of data is less common in the academic research laboratory. A funding agency might require this for certain types of contract work, for example, clinical testing. However, little thought is given to witnessing the data books produced during the course of most basic research projects that constitute thesis or dissertation research. Investigators performing fundamental experiments often do not think about their work leading directly or indirectly to a discovery of a commercial application, requiring the protection of a patent. However, the unexpected bridging of basic and applied research is becoming commonplace today in the biomedical sciences. Witnessing of data is necessary if the work may lead to a patentable discovery or invention. In the academic or research institute setting, where rules for witnessing do not usually exist, establishment and enforcement of such a policy reside with the laboratory director. In deciding to put such a policy in place, the investigator must consider the requirements (if any) of funding agencies and the possibility that applied science may emerge from the research.

Where it is a standard practice in research laboratories, each and every page of the data book is witnessed. The witness signs and dates the page of the data book being examined. The witness must be able to understand the work. The signature may be accompanied by a declaration that says "witnessed and understood." Many commercially available data books have this declaration and a line or box for signature and date printed on each page. The witness must not be a coinventor. In patent prosecution, coinventors are not allowed to corroborate each other's work. Thus, selection of a neutral party who is able to understand the work is needed for appropriate witnessing of scientific data. Consider, for example, a discovery that grew from a predoctoral research project. The trainee's mentor would likely be considered a coinventor and, thus, should not sign as a witness to the data. Another worker in the same lab could sign, assuming he or she understood the work but was not involved in it.

It is desirable to record in the data book discussions with others about the research. Such notes should list the times, names of the individuals talked to, and relevant points of the discussion. This is a good record-keeping habit that will help trace the investigator's thinking processes and provide a prompt when it is time to attribute credit. In addition, should corroboration of data be needed at some point, tracking down individuals who can talk about certain experiments is the next best thing to a witnessed data book page. Correspondence to and from colleagues about your experiments should be recorded in the data book as well. Letters can be photocopied on high-quality paper and then fixed in the data book using archival tape or glue. Alternatively, it may be appropriate to make notes from such correspondence in the data book and then refer to the location of the letter in a file that can be easily found by someone reading the data book.

Finally, names of individuals who have played any role in your research need to be entered in the data book along with a description of their contributions. Collaborative researchers fall into this category. Agreements with collaborators pertaining to research contributions, expenditures on grants, personnel involvement, and perhaps, most important, authorship on papers, should be recorded in the data book. People who have participated in your research, even on a fee-for-service basis, should be noted in your writing. Technicians working in institutional core laboratories are especially important. Who made the oligonucleotide or the hybridoma or ran the automated DNA sequencer for your particular experiment? These notations represent a record of quality control. They can help you in troubleshooting problems and can provide a source of independent corroboration in matters of intellectual property.

Electronic Record Keeping

The electronic data book

Free-standing or networked personal computers are playing an increasing role in research laboratory record keeping. There are several advantages to such record keeping techniques. Clarity and organization are facilitated by keyboard input, and the entry of graphs, tables, and photographic material can be accomplished using software (drawing, graphing, and data analysis programs) as well as scanners that can digitize hand-drawn materials. Access to electronic data can be managed by using passwords or controlled accounts, and this provides a security advantage while allowing more efficient access to data by authorized individuals. For example, mentors could check data of trainees stored on a central computer hard drive. Enormous amounts of electronic data can be stored in a very small space compared with traditional lab data books, and accessing experimental protocols and data by electronic searching is powerful and rapid.

Disadvantages to electronic record keeping exist as well. Witnessing an electronic data book "page" is open to question. Although scanning technology and automated handwriting digitizers can provide the means to do this, the legal acceptance of such methods has not yet been tested. Another disadvantage that will likely be overcome is the inclusion of an unalterable time and date record associated with the data. Techniques to do this are being developed and applied. Digital "time-stamping" programs are being used in some cases, but they usually require an independent party to collect and maintain records; this logistical problem makes the utility of such programs impractical for the typical academic research lab.

At the present time, the use of the computer in laboratory record keeping continues to be in transition. "Laboratory data book" programs are coming on the market, but their usefulness has not been extensively tested. The Collaborative Electronic Notebook Systems Association (CENSA) is an international industry association that promotes advancement in automated technologies, including the use of electronic record keeping in scientific research and discovery. CENSA maintains a Web site (http://www.censa.org/) that provides a convenient means of keeping track of developments in electronic record keeping. For example, their site contains links to groups and vendors developing and marketing electronic notebook systems.

Genomics and electronic data

Although electronic data books are not yet commonplace, one biomedical research methodology is dependent on computers for acquisition, analysis, transfer, and storage of data. Nucleotide sequence determination has created an absolute need for computer-based record keeping. Data acquired in almost all DNA sequencing projects are now collected electronically, with raw data downloaded from automated instruments and provided to the investigator in electronic format. Primary data may be stored on institutional mainframes as well as on free-standing or networked computer stations. Principal investigators need to consider issues of data storage, organization, access, and retention in relation to nucleotide sequencing results. The implementation of lab policies may be necessary to effectively handle such data sets, especially in research groups where large amounts of data are being generated on a regular basis. Superimposed on these usual considerations of record keeping is the issue of data release. DNA sequence data are commonly uploaded to public databases before formal publication in the scientific literature. Typically, such databases allow the investigator to control access to his or her sequence information until it is published. But such data release may have far-reaching effects. As pointed out in chapter 9, the ownership of data is in reality an exercise of property right. In practice, then, the control of scientific data is tantamount to ownership. Data inappropriately posted to a public DNA database by an ill-informed lab associate could compromise a group's priority over a discovery, with implications for attribution of proper credit, peer-reviewed publication, new grant funds, or intellectual property protection.

Bioinformatics has created a whole new set of challenges in scientific record keeping. The new landscape and language that accompanies this world of the Internet, genome databases, file transfer protocols, and search algorithms are explored in several books, including works by Baxevanis and Ouellette (2) and Bishop (4).

Conclusion

Notes prepared for a talk can be categorized into two general forms. The first includes short phrases, words, or occasional sentences that provide triggers for the speaker. The second form is a verbatim text of the speaker's remarks—a script of every word he or she will speak. There are no abbreviations, cryptic reminders, or short-hand notations. If the speaker were suddenly taken ill, a colleague could easily give the speech. However, it is doubtful that a substitute could successfully deliver the speech using only the abbreviated notes. Use this example in thinking about record keeping. A laboratory data book is inherently more useful as the "verbatim text" of experimental work. In his book *The Cuckoo's Egg* (Doubleday, New York, N.Y., 1989), Clifford Stoll lauds the value of a carefully documented data book. His advice rings true as an axiom of scientific record keeping: "If you don't document it, you might as well not have observed it."

Case Studies

11.1 Sheri, a graduate student colleague about to defend her dissertation and take a postdoctoral position, comes to you for advice. Her mentor has instructed her to take all of her data books when she leaves and keep them in her possession. She quotes him as saying: "Sheri, you know how disorganized I am. I'll probably lose track of your data books before a month goes by. Besides, you have one more paper to write and the books are better off in your hands. If I ever need them, I'll know where to call you." Are Sheri's mentor's actions appropriate? What advice will you give her?

11.2 A predoctoral student working in the laboratory of his mentor is gathering data for a federally funded project on which the mentor serves as principal investigator. The student is, of course, going to use the data for her dissertation work. The student and mentor have a terrible falling out. The student leaves the lab and finds a new advisor. The previous mentor notices that data and materials related to the student's project are missing. The student readily admits to removing the data books, tissue sections, gels, and computer disks but asserts that they are hers — the product of her sweat and blood. What issues of data ownership apply here, and what should be done?

11.3 A graduate student, working on a project that involves extensive DNA sequencing, provides his mentor with a computer-generated sequence of a gene. The student tells his mentor that the sequence determination involved complete analysis of both strands of the DNA molecule. Over the next several months, it is determined that not all of the sequence data reflect analysis of both DNA strands. Indeed, follow-up work by a postdoctoral fellow in the laboratory reveals several mistakes in the sequence. The student in question admits to misleading his mentor and, following appropriate investigation, is convicted of scientific misconduct and dismissed from the graduate program. The mentor realizes that the student presented some of the erroneous data at a regional scientific meeting. Proceedings of the meeting were not published, but abstracts of all the works presented were distributed to approximately 100 meeting participants. In addition, the student, with the mentor's permission, sent the sequence by electronic mail to three other laboratories. What, if any, responsibility does the faculty mentor have with regard to disclosing the above developments? What, if anything, should the mentor do about the prematurely released data? Under these circumstances, what is the potential for harm coming from this incident of scientific fraud? Who might be harmed?

11.4 The research laboratory of a faculty investigator has begun using a new electrophoresis technique. The technique works well in the hands of the laboratory investigators. A field service representative from the company that manufactures the apparatus asks several of the workers in the laboratory if he may borrow some of the photographs of their results to show them to potential clients. In return, he offers to take the whole lab to dinner at an expensive restaurant. The lab members comply, and the whole group goes to dinner a few weeks later. You, as laboratory director, are told of this whole series of events after the fact. Comment on the implications of this scenario for data ownership and laboratory record keeping. What action, if any, will you take?

11.5 Jim, a new assistant professor, is getting ready to submit his first paper since joining the faculty. He reviews one of the figures for his paper, which is a photograph of an ethidium bromide-stained agarose gel. The gel contains the products of PCR-amplified whole-cell DNA. The photograph displays the predicted 3-kb DNA fragment. Jim comments to you, his faculty colleague, that a second, minor signal was also evident on the original gel. Based on its size, Jim believes that this second fragment represents a very exciting discovery, but it needs considerable additional work. This second fragment cannot be seen in the photograph. Jim discloses that this is because he has deliberately prepared an underexposed print in order to obscure the second fragment. He says he did this because he is worried that competing groups in larger, more established labs will recognize the potential of the second fragment and will "scoop" him. He has prepared a figure legend that says: "a second, minor signal of unexplained origin was present in this experiment but is not visible in the photograph." But the figure legend does not indicate the size of the unexplained fragment. Thus, he argues he will be telling the truth while protecting himself from his competition. Are Jim's actions appropriate? Is he (i) simply playing fairly in the hotly competitive arena of biomedical research, (ii) falling victim to self-deception, or (iii) perpetrating scientific fraud?

11.6 You have submitted a manuscript to a peer-reviewed journal that contains primary nucleotide sequence data for a new gene and its upstream sequences. When you receive the paper back from the editor, it is accompanied by two favorable reviews written by expert ad hoc referees. One of the referees has some suggestions regarding the interpretation of your sequence data. Specifically, the reviewer attaches a new printout of your entire sequence data with some computer-generated structures. These represent predictions of folded mRNA derived from the transcription of your gene. The reviewer's interpretations have implications for translational genetic control of the gene. It is clear to you that the reviewer has made an electronic file of your sequence data and has subjected the data to his or her own analysis. Did the reviewer do anything wrong in your view? Will you discuss this with the editor of the journal? If so, what inquiries, comments, or requests will you direct to the editor?

11.7 Bob, your fellow graduate student, comes to you for advice. Bob's mentor recently has noticed that he keeps his stained, desiccated polyacrylamide gels in sealed plastic bags that are taped to the pages of his data book. Bob considers such gels to be primary data that must be retained in their original form. Bob's mentor has ordered him to stop doing this. Moreover, he tells Bob to remove the gels already in his data book. Bob's mentor says that polyacrylamide is a neurotoxin and should be disposed of properly. Further, he tells Bob to make black-and-white photographs of all his previous gels and to retain both the print and negative for each gel. He says that in the future, this practice should be followed for all acrylamide gel data storage. He says the photographs are to be considered the primary data and retained in Bob's data book. Bob disagrees with his mentor and argues that photographs can be altered and that a desiccated gel is an accurate representation of the original data. He also argues that once the acrylamide is sealed in plastic, there is no danger of exposure to toxic material. Bob's mentor dismisses these arguments and gives him 1 month to photograph the existing gels and to dispose of them. Bob is very upset. He thinks his mentor is acting irresponsibly with respect to data retention. He also feels his mentor is being a bully, by forcing Bob to adopt his personal preferences. What advice do you give Bob?

11.8 You meet a colleague at a national meeting who is working in your field. You exchange information on your research, and he agrees to send you a diskette containing the amino acid sequence of a protein called XYZ. The nucleotide sequence of the gene encoding this protein was determined in his lab but has not been published. You return from the meeting, and the diskette is in your mail as promised. You are fascinated by the sequence and suggest that one of your students analyze it. The student proceeds to analyze XYZ using computer algorithms and prepares several printouts of amino acid sequence comparisons with proteins being studied in your lab. The student's results are exciting, and you ask her to make a presentation at a journal club involving your research group and those of three other faculty members. Following the meeting, one of your faculty colleagues accompanies you back to your office and asks to speak to you privately. He questions how you obtained the data and indicates concern that your student has disclosed privileged information without permission. He claims this is especially problematic because the data have not appeared in print and you did not seek specific permission to have the data discussed in an open forum. Further, he expresses concern that your new findings on the amino acid sequence of XYZ should have been shared immediately with the investigator who provided you with the data. How do you respond to your faculty colleague?

11.9 A predoctoral trainee under your supervision has had several difficult years finishing up his dissertation research. He has needed continual guidance, and his attitude has not been positive. He does not seem motivated about the work, but you press him almost daily until the work is completed and the dissertation is finally written. The student turns in an average defense and informs you that he is leaving science to take a job in biomedical sales. Several areas of the student's dissertation need additional work before the research can be written up in manuscripts for publication. You turn several portions of the dissertation work over to a competent postdoctoral trainee in your laboratory. Over the course of the next several weeks, the postdoctoral trainee pursues these new lines of experimentation. In the process, however, he uncovers several problems with the data in the dissertation. In fact, a number of experiments cannot be repeated. Moreover, some of the results obtained are opposite to those reported in the student's dissertation. These results have serious implications regarding interpretations and conclusions reached by the student in his dissertation. You review the student's data books and are unable to find entries that could have been used to construct some of the tables included in the dissertation. Moreover, other data sets written into the data book have been used selectively to construct some tables in the dissertation; i.e., critical points that would have confused analysis were omitted. After considerable analysis and discussion with the postdoctoral trainee, you decide that the student has at least falsified data and possibly fabricated data presented in his dissertation. You have not yet published any of the work of the student's dissertation in manuscript form. However, one published abstract contains accurate information that has been authenticated by your postdoctoral trainee. All of the student's work was supported by your NIH grant. What actions, if any, will you take in this situation?

11.10 A graduate student is working on an NIH-funded project to develop new strains that can degrade certain environmental contaminants (e.g., pesticides). The graduate student is using a variety of genetic techniques to develop new strains. She isolates a strain that has a number of unique, unexpected properties. The student proposes that these properties reflect a novel global regulatory gene. The mentor is concerned that the "new strain" is a contaminant from another study and asks that several additional tests be conducted to establish the lineage of the strain. A few weeks later, the student presents data confirming that the new strain is derived from the appropriate lineage. The mentor asks to see the raw data and the experimental documentation (i.e., the electrophoretic gels). The graduate student, with some reluctance, says that she does not have them and offers the explanation that housekeeping personnel have discarded them. The mentor takes the "new strain" and soon establishes that it is identical to a culture used in another study conducted by another member of the laboratory. The graduate student continues to deny wrongdoing. The mentor asks the graduate student to withdraw from the laboratory and she does so. Who should the mentor notify of these events? What actions should the mentor or institution take?

11.11 Donna Adkins has collected blood samples from 100 human patient volunteers to test antibody levels against two different viruses. Relevant clinical histories of these patients, corresponding to the individual samples, are noted in her data book. She has carefully tagged the tubes with self-adhesive labels and stored them in racks of 20 in the freezer. She assays the samples in three of the five racks and obtains interesting results. She records her results meticulously in her lab data book, cross-referencing the antibody values to the clinical patient data. Donna asks you to witness these data book pages because the results have implications for the development of an important diagnostic test. You sign her data book pages as requested. When she opens the freezer to retrieve the sera in the fourth rack, she makes a disturbing discovery. All the labels have fallen off the tubes in racks 1 and 2. (She later finds out she used the wrong kind of self-sticking labels on these tubes, resulting in their failure to adhere at $-70°C$.) Donna proceeds to number all the tubes in racks 1 and 2 by order of their rack location. Then she repeats the antibody assays on these samples. She arranges her resulting data into a summary table that she compares with her original assays of these samples. She is relieved that the data compare favorably, and she relabels the tubes consistent with their original designations. She comes to you for advice on her actions and asks how, if at all, she should record these events in her data book. What do you tell her?

11.12 Dr. Brown's research group recently published an important paper in a leading physiology journal. Four months after the publication of the manuscript, Dr. Brown is contacted by a European colleague who has been unable to reproduce the results presented in two figures of the paper. Dr. Brown faxes copies of the pertinent laboratory protocols and recipes to his colleague and thinks no more of the discrepancy. Two months later, a graduate student in a competitor's laboratory contacts Dr. Brown and reports that he too was unable to reproduce the results. After this second call, Dr. Brown meets with Adam Green, the postdoctoral fellow who did the experiments in question. He asks Adam to bring his data book to the meeting so they can review the results together. Once in Dr. Brown's office, Adam confesses that he has been remiss in keeping his data book. He says that all of his electrophysiology experiments were recorded on VHS tapes with a live microphone into which he reported the experimental proceedings and observations. Adam transcribed these observations into his data book. However, there was a period of several days when his microphone was not working properly. Although Adam replaced the microphone as soon as he found that it was not working, he relied on his memory to transcribe the results of those particular experiments. After completing the figures for the manuscript, Adam was pleased to find that his data supported Dr. Brown's hypothesis. Dr. Brown comes to you for advice on how to handle this situation. What do you suggest?

References

1. **Barker, K.** 1998. *At the Bench: A Laboratory Navigator.* Cold Spring Harbor Laboratory Press, Plainview, N.Y.

2. **Baxevanis, A. D., and B. F. F. Ouellette (ed.).** 1998. *Bioinformatics: A Practical Guide to the Analysis of Genes and Proteins.* Wiley-Interscience, New York, N.Y.

3. **Becker, J., G. A. Caldwell, and E. A. Zachgo.** 1990. *Biotechnology — A Laboratory Course.* Academic Press, Inc., San Diego, Calif.

4. **Bishop, M. J. (ed.).** 1998. *Guide to Human Genome Computing.* Academic Press, Inc., San Diego, Calif.

5. **Grisson, F., and D. Pressman.** 1996. *The Inventor's Notebook.* Nolo Press, Berkeley, Calif.

6. **Hittelman, K. J., and B. Flynn.** 1995. *Investigators' Handbook.* University of California, San Francisco. (Available on-line at http://www.library.ucsf.edu/ih/)

7. **Kanare, H. M.** 1985. *Writing the Laboratory Notebook.* American Chemical Society, Washington, D.C.

8. **U.S. Department of Health and Human Services.** 1998. *National Institutes of Health Grants Policy Statement.* (Available on-line at http://www.nih.gov/grants/policy/nihgps/)

Resources

University bookstores and office supply companies typically sell bound data books suitable for laboratory record keeping like those manufactured by the Avery-Dennison Company ("computation notebooks"). Such data books come in several standard formats. The paper contained in these products may not be acid-free.

Some companies specialize in data book manufacturing. Products marketed by these companies contain acid-free paper and come in standard or custom-designed formats. These companies sell directly to individual customers but usually require a minimum order. Some of their data books are carried by university bookstores.

For standard-format data books of varying styles:

Scientific Notebook Company
P.O. Box 238
Stevensville, MI 49127
Phone: (616) 429-8285

http://www.snco.com/

For custom-manufactured data books:

Eureka Blank Book Company
P.O. Box 150
Holyoke, MA 01041-0150
Phone: (413) 534-5671

http://www.eurekalabbook.com/

Laboratory Notebook Company
P.O. Box 188
Holyoke, MA 01041-0188
Phone: (413) 532-6287

Materials for archiving, including acid-free glue, archival mending tape, and acid-free boxes of varying styles and sizes, are sold by:

University Products, Inc.
517 Main Street
P.O. Box 101
Holyoke, MA 01041-0101
Phone: (800) 762-1165

http://www.universityproducts.com/

appendix I

Survey Tools for Training in Scientific Integrity

Michael W. Kalichman

SURVEYS, LIKE THE SCENARIOS USED FOR CASE STUDY DISCUSSION, require trainees to examine their own perceptions and assumptions. Through the process of this reflection, it is possible to refine existing standards, identify new standards, and develop strategies for responding to difficult questions. The characteristic that distinguishes surveys from case study discussions is that answers are not open-ended. Instead, the respondents are asked for answers that are either categorical (e.g., yes/no) or quantitative (e.g., the degree of agreement or disagreement). These answers then can be reduced and summarized for ready discussion of patterns and correlations within a particular group (e.g., this year's students) or between groups (e.g., those who have had training in scientific integrity versus those who have not). Although the subtleties in complex cases may not always emerge in discussion, it is often possible with surveys to elicit information about common trends and attitudes that would otherwise be lost. The following includes both general observations about the use of the 10 surveys presented here and some specific comments about the use of each.

The surveys included below parallel most of the topic areas of this book. The following points should be kept in mind when employing them as teaching tools. First, because some surveys overlap, their selection is at the discretion of the instructor. Further, not all surveys will be appropriate to meet the needs of a specific course, instructor, or group of students. Second, these surveys should not be viewed as definitive; instructors may want to develop new surveys to meet specific instructional objectives. Third, nearly all of these surveys are suitable for administration during a class or workshop, but some of the longer forms are probably more appropriate as homework assignments. For purposes of homework or distribution, these forms may be copied from this book, or suitable response sheets may be prepared by the instructor. Fourth, simply completing these surveys can

257

have value in stimulating reflection on personal values and the normative conduct of science. However, analysis and discussion of survey results is a key part of this exercise. The instructor could do the analyses, but it is instructive for everyone to have trainees summarize the data, select results of interest, present their findings, and lead class discussion about interesting results. Usually, it is not necessary for the survey discussants to focus on the responses to each and every item in the survey. Instead, identifying questions that reveal differing attitudes and perceptions on the part of the respondents is desirable. These should be used to stimulate class discussion, allowing the discussants to state their positions and the rationales underlying their responses. Such discussions allow students to invoke their critical thinking skills in articulating their arguments. Equally important, these discussions frequently uncover multiple points of view, many of which have merit and can be appropriately defended.

Survey Form I—General Survey.   Survey Form I is modified from one originally used to study perceptions about research misconduct at the University of California, San Diego (3). It has potential value as an introductory exercise for a course in scientific integrity. Ideally, students would be asked to complete the questionnaire immediately before or at the beginning of a workshop or the first meeting of the course. In addition to the raw data being of interest (e.g., what percentage of respondents believe they have firsthand knowledge of plagiarism), secondary analyses and discussion about the meaning of the answers are at least as important. A couple of specific analyses that might be of interest are (i) the correlation between position (question #1), years of experience (#2), or experience as an author (#3) and the answers to questions about misconduct experience or (ii) discussion of the various possible interpretations of, for example, 10% of respondents reporting firsthand knowledge of data fabrication or falsification. It can also be of interest to compare results of this survey with those that are published (2, 3).

Survey Form II—Standards and Options.   Survey Form II is also useful as an introductory survey. This form is modified from a survey published by Brown and Kalichman (1). Answers to these questions can help both the instructor and the students appreciate initial perceptions about differing standards of conduct (questions #1 to 6). The questions about options (#7 to 14) can be a valuable assay of the perceived ability to deal with a variety of problems faced by many scientists. Following analysis, the resulting data can then be compared with those already published for many of the same questions (1).

Survey Form III—Predoctoral Mentoring.   The results of Survey Form III can provide the basis for discussions on the responsibilities of mentoring. It is recommended that students and their mentors complete this survey, followed by a presentation and discussion of the results in class.

Survey IV — Authorship and Data Selection. Survey Form IV is in the form of short cases representing three different dilemmas faced by research scientists. By asking these questions at the beginning of a course, it is possible to obtain a quick assessment of the difference between knowing what you should do and the willingness to do something different. Particularly because of the need to ensure confidentiality (i.e., for those who would do something that they believe is unethical), this survey is best done with students anonymously completing the questionnaire and then returning the pages to the instructor. The instructor or a group of students could then tabulate the results for class discussion of whether or not the proposed actions are unethical and what pressures, if any, justify taking these actions. Alternatively, this form of survey can be used as a pre- and post-test evaluative tool to assess whether there has been a shift in attitudes or moral resolve between the beginning and end of a course or workshop.

Survey Forms V — Publication, and VI — Authorship. The object of Survey Forms V and VI is to elicit opinions about a variety of aspects of the publishing of papers. Publication, particularly authorship, is among the most frequent sources of conflict between scientists. These surveys are short and simple enough for presentation in a course or workshop. Analysis can be accomplished either by assignment to small groups for tabulation or simply by asking for a show of hands (e.g., in Survey V, how many people chose #5 as the most important reason for publishing research findings? How many chose #5 as the least important? In each case, explain why.).

Survey Form VII — Confidentiality of Review. Survey Form VII is specifically designed as a homework assignment. The goals are threefold. First, trainees are asked to think about different uses of manuscripts under review and what is or is not acceptable. Second, trainees will engage in discussion about these specific questions with at least one active investigator. Third, it is of interest to see the difference between self-reported standards and those that are presumed to be characteristic of other scientists. The results of this survey provide the opportunity to discuss such differences as well as the need to ensure confidentiality of reviews.

Survey Form VIII — Use of Animal Subjects. Survey Form VIII is designed to encourage trainees to think about their personal criteria for accepting or not accepting the use of animals in research. Although the questions are fairly straightforward, many will require considerable reflection. For this reason, it is probably best that this survey be given as a homework assignment, thereby encouraging more thoughtful answers. For the purposes of analysis, it should be of value to compare the relative importance placed on species (where is the cut-off and why), the adverse consequences of the experiment (pain, distress, or discomfort), and the utilitarian value of the studies (e.g., cure for AIDS versus cosmetic safety). For the purposes of class discussion, it is to be expected that opinions will vary widely, even within the biomedical research community.

Survey IX — Data Sharing. Survey Form IX deals with collegiality and sharing between research scientists. These questions can be completed readily during a class or workshop, but the answers to questions #5 and #6 might warrant a homework assignment. After discussing the results for answers to questions #1 to 4, this survey can be a starting point for discussion about whether standards for sharing are changing, what pressures keep people from sharing, are these pressures real or imagined, are standards for sharing different in different areas of biomedical research, and how can the barriers to sharing and collegiality be decreased.

Survey Form X — Case of the Mystery Medium. Survey X demonstrates the use of a case discussion scenario in a survey format. This survey is clearly for use in a class or workshop rather than as a homework assignment. One use is to present the case (distribute copies or display as a slide or overhead transparency), allow a moment for reflection about the goals and disadvantages of sharing data, and then present each question. By asking for a show of hands, the instructor and class can then assess the relative importance of each rationale and elicit discussion about when, if at all, withholding the data would be justified.

References

1. **Brown, S., and M. W. Kalichman.** 1998. Effects of training in the responsible conduct of research: a survey of graduate students in experimental sciences. *Sci. Eng. Ethics* **4**:487–498.
2. **Eastwood, S., P. Derish, E. Leash, and S. Ordway.** 1996. Ethical issues in biomedical research: perceptions and practices of postdoctoral research fellows responding to a survey. *Sci. Eng. Ethics* **2**:89–114.
3. **Kalichman, M. W., and P. J. Friedman.** 1992. A pilot study of biomedical trainees' perceptions concerning research ethics. *Acad. Med.* **67**:769–775.

Survey Form I: General Survey
(Chapters 1 and 2)

Background information
1. Which of the following best describes your position?
___ Grad student   ___ Postdoc   ___ Faculty   ___ Staff

2. Which of the following best describes your experience in research?
___ None ___ <1 year ___ 2–5 years ___ >5 years

3. Have you ever been the author of a published paper or abstract?
___ Yes ___ No

Data reporting — past

4. Has your name been omitted from a paper for which you made a substantial contribution?
___ Yes ___ No

5. Have you been an author on a paper for which any of the authors had not made a sufficient contribution to warrant credit for the work?
___ Yes ___ No

6. Do you have firsthand knowledge of scientists plagiarizing the work of someone else?
___ Yes ___ No

7. Have you ever plagiarized the work of someone else?
___ Yes ___ No

8. Do you have firsthand knowledge of scientists intentionally altering research or experimental results for the purpose of publication?
___ Yes ___ No

9. Do you have firsthand knowledge of scientists intentionally altering research or experimental results to enhance a grant application?
___ Yes ___ No

10. Have you ever intentionally altered research or experimental results for the purpose of publication or a grant application?
___ Yes ___ No

11. Do you have firsthand knowledge of scientists intentionally falsifying or fabricating research or experimental results for the purpose of publication?
___ Yes ___ No

12. Do you have firsthand knowledge of scientists intentionally falsifying or fabricating research or experimental results to enhance a grant application?
___ Yes ___ No

13. Have you ever falsified or fabricated research or experimental results for the purpose of publication or a grant application?
___ Yes ___ No

14. Have you ever reported research or experimental results that you knew to be untrue?
___ Yes ___ No

Data reporting—future

15. If you were confident of your findings, would you selectively omit contradictory results to expedite publication?
___ Yes ___ No

16. If you were confident of your findings, would you falsify or fabricate data to expedite publication?
___ Yes ___ No

17. If you were confident of your findings, would you selectively omit contradictory results to enhance a grant application?
___ Yes ___ No

18. If you were confident of your findings, would you falsify or fabricate data to enhance a grant application?
___ Yes ___ No

Reporting misconduct

19. Would you report a coworker who you believe has violated scientific integrity standards?
___ Yes ___ No

20. Would you report your supervisor/advisor who you believe has violated scientific integrity standards?
___ Yes ___ No

Data sharing and ownership

21. Have you ever been advised that sharing of research data constitutes good science?
___ Yes ___ No

22. Have you ever been advised that sharing of research data is not in your best interest?
___ Yes ___ No

23. In general, before publication, would you be willing to share your results with:

___ a colleague from your department?

___ a member of another department at your institution?

___ a scientist from another institution?

___ a friend working in your field of research?

___ a competitor in your field of research?

24. At what time should research data reasonably be made available to anyone who requests them?

___ while the project is in progress

___ when data collection is complete

___ when data analysis is complete

___ when the manuscript has been written

___ when the manuscript has been accepted for publication

___ when the paper is published

25. Who has the final approval as to what will be done with your data (research notebooks, details of methods, raw data)?

___ you

___ your mentor, principal investigator of project

___ granting agency

___ your institution

___ don't know

Survey Form II: Standards and Options
(Chapters 1 and 2)

Based on your impressions of others working in your area of research, how would you rate their standards of conduct relative to your own standards in each of the following areas? Use the following scale.

0 Don't know

1 Much lower

2 Lower

3 Similar

4 Higher

1. Handling financial conflicts of interest   ___

2. Care of animal or human subjects   ___

3. Data ownership   ___

4. Reporting of data   ___

5. Disputes about who should be an author on a paper   ___

6. Reporting research misconduct   ___

If faced with problems in the following areas, how would you rate your knowledge of the options available to you? Use the following scale.

0 Don't know
1 Low
2 Average
3 High
4 Very high

7. Financial conflicts of interest _____
8. Care of animal or human subjects _____
9. Data ownership _____
10. Reporting of data _____
11. Disputes about who should be an author on a paper _____
12. Research misconduct by a graduate student _____
13. Research misconduct by a postdoctoral researcher _____
14. Research misconduct by a faculty member _____

Survey Form III Predoctoral Mentoring (Chapter 3)

Indicate the degree to which you agree or disagree with the following issues related to the duties and responsibilities of predoctoral mentoring.

Strongly disagree 0
Disagree 1
Agree 2
Strongly agree 3

1. Prior to accepting advisees into their laboratories, mentors must be prepared to financially support all aspects of the advisee's graduate training (stipend, tuition and fees, all research expenses, travel).

2. Mentors should not accept an advisee into their laboratory without the student's first spending a brief rotation period working at the bench in that laboratory.

3. Mentors should consider the following in deciding whether to accept a potential student in their laboratories:
   A. academic performance prior to graduate school
   B. academic performance in graduate school
   C. motivation
   D. personality, especially in regard to the student's interaction with other lab members
   E. recommendations from teachers or colleagues who have known the potential trainee

4. Mentors should set a limit on the number of trainees they accept into their laboratories; this limit should be based on financial and physical resources as well as on supervisory considerations.

5. Mentors should set a limit on the overall size of their research groups (trainees, technicians, support personnel) based on financial and physical resources, and on supervisory and management considerations.

6. Mentors should provide specific instruction to their advisees on the organization of data books, including issues related to format, collection and recording of data, retention of data, and ownership of data.

7. Mentors should regularly check that their advisees are adhering to the laboratory standards of good record keeping.

8. Mentors and advisees together should write a detailed plan of the advisee's research goals at a very early point in the advisee's training program.

9. Mentors should personally educate their advisees in all matters related to laboratory safety, use of hazardous materials (including isotopes), and good laboratory practice.

10. Mentors should personally educate their advisees in all matters related to the use of animals in research.

11. Mentors should personally educate their advisees in all matters related to the use of humans in research.

12. Mentors should prepare, distribute, and update, as needed, the supervisory and responsibility structure of their research laboratories.

13. Mentors should have a defined policy with regard to scientific publication, manuscript preparation, and authorship attribution; this should be formally communicated to advisees early in their training program.

14. Mentors should meet privately and regularly (once every 7 to 10 days) with each of their advisees to discuss the advisee's research progress, analyze data, plan experiments, and set goals as appropriate.

15. Mentors should hold regularly scheduled meetings of their whole research group to review individual projects.

16. Mentors should regularly talk to their advisees about the people and politics of science.

17. Mentors should encourage their advisees to present their research results at local, regional, and national meetings.

18. Mentors should financially support their advisees' participation in regional and national meetings, whether or not they present their results at such meetings.

19. Mentors should be active in introducing their advisees to other scientists, e.g., visiting seminar speakers, other scientists at meetings.

20. Mentors should involve their advisees in the preparation of grant proposals, including such activities as:
    A. presentation and analysis of raw data for inclusion in the proposal
    B. development and critique of experimental design
    C. proofreading and editing

21. Mentors should share financial information about their grants with their advisees, including:
    A. how budgets are developed
    B. periodic expenditure reports of grant funds

22. Mentors should monitor advisees closely enough to be able to spot behavioral changes in trainees that might indicate unusual stress.

23. Mentors should provide close supervision and counseling to advisees whom they identify as suffering from inordinate stress of any type.

24. Mentors should provide career counseling throughout the training program, but especially in the latter stages of the advisee's program.

25. Mentors should encourage healthy competition among advisees in their laboratories.

26. Mentors should explain the benefits of professional society membership to their advisees and encourage them to join appropriate societies as student members.

27. Mentors should provide critiques of any verbal scientific presentation a trainee makes.

28. Mentors should provide trainees with assistance and instruction in classroom teaching skills.

29. Mentors should provide trainees with assistance and instruction in how to write a scientific paper.

30. Mentors should provide trainees with assistance and instruction in how to read a scientific paper.

Survey Form IV: Authorship and Data Selection (Chapter 4)

1. Two months after joining a new research group, you are preparing to submit a manuscript based on work you completed while in your previous position. Dr. Helix, one of your new colleagues, has just recommended that you include Dr. Spiral, the head of the new research group, as an author on the paper. When you point out that Dr. Spiral has made no contributions to the work, Dr. Helix observes that adding Spiral's name will improve the chances for publication and increase your prospects for advancement within Spiral's research group.

   Based on this information, should you add Dr. Spiral's name to the manuscript?
   ___ Yes ___ No

   Would adding Dr. Spiral's name as an author be unethical?
   ___ Yes ___ No
   Why or why not?

2. You are a postdoctoral fellow in the laboratory of Dr. Strauss. Dr. Strauss has been asked to review a manuscript, which she has now handed to you for your comments. When you ask if she has notified the journal editor that you will be reviewing the manuscript, she replies that there is no need to do so because sharing the responsibility of manuscript review is common practice.

   Based on this information, should you agree to review the manuscript?
   ___ Yes ___ No

   Would reviewing the manuscript without notifying the journal editor be unethical?
   ___ Yes ___ No
   Why or why not?

3. You have completed a small clinical study in which the drug appears to have worked, but the result just misses statistical significance. It occurs

to you that by randomly selecting values from your previously published study, you could increase the size of your control group and thereby demonstrate a significant effect.

Based on this information, should you supplement your data with numbers from the previously published experiment?
___ Yes   ___ No

Would using the previously published data be unethical?
___ Yes   ___ No
Why or why not?

Survey Form V: Publication (Chapter 4)

Indicate the degree to which you agree or disagree with the following statements about publication of scientific papers. Using the scale below, write the number corresponding to your response on the line next to each statement.

    0   Strongly disagree
    1   Disagree
    2   Agree
    3   Strongly agree

We publish our research findings to
1. contribute to the body of scientific knowledge ___
2. improve our experimental work ___
3. meet research funding requirements ___
4. advertise our work to future trainees and lab associates ___
5. promote our careers ___

It is acceptable to
6. omit contradictory results from a paper ___
7. publish the same paper in two very different journals ___
8. republish data in a second, very different journal ___
9. republish data with citation of the earlier work ___
10. use words written by a colleague without citing the source ___
11. use data from a colleague without citing the source ___

Without notifying the journal editor, a reviewer should never
12. get help with the manuscript from a graduate student, postdoctoral fellow, or faculty member ___
13. use the manuscript review as a tool for training his or her students ___
14. use the manuscript to keep his or her trainees and colleagues up to date ___
15. make scientific use of the manuscript prior to its publication ___

Survey Form VI: Authorship (Chapter 4)

Indicate the degree to which you agree or disagree with the following statements about authorship of scientific papers. Using the scale below, write the number corresponding to your response on the line following each statement.

    0   Strongly disagree
    1   Disagree
    2   Agree
    3   Strongly agree

Authorship is appropriate for someone who has approved the final manuscript and

1. provided the idea for a critical experiment   ___
2. performed an experiment using specialized equipment   ___
3. provided unique materials, critical to the experiments reported in the paper   ___
4. provided unpublished data to augment data obtained for the paper   ___
5. provided statistical analysis of data presented in the paper   ___
6. provided large amounts of unskilled work needed to complete the project   ___
7. organized results and wrote the first draft of the paper   ___

Authorship is not appropriate

8. for someone who contributed to the work only on a fee-for-service basis   ___
9. for someone who cannot scientifically defend all data presented in the paper   ___
10. for someone who has not read and approved the final manuscript   ___
11. solely to advance a student's career   ___
12. solely to recognize leadership of the research group   ___
13. solely to increase chances for publication because of name association   ___

If fabricated data are discovered in a published paper, all coauthors

14. must equally share in the blame   ___
15. must receive the same punishment   ___

Survey Form VII: Confidentiality of Review

Find at least one investigator who is willing to give you a few minutes to talk about the process of manuscript review. Ask the person about his or her own practice as well as his or her impression of what constitutes common practice for each of the following. Use the following scale:

0 never
1 rarely
2 often
3 most of the time
4 always

Without notifying the editor of the journal, the reviewer:	Investigator	Common practice
1. shares the manuscript with a student or colleague to obtain additional help with the review.	____	____
2. shares the manuscript and review with others as a means of training about the process of manuscript review.	____	____
3. shares the manuscript with others to keep them apprised of the latest research.	____	____
4. makes use of the contents of the submitted manuscript in his or her own research prior to publication of the article.	____	____

Survey Form VIII: Use of Animal Subjects (Chapter 6)

For the purpose of the following questions, "animal" is defined as all animal species other than human. For each of the following categories of animal experiments, use this scale to indicate your judgment of the extent to which the experiment is justifiable:

0 typically NOT justifiable
1 may be justifiable
2 typically justifiable

For experiments designed to test a possible cure for AIDS, and

1. procedures likely to be perceived as painful in:

a. human subjects ____	d. pigs ____	g. cockroaches ____
b. nonhuman primates ____	e. rats ____	h. jellyfish ____
c. dogs ____	f. frogs ____	i. bacteria ____

2. procedures likely to cause distress or discomfort in:

a. human subjects ____	d. pigs ____	g. cockroaches ____
b. nonhuman primates ____	e. rats ____	h. jellyfish ____
c. dogs ____	f. frogs ____	i. bacteria ____

For experiments designed to test new therapies to treat chronic pain, and

3. procedures likely to be perceived as painful in:

    a. human subjects   ____   d. pigs ___   g. cockroaches ___
    b. nonhuman primates ___   e. rats   ___   h. jellyfish   ___
    c. dogs   ___   f. frogs ___   i. bacteria   ___

4. procedures likely to cause distress or discomfort in:

    a. human subjects   ____   d. pigs ___   g. cockroaches ___
    b. nonhuman primates ___   e. rats   ___   h. jellyfish   ___
    c. dogs   ___   f. frogs ___   i. bacteria   ___

For experiments designed to better understand the mechanisms of cancer, and

5. procedures likely to be perceived as painful in:

    a. human subjects   ___   d. pigs ___   g. cockroaches ___
    b. nonhuman primates ___   e. rats   ___   h. jellyfish   ___
    c. dogs   ___   f. frogs ___   i. bacteria   ___

6. procedures likely to cause distress or discomfort in:

    a. human subjects   ___   d. pigs ___   g. cockroaches ___
    b. nonhuman primates ___   e. rats   ___   h. jellyfish   ___
    c. dogs   ___   f. frogs ___   i. bacteria   ___

For experiments designed to test cosmetic safety, and

7. procedures likely to be perceived as painful in:

    a. human subjects   ___   d. pigs ___   g. cockroaches ___
    b. nonhuman primates ___   e. rats   ___   h. jellyfish   ___
    c. dogs   ___   f. frogs ___   i. bacteria   ___

8. procedures likely to cause distress or discomfort in:

    a. human subjects   ___   d. pigs ___   g. cockroaches ___
    b. nonhuman primates ___   e. rats   ___   h. jellyfish   ___
    c. dogs   ___   f. frogs ___   i. bacteria   ___

Survey Form IX: Data Sharing (Chapters 8 and 9)

For our discussion of data analysis and sharing, please use a separate sheet of paper to answer the following questions before our next meeting:

1. *Consider your current primary research project. Using the following scale, answer questions 1A to 1D to indicate your willingness to share with someone you do not know from another university.*

0   never
1   only after the paper has been accepted for publication
2   only after the paper has been submitted for publication
3   only after it is possible to begin writing the paper
4   at any time

1A. share your raw data   ____
1B. share your methods   ____
1C. share your reagents   ____
1D. share your relatively rare (or expensive) reagents   ____

2. *What form of sharing would warrant the following costs:*
2A. reimbursement for costs of purchase   ____
2B. acknowledgment in the paper   ____
2C. shared authorship of the paper   ____

3. *Have you ever received the advice that sharing of data or reagents constitutes good science?*
    (Yes/No)

4. *Have you ever received the advice that sharing of data or reagents is not a good idea?*
    (Yes/No)

5. *What are the possible advantages of sharing data or reagents?*

6. *What are the possible disadvantages of sharing data or reagents?*

Survey Form X: The Case of the Mystery Medium (Chapters 8 and 9)

A junior-level scientist presents a poster at a national meeting. She reports being able to cultivate a novel tumor cell line using a new culture medium that she has invented. A cell line for this tumor has never before been established in vitro. Neither her poster nor her published abstract discloses the composition of her new culture medium, and she refuses all requests to reveal its contents. She has a small lab (one technician, a part-time student, and herself) and is struggling to win federal grant support and tenure. Indicate on the following scale the degree to which each of the following reasons (A through E) justifies her not sharing data.

0   not justifiable
1   partially justifiable
2   justifiable
3   mostly justifiable
4   totally justifiable

A. She has discovered that this medium supports the growth in vitro of several other previously noncultivable tumors. Release of the contents of the medium will compromise her ability to protect her invention under intellectual property law (i.e., patent protection).   ___

B. She wants the recognition of being the first to publish the discovery of this culture medium in the peer-reviewed literature so as to establish priority for credit that in turn will help her professional advancement.   ___

C. She fears that availability of the medium will enable larger, established labs to gain a decisive advantage in the field. She views her actions as fair competition.   ___

D. She wants to be the first to report her exciting finding, and release of the medium's contents will compromise her chances of doing so.   ___

E. The transfer of this technology would be too expensive and time-consuming to be effective. The composition and preparation of the medium are complex, requiring several specialty chemicals from foreign distributors as well as the custom preparation of animal tissue extracts that are added to the medium.   ___

Extended Case Studies

The following cases may be given as writing assignments or used for in-class discussion.

Case 1. Electronic Archiving of Research Grant Results (Chapters 1 and 11)

Case information

Assume the following memo reflects a future NIH policy, then respond to questions A through E below.

* * *

TO: Appropriate Program Officer, NIH-NCI
FROM: Principal Investigator
RE: Final electronic report: CA12345-05-10

Attached you will find a copy of a CD-ROM labeled: "CA12345-05-10." The information contained on this diskette corresponds to my NIH grant award, "Antisense control of human leukemias." The funding period corresponds to years -05 (1 May *XXXX*) through -10 (30 April *XXXX*) of this project. A competing renewal of this program (years -11 through -15) was approved for funding earlier this year. I understand that these new funds will not be released until my attached electronic report is filed with and approved by the NIH.

The diskette contains the following information:

1. Names and up-to-date curricula vitae of all professional personnel, trainees, and technical staff who worked on the project during this reporting period.

2. All pages of all data books and original records pertaining to this project. This includes audioradiographic films, photographs, and micrographs. Such photographic

data have been cross-indexed to appropriate experiments in the various data books. These materials were digitized and entered onto the diskette by use of a scanner and appropriate software using "WORM" ("write once–read many") technology as prescribed by the NIH. All materials were entered on a weekly basis and were electronically dated using a digital time-stamping program. The electronic time-stamping data have been submitted to the prescribed independent vendor. The signatures of the principal investigator and the project worker(s) were also entered and dated via scanning on a weekly basis to certify the accuracy and authenticity of the recorded data. Where appropriate, signatures of witnesses to certain results were scanned into the file as part of the page on which they appeared.

3. All published manuscripts and preprints of submitted or in-press manuscripts also are included on this diskette. The published papers were loaded onto this diskette by scanning and digitization. Note that one of the manuscripts contains several color photographs. These have been scanned and processed with the appropriate software so as to be reproduced electronically or in hard copy as full-color illustrations. The preprints and reprints were entered as standard word-processing files, including picture files of all manuscript figures.

4. The original grant proposal and all appendix material have been entered onto the disk using a combination of direct word-processing file transfer and scanning of all halftone and graphic materials. Also included are the progress reports for the four noncompeting renewal applications filed in connection with this grant award.

5. A file designated "Aims — Progress" is also included on this diskette. As prescribed by NIH, this file contains all original, additional, and modified Specific Aims for this project. These are cross-referenced with respect to relevant data books and materials, or with publications associated with this project.

I understand that I am to retain a copy of this disk for 15 years from the date of this memo. I further understand that NIH will retain the attached diskette indefinitely.

Student assignment

A. Is such an electronic filing likely to have an effect the incidence or occurrence of scientific misconduct? Can you envision it deterring misconduct in science? If so, how? If not, why not? In presenting your arguments, consider the effects on two kinds of personalities: the unintentional self-deceiver and the deliberate perpetrator of misconduct.

B. Can such a system create a situation(s) that could lead to even more kinds of scientific misconduct? Explain.

C. Who should have access to this information, under what conditions and by what means? Defend your answer.

D. What implications would such a reporting system have for requests made under the Freedom of Information Act? (see chapter 9)

E. What implications would such a reporting system have for invention disclosures and for the filing of patents? (see chapter 9)

Case 2. Blastomere Analysis before Implantation: Medical Ethics (Chapter 10)

Case information

You practice in a large in vitro fertilization clinic, where blastomere analysis before implantation (BABI) is offered. A couple seeking in vitro fertilization requests that an "embryo biopsy" be done because each is a Tay-Sachs carrier. Determination of whether or not the embryo will have Tay-Sachs does not require determination of the embryo's sex. However, the wife states, "This may be our only chance to have a child, and we both want a boy. Check the sex of the embryos, and don't implant any female embryos!" Consider the following questions in determining what you will say to the couple, and how you will determine which embryos to implant.

Student assignment

A. Is selection against male embryos on the basis of potential sex-linked genetic abnormalities different from selection against female embryos on the basis of cultural biases?

B. If the care provider and recipients differ in their stances on sex selection, should the final decision about sex selection be guided by the ethical base of the care provider or of the recipient?

C. Selection of embryos on the basis of sex is not considered unethical by some contemporary Jewish scholars. (However, it is prohibited under the German Embryo Protection Act of October 24, 1990.) Should the religious affiliation of the couple influence the final decision?

D. More generally, who should control what tests will be run on the embryos and which embryos will be selected for implantation? Does the fact that recipients are paying for the service influence the criteria for selection of embryos? How much input should researchers, care providers, potential parents, legislators, and society in general have in determining what is an acceptable implant?

E. Does BABI differ from amniocentesis? If so, how? Does BABI differ from infanticide? If so, how?

Case 3. Authorship, Peer Review, and Sharing Data (Chapters 4 and 9)

Case information

You are the editor-in-chief of a molecular biology journal. Twelve associate editors constitute your editorial board. They receive papers directly and oversee the ad hoc peer review of these submitted manuscripts. You receive a letter from one of your associate editors asking you for advice on how to handle a problem. The details of this issue are as follows:

1. A manuscript handled by the associate editor (Dr. Red) has been in print for about 6 months.

2. Dr. White is the corresponding author and the head of the lab that submitted the paper. Dr. White was contacted in writing by Dr. Blue requesting a series of recombinant plasmids carrying sequences of a newly discovered herpeslike virus that infects and transforms mouse cells. The published manuscript carefully details the uniqueness of the virus using biological techniques and nucleic acid hybridization studies.

3. Dr. Blue, an expert DNA virologist, informs Dr. White that he wants to use the viral fragments to do complementation experiments with a specific group of viruses.

4. Dr. White writes back to Dr. Blue telling him that those experiments have already been done in his own lab and the results were negative (but not published).

5. Dr. Blue then writes Dr. White again, requesting the same recombinant plasmids. He argues that Dr. White's reason is not an appropriate one and that Dr. White is interfering with his right to confirm and verify results in the published literature.

6. Dr. White responds by letter to Dr. Blue strongly stating that he is only obliged to release such materials when the requestor provides reasonable disclosure of intended use. He accuses Dr. Blue of seeking these materials to advance his research in areas that will directly compete with Dr. White's program. Under the circumstances, Dr. White argues, this is both unfair and unethical.

7. The policy of the journal with respect to sharing of materials reads as follows: "Publication of original research reports in this journal indicates that the authors are willing to make available and distribute clones of cells or DNA molecules, antibodies, or other similar materials to interested

researchers associated with academic or not-for-profit institutions. The authors may reserve the right to limit quantities of such distributed material. Materials requested must have been used in the experiments reported in the published paper."

8. Dr. Red, the associate editor, asks you as editor-in-chief to handle this situation. Dr. Red asks you to write him with advice and a suggested course of action.

Student assignment

Compose the requested letter to Dr. Red. Include the following in your letter:

A. Your perspective on this series of events.

B. Any additional questions you would like Dr. White or Dr. Blue to answer.

C. Your suggested course of action. If you suggest that either Dr. White or Dr. Blue be contacted in writing, be explicit as to what information should be included in the letter(s).

Case 4. Sharing of Research Materials (Chapters 4, 8, and 9)

Case information

You are an NIH-funded university faculty member who writes to an investigator requesting a recombinant plasmid that carries a gene encoding a major surface protein of a human pathogenic bacterium. Your letter precisely spells out your plans for using this recombinant plasmid in your research project. This surface protein gene has been cloned by the investigator, and its cloning has been reported in the peer-reviewed literature. The investigator is a staff scientist at a private research institute. You get a prompt response to your request in the form of a letter from the research administrator of the institute. The letter describes the terms under which a bacterial strain containing the requested recombinant plasmid (referred to in the letter as "the material") will be released to you. The letter contains language that pertains to your use of the material being requested. A list of the issues covered includes:

1. The material must not be administered to humans.

2. The material is being provided only for the stated use; permission must be sought from the research institute if other uses of the material are planned.

3. The material may not be released to any other investigators outside of the faculty member's lab.

4. The faculty member must provide the names of any lab staff or trainees working with the material.

5. You certify that you will not hold the research institute legally responsible for any harm or injury that may be caused by the material or its use.

6. You cannot disclose, by any means, any of the work you do with this material without first seeking and obtaining the permission of the research institute.

7. You cannot use the material for any commercial or profit-making purposes.

8. The research institute will be granted an option for exclusive licensing of any inventions coming from the planned work with this material. The license will be negotiated in good faith between you and the research institute.

9. At the conclusion of the work, the material will be returned to the research institute or destroyed.

10. All lab members must be notified in writing of the terms of the release of the material and its use under those terms.

You are asked to countersign this letter and return it to the research institute before the recombinant plasmid can be released to you. This is your first experience with such a letter. You do have some concerns about certain items in the letter.

You show the letter to a colleague, who comments that such agreements aren't worth the paper they're printed on. He advises you to just sign it and return it so you can get the plasmid and move ahead with your work.

You show the letter to your departmental chair, who informs you that the letter must be countersigned by someone authorized to sign on behalf of your university, in this case the director of sponsored programs.

You show the letter to another faculty colleague, who confirms your notions that some of the clauses are too restrictive and inappropriate according to current standards of exchange of biological materials. He suggests you send the letter back and suggest the deletion or modification of such clauses.

Student assignment

Comment on the advice being given by each of these individuals. What, if any, clauses are unacceptable to you? Why? Where might you find "current standards of exchange of biological materials" if such standards really exist? Finally, explain the course of action you would take in this situation.

Case 5. Conflict of Conscience (Chapters 6 and 7)

Case information

You are chair of an institutional advisory committee on the care and use of laboratory animals. This committee reviews and recommends policies and practices on the use of laboratory animals and reviews experimental protocols prepared by investigators. It consists of nine faculty, two members outside the institution, and you as chair. One-third of the faculty members rotate off the committee each year. A professor in a basic science department asks you to nominate him for membership on the committee. Some salient points are as follows:

1. The basic science professor is a theoretical biologist who is internationally known for his computer simulations.

2. The professor has testified before legislative bodies opposing the use of pets in research.

3. The professor has known associations with members of a militant animal rights and animal liberation group.

4. The professor has been very active in developing computer-assisted instructional material.

5. The professor has been very active in university governance.

6. Your institution has a nationally accredited laboratory animal facility managed by a team of veterinarians and trained animal care workers.

7. The faculty of your institution has a number of extramurally funded research projects using rodents, rabbits, cats, dogs, and primates.

8. The extramurally funded research is sponsored by private agencies, federal agencies, and private industry.

9. The animal research encompasses nutrition, immunology, drug testing, biochemistry, neurology, toxicology, infectious diseases, and chemical dependency.

Student assignment

You discuss this with the senior administrator for research who appoints the membership of this committee. She asks you to make a decision based upon your best judgment and to

1. Write a letter to her, advising her of your recommendation, along with reasons for your decision.

2. Write a letter to the professor who asked you to nominate him for committee membership, giving your decision and reasons for your decision.

3. Prepare a draft of your response should members of the institutional advisory committee ask you about the matter.

4. Prepare a draft of your plan of action should a member of the press or an animal rights organization ask you about the matter.

Case 6. Conflict of Effort (Chapter 7)

Case information

You are chair of a renowned chemistry department in a major research-intensive university. One of your outstanding professors is sought after as a member of advisory panels, lecturer, consultant, and leader in professional organizations. This professor continues to publish and accept assignments as coordinator of formal courses. During the past year, you have received complaints and overheard comments that this professor is missing departmental committee meetings and is frequently not prepared to make scheduled presentations. Upon inquiry, you discover the following:

1. The professor is advisor to two graduate students, two postdoctoral research trainees, and three technicians.

2. One graduate student is supervised by a postdoctoral research trainee, who also supervises one of the technicians.

3. The other graduate student and postdoctoral research trainee each supervise one technician.

4. The professor is on an international commission that meets overseas a total of 15 working days per year.

5. The professor has established a consulting company, which he operates from a home office.

6. The professor makes 12 consulting visits per year.

7. The professor is president of a national professional society that requires that he spend 24 days per year at the society's meetings or at its headquarters.

8. The professor is on a grant review panel that meets 9 days per year.

9. The professor is editor-in-chief of a professional journal. He attends meetings with the publisher 6 days per year and spends approximately 1 hour each day on editorial matters.

10. The professor is on the editorial board of another professional journal and reviews about two dozen manuscripts per year.

11. The professor attends two professional meetings each year, each 4 days long.

12. The professor gives one lecture each month at another institution.

13. The professor is asked to testify before legislative committees and regulatory panels a total of 6 days each year.

14. The professor has assigned his two postdoctoral workers to present 30 hours of his assigned 45 hours of lectures.

15. The professor frequently misses departmental and institutional committee meetings.

Student assignment

Compose an annual letter of evaluation of this professor, addressing these issues:

1. Activities to be encouraged

2. Activities to be discouraged

3. Other advice

4. Overall evaluation

Case 7. Conflict of Interest (Chapter 7)

Case information

You are chair of a department in an academic health science center. You have recruited a faculty member whose spouse has a master's degree in the same field. The faculty member has applied for a grant from a federal agency and has just received an award notice. The faculty member comes to you to establish a research assistant position on the grant. She indicates that she plans to hire her spouse for this position. Some salient points follow:

1. The faculty member and spouse worked together to collect preliminary data. The spouse worked as an unpaid volunteer.

2. The faculty member and spouse both contributed to the preparation of the grant application, in which the faculty member was named as the principal investigator and the spouse was named as the research assistant.

3. The faculty member and spouse have different professional names.

4. The thesis work of the spouse directly relates to the goals and methods described in the grant.

5. The faculty member and the spouse have established that they can work together effectively on this project, so there would be no delay in pursuing the grant.

6. The faculty member and the spouse have published a preliminary report as coauthors.

Student assignment

You discuss this matter with the personnel officer of the academic health sciences center. Several issues are raised, and you are asked to write a memorandum to the faculty member and one to the personnel officer, giving your recommended action and the reasons for your decision. The issues to be addressed include:

1. Does this constitute preselection of a job candidate, thereby violating the academic health sciences center's commitment to equal employment opportunity?

2. Does this involve direct supervision of a member of the immediate family and thereby violate the academic health sciences center's antinepotism rules?

3. Is this an acceptable or desirable arrangement, and if so, how can the arrangement be set up to meet legal requirements?

4. Are there additional issues that might arise in the future that should be resolved at this time; for example, supervision of graduate students or additional technicians?

Case 8. Publication of Data (Chapters 4 and 11)

Case information

A postdoctoral fellow presents her research results at an informal laboratory meeting. Her work involves the molecular cloning of a relatively large genomic fragment isolated from a bacterium. A gene encoding a 200-kDa protein has been mapped on this genomic fragment. The entire fragment was cloned in a bacteriophage lambda vector and is approximately 20 kb in size. The nucleotide sequence of the gene in question has been determined, and the gene product has been biochemically characterized. There is some reason to believe that this gene is part of an operon. The postdoctoral fellow indicates that this portion of the work, including the sequencing and characterization of the gene product, is in advanced draft form for submission to a peer-reviewed journal. The 20-kb lambda insert consists of seven smaller restriction fragments, all of which have been subcloned with the exception of a 2-kb fragment at one of the termini of the original insert. The draft manuscript focuses heavily on the cloning of the 200-kDa protein gene, its mapping on the 20-kb insert, and its sequence determination. The postdoctoral fellow points out that the original lambda clone carrying the 20-kb insert has been lost. The only sample of the recombinant phage that was being stored for future use no longer gives rise to phage plaques. She contends that this is not a significant issue in relation to the manuscript, as most of the fragment has been subcloned and all the clones involving the 200-kDa protein gene are available as plasmid subclones. A discussion among the group members subsequently ensues regarding the handling of this situation. The laboratory head actively solicits the opinions of some of the individuals (all postdoctoral fellows) seated around the table.

Gene Jones suggests that the paper describe the original lambda phage clone and not mention that the clone has been lost. Since 90% of the original clone is available as subcloned fragments, Dr. Jones contends that the point is moot and that there is no need to mention the status of the original clone.

Bill Thomas suggests a modification of Dr. Jones's proposal. He proposes that the missing restriction fragment be isolated from a plasmid library of the organism. This would be far easier than attempting to reclone the entire 20-kb fragment. The cloning of the missing fragment would thus result in the entire 20-kb sequence being available and would preclude problems with supplying any of the fragments of the original clone to other interested investigators. Accordingly, Dr. Thomas argues that with this goal accomplished there won't be any need to mention that the original phage clone is no longer available. He argues that in practice, his solution makes the entire 20-kb sequence available in an appropriate context.

Gail Smith suggests that the manuscript be submitted as soon as possible but that the paper be modified to disclose that the original phage clone is no longer available. She suggests that the clone can be described and its map presented based on experiments that have been performed, but an indication that the clone is no longer available in the laboratory should be made in the manuscript. Since the subclones of the sequenced and genetically characterized area are available, Dr. Smith contends that this solution is practical and is not likely to lead to confusion or to possible misinterpretation in terms of what is being presented in the paper.

Dan Williams states that if the original phage clone is to be described in the paper, it must be available. Accordingly, he argues strongly that the clone must be reisolated and present in the laboratory stock collection before the paper can be submitted.

Kathy Steward favors the idea forwarded by Dr. Williams but suggests that the paper be submitted before the clone is available. The paper is likely to be returned with reviewers' comments, and submission of a revised manuscript can always be delayed at that point if the clone is still not available (which, she contends, is not very likely).

Student assignment

Imagine that you are the laboratory supervisor at this meeting. Respond to the various arguments being presented by these postdoctoral fellows. What proposed solutions or combination of solutions will you embrace as being acceptable? What acceptable solutions of your own, if any, do you have to offer?

Standards of Conduct

A COLLECTION OF GUIDELINES FOR THE CONDUCT OF RESEARCH, specific research policies and practices, and policies and procedures for handling of misconduct was published in 1993 by the U.S. National Academy of Sciences: *Responsible Science*, vol. II, *Ensuring the Integrity of the Research Process* (National Academy Press, 2101 Constitution Avenue, N.W., Washington, D.C. 20418). A collection of professional codes of conduct is also published periodically (Gorlin, R. A. [ed.]. 1999. *Codes of Professional Responsibility: Ethics Standards in Business, Health, and Law.* BNA Books, Washington, D.C.). However, the Internet has become the medium of choice for the dissemination of such information. The Web sites of academic and research institutions, scholarly societies, journal publishers, and government agencies now provide easy access to information relating to responsible research conduct.

The following provides general information on the location of other written documents that deal with standards of conduct in the research and academic settings.

Federal Agency Documents. Federal agency documents are concerned with such things as procedures and regulations related to the identification and prosecution of scientific misconduct. They also deal with other specific issues related to scientific integrity and responsible conduct. They are usually available at the institutional level or can be found directly in the *Federal Register* or the *NIH Guide for Grants and Contracts*, both of which are available on-line.

http://www.access.gpo.gov/ (provides access to the *Federal Register* through the U.S. Government Printing Office home page)

http://www.nih.gov/grants/guide/index.html

Often the subject matter published in the *Federal Register* is under discussion, and subsequent publication occurs. When implemented as policy, the phrase "Final Rule" is included in the title of the article. These documents usually reflect the activities and authority of the Office of Research Integrity of the U.S. Department of Health and Human Services (which encompasses NIH) or the Office of the Inspector General of the National Science Foundation.

Conflict-of-interest documents are a good example of federal policy documents. The National Science Foundation's notice, "Investigator Financial Disclosure Policy" (*Federal Register* **59**[No. 123]:33308–33312, June 28, 1994), and the Department of Health and Human Services's proposed rules, "Objectivity in Research" (*Federal Register* **59**[No. 123]:33242–33251, June 28, 1994), can be accessed on the *Federal Register* Web site given above.

Federal and Institutional Guidelines for the Use and Protection of Human Research Subjects and of Animals.   Guidelines for human and animal experimentation can usually be located at institutional sponsored programs offices or the institutional offices of the federally mandated Investigational Review Board (IRB). They are frequently found on-line at institutional home pages, usually under the heading of "Research." Specific URLs can be found in the resource sections of chapters 5 and 6. Federal guidelines prevail, but special institutional guidelines may augment or supplement them.

Guidelines for Scholarly Publication.   Scientific journals regularly publish guidelines for contributors. They may appear in every issue of the publication, at the beginning or end of volume sequences, or at the beginning or end of the calendar year. Such guidelines vary in scope and content and may cover such things as authorship attribution, sharing of research materials, conflict-of-interest disclosure, and communication of results to the media before manuscript acceptance (see chapter 4). The Web sites of journal publishers almost always contain these guidelines. Investigators should be familiar with the publication guidelines of any journal to which they intend to submit a scientific manuscript.

Appropriate Professional Society Code of Ethics or Standards for Scientific Conduct.   Professional scientific societies have conduct and ethics codes, which may be published from time to time, usually in society-sponsored journals or publications. The central administrative offices of the relevant society may be contacted to get these documents. Alternatively, the Center for the Study of Ethics in the Professions at the Illinois Institute of Technology has a Web site that contains many codes of ethics of professional societies, corporations, government, and academic institutions. The Codes of Ethics Online Project may be found at http://216.47.152.67/codes.

Institutional Policies Document for Conduct of Research. A growing number of academic and research institutions have developed policy documents dealing with the responsible conduct of research. These documents are also frequently found on-line at institutional home pages under the heading of "Research." Specific examples of such documents may also be found in *Responsible Science*, vol. II. Other things to look for at the institutional level, either in print or on-line, include computer ethics policies, copyright and intellectual property policies, conflict-of-interest policies, Worker's Right to Know and Hazard Communication Documentation, and Institutional Academic Honor Code Document. These documents are usually distributed periodically, or faculty and trainees are reminded of their location on the institutional Web site.

The *Guidelines for the Conduct of Research at the National Institutes of Health* (http://www.nih.gov/campus/irnews/guidelines.htm) are now in their third edition and are provided below as a relevant example.

Guidelines for the Conduct of Research at the National Institutes of Health

Preface

The Guidelines for the Conduct of Research expound the general principles governing the conduct of good science as practiced in the Intramural Research Programs at the National Institutes of Health. They address a need arising from the rapid growth of scientific knowledge, the increasing complexity and pace of research, and the influx of scientific trainees with diverse backgrounds. Accordingly, the Guidelines should assist both new and experienced investigators as they strive to safeguard the integrity of the research process.

The Guidelines were developed by the Scientific Directors of the Intramural Research Programs at the NIH and revised this year by the intramural scientists on the NIH Committee on Scientific Conduct and Ethics. General principles are set forth concerning the responsibilities of the research staff in the collection and recording of data, publication practices, authorship determination, peer review, confidentiality of information, collaborations, human subjects research, and financial conflicts of interest.

It is important that every investigator involved in research at NIH read, understand, and incorporate the Guidelines into everyday practice. The progress and excellence of NIH research is dependent on our vigilance in maintaining the highest quality of conduct in every aspect of science.

Michael M. Gottesman, M.D.
Deputy Director for Intramural Research
3rd Edition, January, 1997
The National Institutes of Health

Introduction

Scientists in the Intramural Research Program at the National Institutes of Health generally are responsible for conducting original research consonant with the goals of their individual Institutes and Divisions. Intramural scientists at NIH, as all scientists, should be committed to the responsible use of the process known as the scientific method to seek new knowledge. While the general principles of the scientific method — formulation and testing of hypotheses, controlled observations or experiments, analysis and interpretation of data, and oral and written presentation of all of these components to scientific colleagues for discussion and further conclusions — are universal, their detailed application may differ in different scientific disciplines and in varying circumstances. All research staff in the Intramural Research Programs should maintain exemplary standards of intellectual honesty in formulating, conducting and presenting research, as befits the leadership role of the NIH.

These Guidelines were developed to promote high ethical standards in the conduct of research by intramural scientists at the NIH. It is the responsibility of each Laboratory or Branch Chief, and successive levels of supervisory individuals (especially Institute, Center and Division Intramural Research Directors), to ensure that each NIH scientist is cognizant of these Guidelines and to resolve issues that may arise in their implementation.

These Guidelines complement, but are independent of, existing NIH regulations for the conduct of research such as those governing human subjects research, animal use, radiation, chemical and other safety issues, transgenic animals, and the Standards of Conduct that apply to all federal employees.

The formulation of these Guidelines is not meant to codify a set of rules, but rather to elucidate, increase awareness and stimulate discussion of patterns of scientific practice that have developed over many years and are followed by the vast majority of scientists, and to provide benchmarks when problems arise. Although no set of guidelines, or even explicit rules, can prevent willful scientific misconduct, it is hoped that formulation of these Guidelines will contribute to the continued clarification of the application of the scientific method in changing circumstances.

The public will ultimately judge the NIH by its adherence to high intellectual and ethical standards, as well as by its development and application of important new knowledge through scientific creativity.

Responsibilities of Research Supervisors and Trainees

Research training is a complex process, the central aspect of which is an extended period of research carried out under the supervision of an experienced scientist. This supervised research experience represents not merely performance of tasks assigned by the supervisor, but rather a process wherein the trainee takes on an increasingly independent role in the selection, conceptualization and execution of research projects. To prepare a young scientist for a successful career as a research investigator, the trainee should be provided with training in the necessary skills. It should be recognized that the trainee has unique needs relevant to career development.

In general a trainee will have a single primary supervisor but may also have other individuals who function as mentors for specific aspects of career development. It is the responsibility of the primary supervisor to provide a research environment in which the trainee has the opportunity to acquire both the conceptual and technical skills of the field. In this setting, the trainee should undertake a significant piece of research, chosen usually as the result of discussions between the mentor and the trainee, which has the potential to yield new knowledge of importance in that field. The mentor should supervise the trainee's progress closely and interact personally with the trainee on a regular basis to make the training experience meaningful. Supervisors and mentors should limit the number of trainees in their laboratory to the number for whom they can provide an appropriate experience.

There are certain specific aspects of the mentor-trainee relationship that deserve emphasis. First, training should impart to the trainee appropriate standards of scientific conduct both by instruction and by example. Second, mentors should be particularly diligent to involve trainees in research activities that contribute to their career development. Third, mentors should provide trainees with realistic appraisals of their performance and with advice about career development and opportunities.

Conversely, trainees have responsibilities to their supervisors and to their institutions. These responsibilities include adherence to these Guidelines, applicable rules, and programmatic constraints related to the needs of the laboratory and institute. The same standards of professionalism and collegiality apply to trainees as to their supervisors and mentors.

Data Management

Research data, including detailed experimental protocols, all primary data, and procedures of reduction and analysis are the essential components of scientific progress. Scientific integrity is inseparable from meticulous attention to the acquisition and maintenance of these research data.

The results of research should be carefully recorded in a form that will allow continuous access for analysis and review. Attention should be given to annotating and indexing notebooks and documenting computerized information to facilitate detailed review of data. All data, even from observations and experiments not directly leading to publication, should be treated comparably. All research data should be available to scientific collaborators and supervisors for immediate review, consistent with requirements of confidentiality. Investigators should be aware that research data are legal documents for purposes such as establishing patent rights or when the veracity of published results is challenged and the data are subject to subpoena by congressional committees and the courts.

Research data, including the primary experimental results, should be retained for a sufficient period to allow analysis and repetition by others of published material resulting from those data. In general, five to seven years is specified as the minimum period of retention but this may vary under different circumstances.

Notebooks, other research data, and supporting materials, such as unique reagents, belong to the National Institutes of Health, and should be maintained and made available, in general, by the Laboratory in which they were developed. Departing investigators may take copies of notebooks or other data for further work.

Under special circumstances, such as when required for continuation of research, departing investigators may take primary data or unique reagents with them if adequate arrangements for their safekeeping and availability to others are documented by the appropriate Institute, Center or Division official.

Data management, including the decision to publish, is the responsibility of the principal investigator. After publication, the research data and any unique reagents that form the basis of that communication should be made available promptly and completely to all responsible scientists seeking further information. Exceptions may be necessary to maintain confidentiality of clinical data or if unique materials were obtained under agreements that preclude their dissemination.

Publication Practices

Publication of results is an integral and essential component of research. Other than presentation at scientific meetings, publication in a scientific journal should normally be the mechanism for the first public disclosure of new findings. Exceptions may be appropriate when serious public health or safety issues are involved. Although appropriately considered the end point of a particular research project, publication is also the beginning of a process in which the scientific community at large can assess, correct and further develop any particular set of results.

Timely publication of new and significant results is important for the progress of science, but fragmentary publication of the results of a scientific investigation or multiple publications of the same or similar data are inappropriate. Each publication should make a substantial contribution to its field. As a corollary to this principle, tenure appointments and promotions should be based on the importance of the scientific accomplishments and not on the number of publications in which those accomplishments were reported.

Each paper should contain sufficient information for the informed reader to assess its validity. The principal method of scientific verification, however, is not review of submitted or published papers, but the ability of others to replicate the results. Therefore, each paper should contain all the information that would be necessary for scientific peers of the authors to repeat the experiments. Essential data that are not normally included in the published paper, e.g. nucleic acid and protein sequences and crystallographic information, should be deposited in the appropriate public data base. This principle also requires that any unique materials (e.g. monoclonal antibodies, bacterial strains, mutant cell lines), analytical amounts of scarce reagents and unpublished data (e.g. protein or nucleic acid sequences) that are essential for repetition of the published experiments be made available to other qualified scientists. It is not necessary to provide materials (such as proteins) that others can prepare by published procedures, or materials (such as polyclonal antisera) that may be in limited supply.

Authorship

Authorship refers to the listing of names of participants in all communications, oral and written, of experimental results and their interpretation to scientific colleagues. Authorship is the fulfillment of the responsibility to communicate research results to the scientific community for external evaluation.

Authorship is also the primary mechanism for determining the allocation of credit for scientific advances and thus the primary basis for assessing a scientist's contributions to developing new knowledge. As such, it potentially conveys great benefit, as well as responsibility. For each individual the privilege of authorship should be based on a significant contribution to the conceptualization, design, execution, and/or interpretation of the research study, as well as a willingness to assume responsibility for the study. Individuals who do not meet these criteria but who have assisted the research by their encouragement and advice or by providing space, financial support, reagents, occasional analyses or patient material should be acknowledged in the text but not be authors.

Because of the variation in detailed practices among disciplines, no universal set of standards can easily be formulated. It is expected, however, that each research group and Laboratory or Branch will freely discuss and resolve questions of authorship before and during the course of a study. Further, each author should review fully material that is to be presented in public forums or submitted (originally or in revision) for publication. Each author should be willing to support the general conclusions of the study.

The submitting author should be considered the primary author with the additional responsibilities of coordinating the completion and submission of the work, satisfying pertinent rules of submission, and coordinating responses of the group to inquiries or challenges. The submitting author should assure that the contributions of all collaborators are appropriately recognized and that each author has reviewed and authorized the submission of the manuscript in its original and revised forms. The recent practice of some journals of requiring approval signatures from each author before publication is an indication of the importance of fulfilling the above.

Peer Review and Privileged Information

Peer review can be defined as expert critique of either a scientific treatise, such as an article prepared or submitted for publication, a research grant proposal, a clinical research protocol, or of an investigator's research program, as in a site visit. Peer review is an essential component of the conduct of science. Decisions on the funding of research proposals and on the publication of experimental results must be based on thorough, fair and objective evaluations by recognized experts. Therefore, although it is often difficult and time-consuming, scientists have an obligation to participate in the peer review process and, in doing so, they make an important contribution to science.

Peer review requires that the reviewer be expert in the subject under review. The reviewer, however, should avoid any real or perceived conflict of interest that might arise because of a direct competitive, collaborative or other close relationship with one or more of the authors of the material under review. Normally, such a conflict of interest would require a decision not to participate in the review process and to return any material unread.

The review must be objective. It should thus be based solely on scientific evaluation of the material under review within the context of published information and should not be influenced by scientific information not publicly available.

All material under review is privileged information. It should not be used to the benefit of the reviewer unless it previously has been made public. It should not be shared with anyone unless necessary to the review process, in which case the names of those with whom the information was shared should be made known to those managing the review process. Material under review should not be copied and retained or used in any manner by the reviewer unless specifically permitted by the journal or reviewing organization and the author.

Collaborations

Research collaborations frequently facilitate progress and generally should be encouraged. It is advisable that the ground rules for collaborations, including eventual authorship issues, be discussed openly among all participants from the beginning. Whenever collaborations involve the exchange of materials between NIH scientists and scientists external to NIH, a Material Transfer Agreement (MTA) or other formal written agreements may be necessary. Information about such agreements and other relevant mechanisms, such as licensing or patenting discoveries, may be obtained from each ICD's Technology Development Coordinator or the NIH Office of Technology Transfer.

Human Subjects Research

Clinical research, for the purposes of these Guidelines, is defined as research performed on human subjects or on material or information obtained from human subjects as a part of human experimentation. All of the topics covered in the Guidelines apply to the conduct of clinical research; clinical research, however, entails further responsibilities for investigators.

The preparation of a written research protocol ("Clinical Research Protocol") according to existing guidelines prior to commencing studies is almost always required. By virtue of its various sections governing background; patient eligibility and confidentiality; data to be collected; mechanism of data storage, retrieval, statistical analysis and reporting; and identification of the principal and associate investigators, the Clinical Research Protocol provides a highly codified mechanism covering most of the topics covered elsewhere in the Guidelines. The Clinical Research Protocol is generally widely circulated for comment, review and approval. It should be scrupulously adhered to in the conduct of the research. The ideas of the investigators who prepared the protocol should be protected by all who review the document.

Those using materials obtained by others from patients or volunteers are responsible for assuring themselves that the materials have been collected with due regard for principles of informed consent and protection of human subjects from research risk. Normally, this is satisfied by a protocol approved by a human subjects committee of the institution at which the materials were obtained.

The supervision of trainees in the conduct of clinical investigation is complex. Often the trainees are in fellowship training programs leading to specialty or subspecialty certifications as well as in research training programs. Thus, they should be educated in general and specific medical management issues as well as in the conduct of research. The process of data gathering, storage, and retention can also be complex in clinical research which sometimes cannot easily be repeated. The principal investigator is responsible for the quality and maintenance of the records and for the training and oversight of all personnel involved in data collection.

Epidemiologic research involves the study of the presence or absence of disease in groups of individuals. Certain aspects of epidemiologic research deserve special mention. Although an epidemiologist does not normally assume responsibility for a patient's care, it is the responsibility of the epidemiologist to ensure that the investigation does not interfere with the clinical care of any patient. Also, data on diseases, habits or behavior should not be published or presented in a way that allows identification of any particular individual, family or community. In addition, even though it is the practice of some journals not to publish research findings that have been partially released to the public, it may be necessary for reasons of immediate public health concerns to report the findings of epidemiologic research to the study participants and to health officials before the study has been completed; the health and safety of the public has precedence.

Development and review of detailed protocols are as important in epidemiologic research as in clinical research and any other health science. However, the time for protocol development and review may be appropriately shortened in circumstances such as the investigation of acute epidemic or outbreak situations where the epidemiologic investigation may provide data of crucial importance to the identification and mitigation of a threat to public health. Nevertheless, even in these situations, systematic planning is of great importance and the investigator should make every attempt to formalize the study design in a written document and have it peer-reviewed before the research is begun.*

Financial Conflicts of Interest

Potential conflicts of interest due to financial involvements with commercial institutions may not be recognized by others unless specific information is provided. Therefore, the scientist should disclose all relevant financial relationships, including those of the scientist's immediate family, to the Institute, Center or Division during the planning, conducting and reporting of research studies, to funding agencies before participating in peer review of applications for research support, to meeting organizers before presentation of results, to journal editors when submitting or refereeing any material for publication, and in all written communications and oral presentations.

Concluding Statement

These Guidelines are not intended to address issues of misconduct nor to establish rules or regulations. Rather, their purpose is to provide a framework for the fair and open conduct of research without inhibiting scientific freedom and creativity.

These Guidelines were originally prepared by a Committee appointed by the NIH Scientific Directors. This third edition was prepared by the NIH Committee on Scientific Conduct and Ethics and approved by the NIH Scientific Directors.

*The section on epidemiological research is adapted from the *Guidelines for the Conduct of Research within the Public Health Service*, January 1, 1992.

Sample Protocols for Human and Animal Experimentation

This appendix has three parts:

1. Abridged Human Subject Protocol. An abridged version of a clinical research trial protocol involving human subjects, condensed from an approved protocol document, is presented. The following modifications have been made to simplify and shorten the presentation.

- Under the heading of Rationale, section 1.1 was deleted. This was a technical background review of previous preclinical and clinical studies ($\sim 2\frac{1}{2}$ typewritten pages).
- The table presenting the schedule of procedures as a matrix was deleted.
- The nine references to the literature were deleted.
- All of the 10 appendixes were deleted.

The resulting abridged protocol document provides an example of the style and degree of detail required to write an experimental plan suitable for consideration by an institutional review board (IRB).

2. Informed Consent Document. The full text of the informed consent document for the above human subject protocol is included, providing an example of the scope and level of presentation required for such documents.

3. Animal Use Protocol. The full text of an animal use protocol is presented to illustrate the scope and detail of preparation for such documents. Such a document would be submitted for consideration by the institutional animal care and use committee (IACUC).

1. Abridged Human Subject Protocol

Title: **Phase I study of weekly intravenous lometrexol with continuous oral folic acid supplementation.**

Principal Investigator: John D. Roberts, M.D., Massey Cancer Center, Virginia Commonwealth University, P.O. Box 980037, 401 College St., Richmond, VA 23298-0037, (804) 555-*XXXX*, FAX (804) 555-*XXXX*

Co-Investigator: Richard G. Moran, Ph.D. The Massey Cancer Center and the Departments of Internal Medicine and Pharmacology, Virginia Commonwealth University

IND agents: lometrexol NSC # 660025

Index

Schema

Phase I study of weekly intravenous lometrexol with continuous oral folic acid supplementation
John D. Roberts, M.D., and Richard G. Moran, Ph.D.

Objectives:

1. Determine a recommended phase II dose combination of lometrexol administered by weekly short intravenous (i.v.) infusion with concurrent oral folic acid. Describe the toxicity associated with this dose combination.

2. Determine whether weekly lometrexol with concurrent folic acid results in a protracted terminal elimination phase with potentially cytotoxic lometrexol levels.

3. Study the accumulation of lometrexol and its polyglutamate metabolites in peripheral red cells and mononuclear white blood cells (WBC).

4. Measure the effects of lometrexol treatment upon GAR (glycinamide ribonucleotide) transformylase activity in mononuclear blood cells and serum purines.

5. Identify lometrexol antitumor activity that occurs within the context of the clinical trial.

1 Rationale
1.1 Preclinical and clinical studies: ***NOT INCLUDED IN THIS PRESENTATION***

1.2 Study design

This is a phase I study of lometrexol administered weekly with concurrent folic acid supplementation in the expectation that this schedule will be feasible and optimal for the demonstration of antitumor activity in phase II trials.

1.2.1 Lometrexol dose escalation; folic acid dose de-escalation.
Both animal studies and phase I experience to date suggest that when lometrexol is to be used in combination with folic acid, there is a dose range for folic acid which is sufficient to eliminate lometrexol cumulative toxicity but which does not ablate lometrexol antitumor efficacy. The lower limit of this range in humans probably is greater than 1 mg every day by mouth; the upper limit is greater than 5 mg every day by mouth. Although it is apparent that the maximum tolerated dose (MTD) of lometrexol increases with concurrent folic acid, there is little quantitative information concerning the impact of increases in folic acid upon the lometrexol MTD. The experience in an ongoing phase I trial of lometrexol for 3 weeks with folic acid 5 mg administered by mouth 1 week before and one week following each lometrexol dose, in which the MTD is much greater than originally anticipated, suggests that the effect may be large.

It is likely that there are many dose combinations for lometrexol and folic acid which would be associated with both dose-limiting toxicity and antitumor efficacy. The primary objective of this study is to identify one such dose combination in an efficient manner. In order to do this, the study will begin with a relatively high dose of folic acid, in order to increase the probability that cumulative toxicity is eliminated to the greatest extent possible, and lometrexol dose will be escalated in successive patient cohorts. If an MTD for lometrexol is not identified after a maximum of four dose escalations, then the lometrexol dose will be fixed at the highest dose administered, and the dose of folic acid will be de-escalated in subsequent patient cohorts. In the absence of toxicity, lometrexol dose will be escalated by increments of approximately 60%; in the presence of toxicity, lometrexol dose will be escalated by increments of approximately 30%. Folic acid will be de-escalated by decrements of approximately 35%.

1.2.2 Folic acid schedule and starting dose.
In order to ensure that patients are folate replete, folate will be administered on a once-per-day basis beginning 1 week before lometrexol and extending 1 week beyond the last lometrexol dose. This can be economically accomplished with folic acid administered by mouth. Cumulative toxicity observed in a study of lometrexol twice weekly × 2, every 4 weeks with folic acid 1 mg every day by mouth suggests that this folic acid dose is too low. In another study, lometrexol 5 mg weekly with folic acid 1 mg every day by mouth was toxic, whereas the same dose of lometrexol with folic acid 2 mg every day by mouth was associated only with anemia. In an ongoing study of lometrexol administered every 3 or 4 weeks with folic acid 5 mg every day by mouth 1 week before and 1 week following each lometrexol dose, significant cumulative toxicity has yet to be observed, and antitumor responses have been observed (Hilary Calvert, personal communication). These observations suggest that the optimal folic acid dose is greater than 1 mg and not exceeded by 5 mg every day by mouth.

If the window for folate repletion between elimination of toxicity and elimination of efficacy is narrow, then interindividual precision in folate dosing will be important. In this study of weekly lometrexol, the initial folic acid dose will be 3 mg/m^2 every day by mouth.

1.2.3 Lometrexol starting dose. In previous phase I studies of various schedules with and without scheduled folate cumulative effects generally were apparent after from one to six doses. In order to determine a lometrexol dose which can be administered weekly for a protracted period of time (ideally, for responding patients, indefinitely), patients will receive lometrexol weekly × 8. Dose-limiting toxicity will be identified not only by severe toxicity but also by inability to adhere to this schedule.

The prior phase I experience has been considered in order to select a starting dose intended to be both efficient and safe. In a study of lometrexol 5 mg/m^2 weekly × 3 with concurrent folic acid 2 mg every day by mouth, one of two patients experienced grade 2 or greater hemoglobin (Hgb) toxicity that required transfusion, and no other toxic effects were reported. In a study of lometrexol twice weekly × 2, every 4 weeks (a dose rate equivalent to this study) with concurrent folic acid 1 mg every day by mouth, a lometrexol dose of 5.0 mg/m^2 was safe and 6.4 mg/m^2 was above the MTD. The higher dose of folic acid in the present study provides a margin of safety so that 5.0 mg/m^2 is an appropriate starting dose. This dose may be considerably below the MTD. In an ongoing study of lometrexol administered for 3 weeks with folic acid 5 mg every day by mouth 1 week before and 1 week following each lometrexol dose, the MTD is more than 60 mg/m^2, a dose rate of 20 mg/m^2/wk (Hilary Calvert, personal communication). On the other hand, this extrapolation almost certainly overestimates the MTD on a weekly schedule as lometrexol fractional urinary excretion increases with dose.

1.2.4 Anemia as a toxic effect. Anemia has been a significant toxic effect in previous lometrexol studies. Evaluation of anemia is complicated by the frequency of abnormal red cell production in patients with cancer, the prolonged time course over which inadequate red cell production becomes apparent, and repeated phlebotomy associated with a phase I study, especially for patients in which ancillary studies are performed. The advent of erythropoietin for chemotherapy-induced anemia further complicates matters as anemia in the absence of other significant adverse effects may be treatable without blood transfusion. In this study, eligible patients will be transfusion independent. Transfusion of packed red blood cells within defined limits is permissible; patients transfused in excess of these limits will be scored as having experienced dose-limiting toxicity. Use of erythropoietin in order to continue lometrexol administration is prohibited (exceptions may be made in the case of patients eligible for continuation therapy by virtue of a disease response). If anemia proves to be the only significant toxicity at the MTD, then further dose escalation with erythropoietin support may be indicated, which could be pursued through an amendment to this protocol.

1.2.5 Continuation therapy. Patients experiencing a response may continue lometrexol until there is progression of disease or unacceptable toxicity. Continuation therapy will stop after 1 year.

2 Objectives
Primary Objective
2.1 Determine a recommended phase II dose combination for lometrexol administered by weekly short i.v. infusion with concurrent oral folic acid. Describe the toxicity associated with this dose combination.

Secondary Objectives

2.2 Determine whether weekly lometrexol with concurrent folic acid results in a protracted terminal elimination phase with potentially cytotoxic lometrexol levels.

2.3 Study the accumulation of lometrexol and its polyglutamate metabolites in peripheral red cells and mononuclear WBC.

2.4 Measure the effects of lometrexol treatment upon GAR transformylase activity in mononuclear blood cells and serum purines.

2.5 Identify lometrexol antitumor activity that occurs within the context of the clinical trial.

3 Agents
3.1 Lometrexol (NSC # 660025)
3.1.1 Lometrexol is formulated as a disodium salt mixed with mannitol and hydrochloric acid or sodium hydroxide (to adjust pH) in water and lyophilized; it is available in vials containing lometrexol 50 or 100 mg. It is reconstituted with 0.9% Sodium Chloride Injection, USP (nonbacteriostatic) to yield a solution containing lometrexol 10 mg/ml. Upon reconstitution, it should be used without delay. It is manufactured by Eli Lilly and Company and will be provided by the Cancer Treatment Evaluation Program (CTEP), National Cancer Institute (NCI).

3.1.2 (Storage and stability information is not available at this time; it will be obtained from Eli Lilly and Company and the protocol amended.)

3.1.3 Lometrexol may be requested by completing a Clinical Drug Request (NIH-986) and mailing it to Drug Management and Authorization Section, DCT, NCI, EPN Room 707, Bethesda, MD 20892 or through the DMAS Electronic Clinical Drug Request System (ECDR). The MCV Hospitals Investigational Pharmacy will maintain a record of the inventory and disposition of all lometrexol using the NCI Drug Accountability Record Form (Appendix 5).

3.1.4 Expected toxicities of lometrexol are: mucositis, anemia, leucopenia, and thrombocytopenia. Patients treated with lometrexol also have experienced myocardial infarction, but a causal relationship has not been determined.

3.2 Folic acid
Folic acid is formulated for oral administration only as a 1 mg tablet and is available commercially from a number of suppliers. Folic acid for this study will be purchased through the MCV Hospitals pharmacy. Currently, the pharmacy changes suppliers irregularly according to need and market conditions. At the initiation of this study, a supply sufficient for the first year will be ordered through the pharmacy and sequestered for exclusive use in this study. This procedure will be repeated at the beginning of the second year.

4 Patient Eligibility
Patients will be defined as ELIGIBLE according to the following criteria.
4.1 Histologically or cytologically confirmed lymphoma or solid tumor malignancy.

4.2 No reasonable prospect for benefit from any conventional therapy if administered at the time of enrollment.

4.3 Written documentation of informed consent.

4.4 Age ≥18 years.

4.5 Zubrod Performance Status ≤2.

4.6 Life expectancy ≥16 weeks.

4.7 At least 4 weeks from prior chemotherapy (6 weeks in a case of nitrosourea or mitomycin C treatment) and radiation therapy. No planned concurrent chemotherapy, hormonal, or radiation therapy.

4.8 The following must be normal as determined by MCV Hospitals or another certified clinical pathology laboratory: WBC, platelets, creatinine, bilirubin (conjugated and unconjugated), prothrombin time. Aspartate aminotransferase (AST) must be normal, unless an abnormality is presumed due to metastatic disease, in which case it must be ≤3× normal.

4.9 Hgb ≥10.0 without red cell transfusion within 3 weeks.

4.10 A urinalysis for blood, protein, glucose, and ketones must be obtained and any abnormalities evaluated with reference to section 4.15.

4.11 Willing to practice a medically accepted form of contraception (sexual abstinence, birth control pills, IUD, condoms, diaphragm, implant; surgical sterilization; postmenopausal) during and for 3 months following lometrexol therapy.

Patients INELIGIBLE according to the following criteria will not be enrolled.
4.12 Pregnant or nursing.

4.13 A continuing requirement for allopurinol treatment.

4.14 Myocardial infarction within the past 12 months or unstable cardiac disease.

4.15 Disease of the gastrointestinal tract associated with maladsorption such that there would be a risk that oral folic acid would not be adsorbed.

4.16 Any other serious or chronic medical condition which would significantly compromise a patient's ability to tolerate or the investigator's ability to evaluate lometrexol toxicity, especially toxicities of anemia, thrombocytopenia, leucopenia, and stomatitis.

5 Study Design and Patient Treatment
5.1 Folic acid

5.1.1 Patients will begin folic acid every day by mouth 7 days prior to the first scheduled dose of lometrexol and continue for 7 days following the last dose of lometrexol.

5.1.2 The starting dose will be folic acid 3 mg/m^2 every day by mouth.

5.1.3 Body surface area (BSA) will be calculated according to actual height and weight to the nearest 0.1 m^2. Dosing:

Folic acid: 3 mg/m^2 dosing table

BSA (m^2)	dose (mg)
$\leq$1.5	4
1.6–1.8	5
1.9–2.1	6
$\geq$2.2	7

5.1.4 In the event of omission of a dose(s) of folic acid, up to two doses may be taken on any given day in order to make up for missed doses. At treatment initiation and following each dose of lometrexol, a new supply of folic acid will be dispensed in a calendar pill pack. Prior to each lometrexol dose, a pill count will be performed to check monitor compliance. Failure to take at least four doses in any 7-day period will be reason to omit a lometrexol dose. Omission of two lometrexol doses due to failure to take folic acid prior to completion of 8 weeks of treatment will be reason to declare a patient nonevaluable for lometrexol tolerance and to remove the patient from study. Such a patient will be monitored for toxicity according to the usual guidelines.

5.2 Lometrexol
5.2.1 Schedule, starting dose, and dose modifications

5.2.1.1 Patients will receive lometrexol by short ($\leq$2 minutes) i.v. infusion weekly $\times$ 8 weeks.

5.2.1.2 When necessary for reasons other than toxicity (holidays, other schedule conflicts, etc.), a dose may be administered up to 1 day early or up to 2 days late, in which case subsequent doses will be administered weekly according to the last date of administration. No more than two such changes are permitted. Patients requiring a third change should stop treatment and undergo evaluation for response; such patients are not evaluable for lometrexol tolerance.

5.2.1.3 Lometrexol will be stopped if a patient experiences dose-limiting toxicity.

5.2.1.4 Patients experiencing a response may receive continuation therapy.

5.2.2 BSA will be calculated according to section 5.1.3. Dose will be calculated to the nearest 0.1 mg.

5.2.3 Starting dose is lometrexol 5 mg/m^2 i.v.

5.2.4 Doses will not be modified except in continuation therapy.

5.2.5 A dose will be omitted if any of the following is observed on the scheduled day of treatment:

5.2.5.1 Platelets <100,000.

5.2.5.2 Grade 2 or greater toxicity in any NCI CTC category except Hgb, alopecia, local, and weight loss.

5.3 Dose-limiting toxicity

5.3.1 Any grade 3 nonhematologic toxicity except infection, local, and weight loss.

5.3.2 Grade 4 WBC toxicity; grade 3 platelet toxicity.

5.3.3 Omission of three doses due to toxicity.

5.3.4 Omission of both the seventh and eighth scheduled doses due to toxicity, if a scheduled ninth dose would have been omitted due to failure of toxicity to resolve.

5.3.5 Transfusion of more than two units of packed red blood cells in excess of documented cumulative phlebotomy subsequent to initiation of folic acid (calculated as 450 ml whole blood per unit packed red blood cells).

5.4 Lometrexol escalation and folic acid de-escalation

5.4.1 Patient registration, patient cohorts, and dose changes

5.4.1.1 Patients will be registered on study by a Massey Cancer Center (MCC) Clinical Research Associate or Research Nurse of the Office for Clinical Research prior to receiving the first dose of folic acid. Patients receiving any lometrexol will be evaluable for toxicity. Patients discontinuing lometrexol early for reasons other than toxicity (for example, noncompliance, patient preference, tumor progression) are not evaluable for lometrexol tolerance.

5.4.1.2 A patient cohort is a group of patients treated with the same dose combination of lometrexol and folic acid. A patient cohort will consist of not less than three and not more than five patients evaluable for lometrexol tolerance. Within a patient cohort the first three patients will start therapy one at a time and not more frequently than every 2 weeks. A patient cohort will be complete when either (1) three patients have completed therapy without dose-limiting toxicity or (2) three patients have experienced dose-limiting toxicity. A new patient cohort will not be started until the previous patient cohort is complete, except that a new patient cohort may be started if a possible last patient (not the fifth patient) of the previous cohort has actually received four lometrexol doses without dose-limiting toxicity. If three patients of a patient cohort experience dose-limiting toxicity, the dose combination of that cohort is associated with greater than maximum tolerated toxicity, and further patient enrollment will be according to section 5.4.3.2.

5.4.2 Lometrexol dose escalation with folic acid 3 mg/m^2

5.4.2.1 In the absence of grade 1 or greater toxicity, lometrexol will be escalated as follows (mg/m^2): 5, 8, 13, 20.

5.4.1.2 Once grade 1 or greater toxicity is observed in any patient, subsequent lometrexol dose escalation will be by 30% increments.

5.4.3 Folic acid de-escalation

5.4.3.1 If a patient cohort is treated with lometrexol 20 mg/m^2 and concurrent folic acid 3 mg/m^2 without dose-limiting toxicity, the lometrexol dose will be fixed and subsequent patient cohorts will be treated with de-escalated doses of folic acid according to the following schema:

Folic acid: 2 mg/m^2 dosing table

BSA (m^2)	dose (mg)
$\leq$1.7	3
1.8–2.2	4
$\geq$2.3	5

Folic acid: 1.3 mg/m^2 dosing table

BSA (m^2)	dose (mg)
$\leq$1.9	2
$\geq$2.0	3

5.4.3.2 Upon identification of a dose combination associated with dose-limiting toxicity, further patients will be treated at either the previous dose combination or an intermediate dose combination. Up to a total of eight patients will be treated at a previous or new dose combination. Identification of the recommended dose combination will require treatment of five of eight patients at a dose combination without dose-limiting toxicity. It is anticipated that the recommended phase II dose will be this dose combination. If it appears that tolerance of lometrexol is affected by the extent of prior chemotherapy or radiation therapy, the recommended phase II dose combination may include a contingency for lometrexol dose escalation following an initial treatment phase.

5.5 Leucovorin
Patients experiencing grade 4 toxicity will receive leucovorin 15 mg every 6 hours by mouth for at least 3 days. Patients experiencing grade 3 toxicity may be treated with leucovorin at the discretion of the treating physician and principal investigator.

5.6 Toxicity grading
5.6.1 Toxicity will be graded by NCI Common Toxicity Criteria (Appendix 2) using information available according to the schedule of observations (section 7); additional observations and tests may be obtained as clinically indicated.

5.6.2 Prior conditions

5.6.2.1 Conditions documented prior to treatment and stable during the course of treatment will not be scored as toxic events.

5.6.2.2 Liver abnormalities. For patients enrolled with abnormal AST values attributed to metastatic disease, a 2–4× increase from pretreatment will be scored as grade 2 toxicity. A > 4× increase will be scored as grade 3 toxicity.

5.7 Concurrent therapy
5.7.1 Patients on treatment will not receive concurrent chemotherapy, immunotherapy, hormonal, or radiation therapy for malignant disease except short-course

palliative radiation to painful bone metastases with fields involving less than 10% of total bone marrow.

5.7.2 Patients experiencing anemia who are to continue on study will be transfused for Hgb <8.0.

5.7.3 Patients will receive medically appropriate supportive care, including treatment of pain and infection, except that use of hematopoietic growth factors (erythropoietin, G-CSF, GM-CSF) in order to prevent lometrexol toxicity is not permitted; except erythropoietin may be used in continuation therapy. Hematopoietic growth factors may be used in the support of patients for whom lometrexol is discontinued. Medical care will be documented.

5.8 Adverse event reporting
The following adverse events will be reported by telephone or fax to Investigational Drug Branch (IDB) within 24 hours; a written report will follow within 10 working days addressed to:

Investigational Drug Branch
P.O. Box 30012
Bethesda, MD 20824

5.8.1 All nonhematological life-threatening events (grade 4) which may be due to drug administration

5.8.2 All fatal events

5.8.3 A first occurrence of any unexpected toxicity (see section 3.1.4) regardless of grade. With regard to adverse event reporting, myocardial infarction will be considered an unexpected toxicity.

5.9 Continuation therapy
Patients documented to experience a response may continue treatment at the discretion of the principal investigator. A patient eligible for continuation therapy by virtue of disease response who has experienced dose-limiting toxicity may, at the discretion of the principal investigator, be treated on a weekly schedule with dose omissions and modifications as seem appropriate. Erythropoietin may be used, if clinically indicated, in continuation therapy. Patients receiving continuation therapy will undergo tumor response at least every 3 months. Continuation therapy will stop after 1 year.

5.10 Response evaluation
5.10.1 Patients will be assigned to one of three categories prior to lometrexol: measurable disease; evaluable disease; nonevaluable disease. Lesions to be measured will be identified prior to lometrexol. In the case of patients with lesions too numerous to measure, four lesions may be selected for measurement, but in any response evaluation all lesions will be inspected and measured if indicated in order to rule out a 25% increase. Patients with both measurable tumor, and serum tumor

markers will be evaluated on the basis of measurable tumor in which case the serum tumor marker will be considered an evaluable feature. Patients with evaluable or nonevaluable tumor and serum tumor markers will be evaluated on the basis of a serum tumor marker which will be considered a measurable feature. A single serum tumor marker will be selected prior to the first lometrexol dose.

5.10.2 For patients with measurable disease, a complete response will be recognized as the absence of all clinical evidence of persistent malignant disease. Recognition of a complete response will require reevaluation of all previously known sites of disease.

A partial response will be a greater than 50% decrease in the sum of the products of two perpendicular diameters of all measurable lesions without an increase greater than 25% in any single lesion and without the appearance of any new lesions.

A minimal response will be a reduction in the sum of the products of two perpendicular diameters of all measurable lesions without an increase greater than 25% in any single lesion and without the appearance of any new lesions.

Patients not experiencing a complete, partial, or minimal response will be categorized as without a response.

For patients to be evaluated on the basis of a serum tumor marker, a partial response will be recognized as a 50% reduction, and a complete response as normalization.

5.10.3 For patients with evaluable disease, a response will be recognized as absence of clinical evidence of persistent disease.

5.10.4 Patients with nonevaluable disease will not be eligible for continuation therapy.

6 Laboratory studies

6.1 The following will be evaluated:

6.1.1 Urinary excretion by 24-hour urine collection

6.1.2 Pharmacokinetics: terminal elimination phase

6.1.3 Distribution, metabolism, and pharmacodynamics: accumulation of lometrexol and its polyglutamate metabolites in red and mononuclear blood cells; GAR transformylase activity in mononuclear blood cells; serum purines

6.2 At least one patient at each dose combination and at least four patients at the recommended phase II dose combination will be studied following doses 1, 4, and 8; in a case in which dose 4 is omitted, a patient will be studied following the next administered dose.

6.3 For pharmacokinetic, distribution, metabolism, and pharmacodynamic studies, samples will be drawn on days 2, 3, and day 1 of the subsequent week (prior to administration of a next dose). Each sample will consist of 10 ml drawn into a syringe containing 0.1 ml of tetrasodium EDTA 0.15 g/ml, 0.1 ml of deoxycoformycin

1 μg/ml, and 0.1 ml of dipyridamole 300 mg/ml through a gauge 21 or larger bore needle with care taken to avoid hemolysis.

6.4 See Appendix 3 for analytical methods.

7 Schedule
7.1 Schedule of procedures: NOT INCLUDED IN THIS PRESENTATION

7.2 Patients with any persistent lometrexol toxicity will be followed on at least a monthly basis with appropriate clinical and laboratory evaluation until resolution of toxicity.

8 Statistical considerations
The patient cohort design is a modification of a "3 in 5" schema (1).

9 References

2. Informed Consent Document

Massey Cancer Center, Virginia Commonwealth University, Informed Consent Form

Phase I study of weekly intravenous lometrexol with continuous oral folic acid supplementation

You have been asked to participate in an experimental research study being conducted at the Virginia Commonwealth University by Dr. John D. Roberts and colleagues.

Your Situation
Your doctor has asked you to consider joining this research study because you have a cancer (either a lymphoma or a solid tumor malignancy) for which there is no known cure and for which there is no treatment to offer at this time that would likely be of benefit.

The Study
Lometrexol is an unproven, but promising, new drug that may turn out to be useful in the treatment of cancer. The purpose of this study is find the right amount of lometrexol to give to people. In addition to lometrexol, patients in this study will take the vitamin, folic acid. Another purpose of the study is to find the right amount of folic acid to give to people who are receiving lometrexol. Other purposes are to learn about the side effects of lometrexol when given with folic acid and to learn about the chemistry of lometrexol and folic acid in the body. About 11 to 40 patients will be in this study. The doctors running this study are looking for patients with certain types of illness and conditions. Only patients who have these certain types of illness and conditions will be allowed to join the study.

You will take folic acid pills daily for 10 weeks. Starting the second week, you will receive lometrexol weekly for 8 weeks. Lometrexol will be dissolved in water and run into a plastic needle placed in one of your veins. Lometrexol will be run in over 1 or 2 minutes' time. During this time and for at least 3 weeks after, you will see a doctor weekly, and tests, including blood tests, will be taken weekly or more often. You may be asked to stay in the hospital for one or two days following three of the lometrexol doses for further testing, including further blood tests. The total amount of blood taken during the course of the study will not be more than what you would lose if you donated blood (about 1 pint).

After about 8 weeks of lometrexol treatment, your doctor will look at whether your cancer has shrunk. If it has, or if your disease is stable, you will be able to continue lometrexol for up to 1 year, so long as the side effects are not too bad, and the doctors think it is safe and the right thing to do.

Should your doctor or the doctors running this study learn new, important facts about lometrexol while you participate in the study, you will be told.

Benefits

There may not be any direct benefit to you from participating in this study. Your cancer may shrink as a result of lometrexol treatment, and this might allow you to feel better and live longer. Your participation may provide valuable information about the use of lometrexol. However, there are no guarantees this treatment will work.

Alternatives

Your participation is voluntary; you could decide that you do not want to get lometrexol. If so, you should talk with your doctor about what to do next.

Risks and Side Effects

Although designed with your safety in mind, experimental research studies are risky. You probably will experience side effects from lometrexol.

Lometrexol may cause a drop in the numbers of blood cells. In the case of red blood (oxygen-carrying) cells, this could cause shortness of breath or fatigue. In the case of white blood (infection-fighting) cells, this could cause infections. In the case of platelets (blood-clotting cells), this could cause bleeding or bruising.

Lometrexol may cause soreness and sores in the mouth, swallowing tube (esophagus), and intestines, as well as diarrhea.

Lometrexol may cause nausea and vomiting.

Lometrexol may cause heart attacks.

It is possible that you could experience other side effects, including serious side effects. It is possible that the side effects you experience could be long-lasting, life-threatening, or deadly.

For Men and Women Who Are Sexually Active

You should be aware that every effort will be made to have females enter this study on an equal basis with male subjects. You should not become pregnant or father a child before, during, or in the months after you receive lometrexol as it could cause miscarriage, birth defects, or other unforeseen problems. Medically accepted birth control is required to participate in this study. This may include, but is not limited to, not having sex, using birth control pills, IUDs, condoms, diaphragms, implants, being surgically sterile, or being in a postmenopausal state. You should not nurse a baby in the months after you receive lometrexol. If you have questions about these matters, you should ask your doctor or the doctors running this experimental study.

Costs

Lometrexol will be provided without cost to you by the Division of Cancer Treatment of the National Cancer Institute. This may change, however, in which case you might need to pay for lometrexol in order to continue treatment. You <u>may be billed</u> for doctor visits and tests done during the period of lometrexol treatment. If you develop side effects from lometrexol that require additional tests or the help of other doctors, you <u>will be billed</u> for these.

Most insurance policies do not cover experimental research studies. Insurance may or may not cover bills that result from lometrexol treatment or treatment of side effects. In any case, the usual deductions and copayments would apply. If you have questions about this, ask your doctor or one of the doctors running the study.

What if you are hurt as a result of participating in this experimental study?

If you are physically or mentally injured as a result of participating in this experimental study, Virginia Commonwealth University will not provide compensation. Medical treatment will be available at MCV Hospitals. You or your insurance will be billed for this treatment.

Who will know about you in this experimental study?

The results of this study may be published in scientific or regular magazines or papers, but your name will not appear.

The results will be discussed by doctors and others at the Medical College of Virginia Hospitals/Virginia Commonwealth University, the National Cancer Institute, the Lilly Pharmaceutical Company, the Federal Food and Drug Administration, and, possibly, other groups and persons. People of these organizations may look at the records related to this study, including your medical records.

What if you sign up, or even start, and then you want to quit? What if your doctors think things aren't going well?

You can quit at any time. Also, your doctor or the doctors running this study can remove you from the study at any time that is thought to be in your best interest to do so. In either case, your doctor and the doctors running this experimental study will continue to care for you according to good medical practice.

What if you have questions about your rights?

If you have questions about your rights as a research subject, you may contact the Chairman of the Committee on the Conduct of Human Research at (804) 555-XXXX or the doctors listed below.

What does it mean if you sign this form?

By signing this form, you indicate that you have read the form, that your questions have been answered, and that you want to participate in the study. You will be given a copy of this form.

(Patient's Signature)	(Date)
(Physician's Signature)	(Date)
(Witness's Signature)	(Date)
(Principal Investigator's Signature)	(Date)

Principal Investigator

Dr. John Roberts   Office 555-*XXXX*
                          Physician-on-call 555-*XXXX*

3. Animal Use Protocol

IACUC NO.<u>9804-2528</u>

Virginia Commonwealth University Institutional Animal Care and Use Committee (IACUC) Protocol for Use of Vertebrate Animals in Research

PROJECT TITLE: <u>Studies on the Genetics of Oral Microflora</u>

PROJECT STATUS: New_____ Pilot_____ Teaching_____ Renewal__X__ Modified_____

Existing IACUC No. <u>9204-2137</u> (for 3-Year Renewal or Modification)

FUNDING: Department Chair's signature is required prior to initiating work on any protocol which has not received scientific review from a recognized body (i.e. NIH Study Section, VAMC Review Board, NSF Review, American Heart Association, etc.). The Chair's signature indicates that he takes responsibility for scientific review of the protocol.

Intramural _____ Budget No. <u>5-00000</u>

Extramural (Sponsor's Name) <u>NIH-NIDCR (DE04224-23)</u> Proj. Dates <u>5/1/96 to 4/30/01</u>

Deadline for submission: <u>active project</u>

Dept. Chair's Signature (or Dean, if Chair is PI): TYPED NAME: Harvey Schenkein, Asst. Dean

Signature:

PRINCIPAL INVESTIGATOR: <u>Francis L. Macrina</u> Title: <u>Director</u>

Dept: <u>Institute of Oral and Craniofacial Molec. Biol.</u>

Off Ph: <u>555-XXXX</u> Box # <u>980566</u> Home/Emer Ph: <u>555-XXXX</u>

SECOND INVEST: Cindy L. Munro Title: Assoc. Prof. Off Ph: 555-XXXX

Department: Adult Health (School of Nursing) Box # 980567 Home/Emer Ph: 555-XXXX

SENIOR TECH: Kevin R. Jones Off Ph. 555-XXXX Home/Emer Ph. 555-XXXX

TYPE OF STUDY: (Check <u>all</u> applicable categories and complete appropriate pages; remove pages "not applicable."

__X__ COLLECTION OF BLOOD/TISSUE OR NON-INVASIVE STUDY

_____ BEHAVIORAL

__X__ BIOHAZARD <u>Circle</u> all that apply - Infectious, Carcinogens, Mutagens, Toxic chemicals, Radioisotopes

_____ SURGICAL-ACUTE (surgical procedures in which the animal is euthanized prior to recovery from anesthesia)

_____ SURGICAL-SURVIVAL (surgical procedures in which the animal is allowed to recover from anesthesia)

_____ FIELD STUDIES/BIOLOGICAL SURVEYS (complete Appendix B on page 13)

DESCRIPTION OF ANIMAL SUBJECTS:

Protocols are approved for a three-year period. Please specify numbers of animals to be used for the first year and total for three years for each species. Space is provided for three species provided experimental procedures are similar for all three. One protocol form may be used for rodents and rabbits, but separate protocol forms must be completed for higher species.

1. SPECIES A: <u>Rabbit</u> 2. Strain <u>New Zealand White</u> 3. Sex <u>Female</u>

4. Age/Weight <u>adult/4kg</u> 5. #/1st Year <u>15</u> 6. Total # for 3 Years <u>45</u>

Will animals be held more than 12 hours outside the vivarium? YES _____ NO __X__ (If yes, justify in Summary).

INDICATE <u>USDA</u> PAIN CATEGORIES: (SEE INSTRUCTIONS FOR DEFINITION OF PAIN CATEGORIES)

No Distress/No Anesthesia Pain Category A - # 1st Year = _____

Total # for 3 Years = _____

Alleviated Distress Pain Category B - # 1st Year = 15 _____

Total # for 3 Years = 45 _____

Unrelieved Distress Pain Category C* - # 1st Year = _____

Total # for 3 Years = _____

*If Category "C" applies, Appendix A on page 12 must be completed.

EUTHANASIA: Describe method(s) of euthanasia of animals including dose (mg/kg) and route of administration of applicable agent:

5mg/kg-35mg/kg xylazine-ketamine administered IV   (Employed prior to cardiac puncture and exsanguination)

Techniques for euthanasia shall follow current guidelines established by the American Veterinary Medical Association Panel on Euthanasia (1993). Other methods must be reviewed and approved by the Institutional Animal Care and Use Committee. If other than approved methods are needed, include justification in the summary.

SUMMARY: Describe your proposed protocol, emphasizing the use of animals. Do not submit an abstract of your grant proposal. Write in terminology understandable to educated lay persons, not scientific specialists, and avoid or define abbreviations. Describe specifically what will be done with the animals, and indicate the expected results. Discuss the procedures in order, and give time intervals (use tables to indicate uses of animals in complex protocols) occurring between procedures. The Committee needs to understand what happens to each animal. Use additional sheets if necessary.

These animals will be used to generate polyclonal antibodies to purified or partially purified protein prepared from bacterial cells. Protein immunogen preparations will fall into two categories. First we will use extracellular proteins prepared from viridans streptococci or from Bacteroides or Porphyromonas; such preparations will contain few to several proteins depending on the type of purification used. The second class of immunogen will consist of highly purified proteins made from recombinant Escherichia coli. Neither protein preparation will contain whole cells at the time of administration. In all cases the proteins being used will consist of one of more of the following enzymes (or portions thereof genetically linked to the B subunit {binding, non-toxic subunit} of cholera toxin or a portion of the core protein of hepatitis B virus): water soluble synthesizing glucosyltransferase, water insoluble synthesizing glucosyltransferase and/or fructosyltransferase. Proteases or structural proteins from streptotococci, Bacteroides or Porphyromonas will also be similarly used. Antibodies generated will be used to probe enzyme/structural protein structure-function relationships in in vitro enzyme assays or in immunolocalization studies.

Female NZW rabbits (approx 4 kg) will be primed with 0.5 to 5.0 mg of protein antigen preparation. The antigen will be suspended in 0.5 ml of phosphate buffered saline (pH 7.3) and emulsified in an equal volume of complete Freund's adjuvant (CFA). Freund's adjuvant is a well-established adjuvant system which is appropriate for use in these experiments where small amounts of antigen are used, and where immunogenicity of the antigen (although likely) is unknown. Published guidelines for use will be followed, including limiting injection to 0.1 ml at each site, using CFA ONLY for initial immunization dose. This antigen preparation (1 ml total volume) will be injected subdermally in the loose skin on the backside of the rabbit's neck. This injection route is immunologically effective and minimizes the possibility of local inflammation associated with unilateral or bilateral flank injection (such ensuing flank inflammation can impair animal mobility). After resting for 3 weeks, one ml of blood will be removed from the ear artery for a test bleed. Antibodies will be boosted if titers of the desirable antibodies are judged to be too low. Rabbits with adequate antibody levels will be boosted subdermally 1.0 mg of antigen contained in Freund's "incomplete" adjuvant. Boosted animals will be bled after two weeks; i.e., 15 ml of blood will be taken from the ear artery using a heat lamp to dilate the blood vessel. The rabbit will be placed in a commercial restraint, tranquilized with xylazine not more than seven times in total after which the rabbit will be exsanguinated by cardiac puncture following anesthesia using xylazine/ketamine.

PERSONNEL QUALIFICATION: (It is an institutional obligation to ensure that professional and technical personnel and students who perform animal anesthesia, surgery, or other experimental manipulations are qualified through training or experience to accomplish these tasks in a humane and scientifically acceptable manner, Guide, pg. 6, 1985.)

Indicate personnel who will be performing the animal procedures and indicate the training and number of years of experience of each person for the types of animal procedures proposed for each species. Personnel who will be irradiating experimental animals must be trained and have approval from the Radiation Safety Section of the Office of Environment Health & Safety. **Please notify the committee by memorandum of any changes in personnel after approval of protocol.**

The following will be involved in this work:

Francis L. Macrina, Ph.D.	*20 years experience and VCU on-site animal training*
Cindy L. Munro, Ph.D.	*10 years experience and VCU on-site animal training*
Kevin R. Jones, M.S.	*10 years experience and VCU on-site animal training*
Todd Kitten, Ph.D.	*8 years experience and VCU on-site animal training*
Janina Lewis, Ph.D.	*4 years experience and VCU on-site animal training*
Janet Dawson, Ph.D.	*6 years experience and VCU on-site animal training*

As in the past, we will employ the services of the Division of Animal Resources to perform all animal injections and bleedings.

JUSTIFICATION FOR THE USE OF PROPOSED ANIMAL MODEL

1. What are the probable benefits of this work to human or animal health, the advancement of knowledge, or the good of society?

This work is designed to probe the structure-function relationships of proteins of the viridans streptococci including the agents of infective endocarditis and smooth surface dental decay. Similarly we will examine the properties of proteases and structural proteins that are involved in soft-tissue infections of the oral cavity. The proposed work will broaden our understanding of specific protein factors in streptococcal and anaerobic bacteral virulence and is aimed at identifying proteins or peptides that can be tested as vaccinogens or further characterized as targets for interventional chemotherapy (e.g., protease inhibitors).

2. Justify the selection of the proposed animal species, strain, and numbers (**include statistical or other criteria for animal numbers**). Cost is not a valid justification.

Rabbits have been used historically and extensively to generate antisera to purified and partially purified proteins. They are preferred due to their ease of handling and their ability to generate a large volume of usable antisera using a minimal number of animals.

3. What databases or services have you used to determine that alternative methods, such as <u>in vitro</u> studies, would not be acceptable? The Animal Welfare Act dictates that the investigator must provide written documentation that alternatives were not available. An alternative is any procedure which results in the reduction in the numbers of animals used, refinement of techniques, or replacement of animals. If the project involves teaching, explain why films, videotapes, demonstrations, etc. would not be acceptable.

The MEDLINE database was searched (1966-present) to determine if alternative methods were available. None were found.

4. Does this experiment duplicate previous experiments? Yes _____ No __X__
If yes, explain why duplication is necessary for your research.

FOR ALL EXPERIMENTAL PROCEDURES:
Are procedures to be used that are **intended to study pain?** Yes _____ No __X__

If the answer is YES, what criteria will be used to assess pain/discomfort and what will be done to minimize or relieve pain/discomfort? (If analgesics cannot be used and pain/discomfort is going to be minimized by early euthanasia of the animals rather than using analgesics, describe the monitoring schedule and the criteria which will determine the time of euthanasia.)

EXPERIMENTAL PROCEDURES (COLLECTION OF BLOOD/TISSUE OR NON-INVASIVE)
This section includes antibody production, blood/tissue collection, or any non-invasive study.

NOTE: Include in summary: expected rate of growth of tumors or ascites, monitoring schedule, criteria for assessment of distress, and earliest point at which animals in distress will be euthanized.

Materials To Be Administered to Animals as Part of Experimental Protocol (do not include hazardous materials which you have listed and described on page 4):

Species	Antigen/Drug	Dose(mg/kg)	Route	Frequency	No. Animals Used
rabbit	*native protein*	*variable (see summ)*		*8/yr*	*24*
rabbit	*recombinant proteins*	*variable (see summ)*		*7/yr*	*21*

Describe expected results:

Animals will generate antisera to proteins present in the preparations.

How long will individual animals be on the study?
One to three months is the expected time needed to generate a usable antibody titer.

Blood or Tissue Collection:
Describe technique used to collect blood or tissue (include route of collection and anesthetic, sedative or tranquilizing agents administered prior to specimen collection):

1. Ear artery test bleed (1 ml) - three times maximum - with xylazine, 3-6 mg/kg.

2. Ear artery bleed (15 ml) - four times maximum - with xylazine, 3-6 mg/kg

3. Exsanguination by cardiac puncture following anesthesia - with xylazine/ketamine, supplemented by methoxyflurane, if required. Methoxyflurane will be used only in a properly ventilated fume hood as a safety precaution.

Blood/Tissue	Amount/Size	Frequency	No. Animals Used
Blood	*1 ml (1-3 times)*	*2 wk. intervals*	*15 (total)*
Blood	*15 ml (3-4 times)*	*2 wk. intervals*	*15 (total)*
Blood	*Exsanguination by cardiac puncture*		

If the nature of your project makes it difficult to complete the above table, include a bleeding or collection schedule:

Not applicable

Indicate methods for the prevention of anemia:

The volume of blood proposed to be collected at the stated two-week intervals should prevent anemia from occurring.

VIRGINIA COMMONWEALTH UNIVERSITY
INSTITUTIONAL ANIMAL CARE AND USE COMMITTEE

ASSURANCE FOR THE HUMANE CARE AND USE OF ANIMALS
USED FOR TEACHING AND RESEARCH

1. I agree to abide by all the federal, state and local laws and regulations governing the use of animals in research. I understand that emergency veterinary care will be administered to animals showing evidence of pain or illness.

2. I have considered alternatives to the animal models used in this project and found other methods unacceptable.

3. I affirm that the proposed work does not unnecessarily duplicate previous experiments.

4. I affirm that all experiments involving live animals will be performed under my supervision or that of another qualified biomedical scientist. Technicians involved have been trained in proper procedures in animal handling, administration of anesthetics, analgesics and euthanasia to be used in this project.

5. I further affirm the information provided in the accompanying protocol is accurate to the best of my knowledge. Any proposed revisions to the animal care and use procedures will be promptly forwarded in writing to the Committee for approval.

I have read and understand the above statements.

Francis L. Macrina

Typed Name

Signature of Principal Investigator _____ Date

SIGNATURE FOR INSTITUTIONAL ANIMAL CARE AND USE COMMITTEE APPROVAL

Signature of Chairman, IACUC

Example of a U.S. Patent

A S A POSTDOCTORAL TRAINEE at the University of Illinois in the late 1960s, Ananda Chakrabarty was intrigued with the extraordinary nutritional diversity of microorganisms. The pioneering work of Chakrabarty, his postdoctoral mentor I.C. Gunsalus, and their lab colleagues helped define the genetic basis of hydrocarbon degradation in species of the bacterium *Pseudomonas*. Their research revealed that pseudomonad bacteria often carried genes on extrachromosomal elements (plasmids) that encoded enzymes able to degrade complex hydrocarbons. Such gene ensembles were found on specific plasmids: for example, one plasmid encoded the enzymatic pathway for the degradation of *n*-octane, another encoded the instructions for camphor degradation, and so on. Equally important was that many of these plasmids could be moved between bacterial cells by a genetic transfer mechanism called conjugation. So it was possible to construct pseudomonad strains that contained multiple plasmids and, thus, the genetic information encoding multiple degradative pathways.

In 1971 Chakrabarty joined the research and development center of the General Electric Company in Schenectady, N.Y. Although his initial work at GE involved the study of bacterial degradation of lignocellulosic compounds, he continued to think about the hydrocarbon-degrading capabilities of the pseudomonads. Chakrabarty reasoned that it should be possible to bring together by conjugation several different plasmids into the same *Pseudomonas* strain. He further believed that selection of specific plasmids to ensure a variety of degradative capabilities would create a "superstrain" able to degrade components of crude oil, for example. Chakrabarty did the experiments to test these ideas and demonstrated that a multi-plasmid-containing pseudomonad strain which he constructed was able to degrade an oil spill in a fish tank.

On June 7, 1972, GE on behalf of Chakrabarty sought patent protection for the oil-eating pseudomonad and the process of bacterial crude oil degradation. As noted in chapter 9, the eventual issuance of this patent had a major impact on biotechnological intellectual property. However, once submitted, this patent was almost 9 years in the making, finally being granted on March 31, 1981. This extended period was considerably longer than it takes most patent applications to be considered and acted upon. In large part, the reason for this was that the Chakrabarty invention stirred much controversy and debate once it was filed for consideration. In 1974 the Patent and Trademark Office (PTO) rejected the claims related to the oil-eating bacterium but allowed the claims related to the degradation process. GE appealed the rejection, beginning a 6-year legal odyssey. During this period the application moved multiple times between the PTO, the PTO Board of Appeals, the U.S. Court of Customs and Patent Appeals, and the U.S. Supreme Court. At the center of the legal deliberations was the historic conception that living organisms — products of nature — were unpatentable. At times the arguments became sidetracked over issues like the possible threats of genetic engineering research and the deliberate release of engineered organisms into the environment. On June 16, 1980, the U.S. Supreme Court voted 5 to 4 to uphold the patentability of Chakrabarty's oil-eating bacterium. In effect, this decision said that a nonhuman life form could be patented if its creation could be attributed in some way to human intervention. The Court's ruling marked a sea change in the intellectual property and biotechnology worlds.

In the parlance of intellectual property law, Chakrabarty's experiments at GE "reduced his invention to practice." First, he showed that strains with multiple degradative capability could be constructed in the laboratory. A novel, nonhuman life form had been created by a scientist. Second, he demonstrated that a proposed use of this new life form — cleanup of environmental oil spills — was possible in a controlled, laboratory setting.

In doing so, Chakrabarty met important legal criteria critical to obtaining a patent. First, he demonstrated his invention — a genetically modified *Pseudomonas* bacterium — to be a new composition of matter that was useful for oil spill cleanup by biological means. Second, his invention was novel, not having been previously described or known to exist in nature or in the laboratory. Last, he could show that the claimed bacterium was not obvious compared with reports of other bacterial cultures.

The full text of the Chakrabarty patent specification is presented in this appendix. "Specification" is the term used to identify the document describing the invention for which the patent protection is being sought. Attached to the specification are one or more drawings that are necessary to support or help understand the invention. These are like figures in a scientific paper. Another similarity to scientific papers is that a patent specification usually contains data and information presented in the form of tables.

The specification begins with a descriptive title. It then lists the inventor and may list the assignee, if the inventor has assigned or transferred ownership in the invention to another. The inventor is the person or persons who conceived and reduced the invention to practice either by the filing of a patent application or by developing and testing the invention. The assignee is the organization (or person) to which the property rights of the patent are granted. In a university setting, the inventor and the assignee have usually worked out an agreement for the sharing of profits associated with the use or licensing of the patent. The inventor and the assignee could be one and the same. Patents, like scientific publications, have a brief abstract summarizing the invention and how it works.

The Background section of the specification is what we might call an extended version of the Introduction in a scientific paper. This section contains a discussion of the relevant literature to the field or area of "art." A variation of this latter terminology is "prior art," which strictly means references or other documentation that the PTO or a court (or possibly the patent applicant) considers to be a barrier to the patentability of the claimed invention. In the Chakrabarty specification, a glossary was included in the Background section to aid in understanding the rest of the specification. Next there is a brief Summary of the Invention. Then there is a description of the invention, referred to as the "preferred embodiment." This is the section where the inventor explains how the invention works or operates. The description here must be detailed and thorough. The presentation must be precise enough that someone "skilled in the art" can practice the invention. A patent specification must fully disclose everything needed to replicate and use the invention. This section ends with a summation of the invention, emphasizing its advantages and use.

A patent specification concludes with a description of the claims associated with the invention. These numbered sentences or sentence fragments precisely define the nature of the invention and its uses. In other words, the claims are what others will be prevented from making, using, selling, or importing should the patent be granted. Only the inventor (or assignee) will have the right to exclude others from carrying out the actions listed in the claims once patent protection is awarded. Even the inventor and assignee may be excluded from practicing the claimed invention, if it is dominated or considered to be within the claims of a broader invention in a patent to another party.

To fulfill the requirements for a patent application, the patent applicant must also pay a filing fee and provide a signed oath that the inventor(s) is the original and first true inventor of the claimed invention.

United States Patent [19]

Chakrabarty

[11] **4,259,444**

[45] **Mar. 31, 1981**

[54] **MICROORGANISMS HAVING MULTIPLE COMPATIBLE DEGRADATIVE ENERGY-GENERATING PLASMIDS AND PREPARATION THEREOF**

[75] Inventor: **Ananda M. Chakrabarty,** Latham, N.Y.

[73] Assignee: **General Electric Company,** Schenectady, N.Y.

[21] Appl. No.: **260,563**

[22] Filed: **Jun. 7, 1972**

[51] Int. Cl.³ **C12N 15/00**
[52] U.S. Cl. **435/172;** 435/253; 435/264; 435/281; 435/820; 435/875; 435/877
[58] Field of Search .... 195/28 R, 1, 3 H, 3 R, 195/96, 78, 79, 112; 435/172, 253, 264, 820, 281, 875, 877

[56] **References Cited**

PUBLICATIONS

Annual Review of Microbiology vol. 26 Annual Review Inc. 1972 pp. 362–368.
Journal of Bacteriology vol. 106 pp. 468–478 (1971).
Bacteriological Reviews vol. 33 pp. 210–263 (1969).

Primary Examiner—R. B. Penland

Attorney, Agent, or Firm—Leo I. MaLossi; James C. Davis, Jr.

[57] **ABSTRACT**

Unique microorganisms have been developed by the application of genetic engineering techniques. These microorganisms contain at least two stable (compatible) energy-generating plasmids, these plasmids specifying separate degradative pathways. The techniques for preparing such multi-plasmid strains from bacteria of the genus Pseudomonas are described. Living cultures of two strains of Pseudomonas (*P. aeruginosa* [NRRL B-5472] and *P. putida* [NRRL B-5473]) have been deposited with the United States Department of Agriculture, Agricultural Research Service, Northern Marketing and Nutrient Research Division, Peoria, Ill. The *P. aeruginosa* NRRL B-5472 was derived from *Pseudomonas aeruginosa* strain 1c by the genetic transfer thereto, and containment therein, of camphor, octane, salicylate and naphthalene degradative pathways in the form of plasmids. The *P. putida* NRRL B-5473 was derived from *Pseudomonas putida* strain PpG1 by genetic transfer thereto, and containment therein, of camphor, salicylate and naphthalene degradative pathways and drug resistance factor RP-1, all in the form of plasmids.

18 Claims, 2 Drawing Figures

U.S. Patent   Mar. 31, 1981   4,259,444

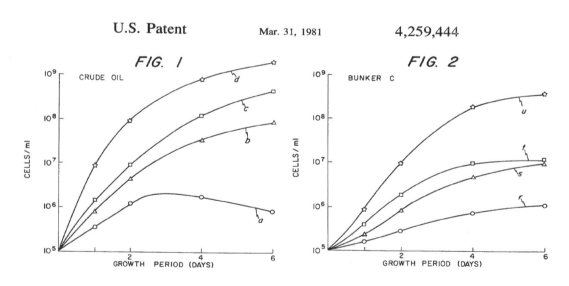

4,259,444

1

MICROORGANISMS HAVING MULTIPLE COMPATIBLE DEGRADATIVE ENERGY-GENERATING PLASMIDS AND PREPARATION THEREOF

BACKGROUND OF THE INVENTION

The terminology of microbial genetics is sufficiently complicated that certain definitions will be particularly useful in the understanding of this invention:

Extrachromosomal element . . . a hereditary unit that is physically separate from the chromosome of the cell; the terms "extrachromosomal element" and "plasmid" are synonymous; when physically separated from the chromosome, some plasmids can be transmitted at high frequency to other cells, the transfer being without associated chromosomal transfer;

Episome . . . a class of plasmids that can exist in a state of integration into the chromosome of their host cell or as an autonomous, independently replicating, cytoplasmic inclusion;

Transmisible plasmid . . . a plasmid that carries genetic determinants for its own intercell transfer via conjugation;

DNA . . . deoxytribonucleic acid;

Bacteriophage . . . a particle composed of a piece of DNA encoded and contained within a protein head portion and having a tail and tail fibers composed of protein;

Transducing phage . . . a bacteriophage that carries fragments of bacterial chromosomal DNA and transfers this DNA on subsequent infection of another bacterium;

Conjugation . . . the process by which a bacterium establishes cellular contact with another bacterium and the transfer of genetic material occurs;

Curing . . . the process by which selective plasmids can be eliminated from the microorganism;

Curing agent . . . a chemical material or a physical treatment that enhances curing;

Genome . . . a combination of genes in some given sequence;

Degradative pathway . . . a sequence of enzymatic reactions (e.g. 5 to 10 enzymes are produced by the microbe) converting the primary substrate to some simple common metabolite, a normal food substance for microorganisms;

(Sole carbon source)$^-$ . . . indicative of a mutant incapable of growing on the given sole carbon source;

(Plasmid)^{del} . . . indicative of cells from which the given plasmid has been completely driven out by curing or in which no portion of the plasmid ever existed;

(Plasmid)$^-$ . . . indicative of cells lacking in the given plasmid; or cells harboring a non-functional derivative of the given plasmid;

(Amino-acid)$^-$ . . . indicative of a strain that cannot manufacture the given amino acid;

(Vitamin)$^-$ . . . indicative of a strain that cannot manufacture the given vitamin and

(Plasmid)$^+$ . . . indicates that the cells contain the given plasmid.

Plasmids are believed to consist of double-stranded DNA molecules. The genetic organization of a plasmid is believed to include at least one replication site and a maintenance site for attachment thereof to a structural component of the host cell. Generally, plasmids are not essential for cell viability.

Much work has been done supporting the existence, functions and genetic organization of plasmids. As is

2

reported in the review by Richard P. Novick "Extrachromosomal Inheritance in Bacteria" (Bacteriological Reviews, June 1969, pp. 210–263, [1969]) on page 229, "DNA corresponding to a number of different plasmids has been isolated by various methods from positive cells, characterized physiochemically and in some cases examined in the electron microscope".

There is no recognition in the Novick review of the existence of energy-generating plasmids specifying degradative pathways. As reported on page 237 of the Novick review, of the known (non energy-generating) plasmids "Combinations of four or five different plasmids in a cell seem to be stable."

Plasmids may be compatible (i.e. they can reside stably in the same host cell) or incompatible (i.e. they are unable to reside stably in a single cell). Among the known plasmids, for example, are sex factor plasmids and drug-resistance plasmids.

Also, as stated on page 240 of the Novick review, "Cells provide specific maintenance systems or sites for plasmids. It is though that attachment of such sites is required for replication and for segregation of replicas. Each plasmid is matched to a particular maintenance site . . . ". Once a plasmid enters a given cell, if there is no maintenance site available, because of prior occupancy by another plasmid, these plasmids will be incompatible.

The biodegradation of aromatic hydrocarbons such as phenol, cresols and salicylate has been studied rather extensively with emphasis on the biochemistry of these processes, notably enzyme characterization, nature of intermediates involved and the regulatory aspects of the enzymic actions. The genetic basis of such biodegradation, on the other hand, has not been as thoroughly studied because of the lack of suitable transducing phages and other genetic tools.

The work of Chakrabarty and Gunsalus (Genetics, 68, No. 1, page S10, [1971]) has showed that the genes governing the synthesis of the enzymes responsible for the degradation of camphor constitute a plasmid. Similarly, this work has shown the plasmid nature of the octane-degradative pathway. However, attempts to provide a microorganism with both CAM and OCT plasmids were unsuccessful, these plasmids being incompatible.

In *Escherichia coli* artificial, transmissible plasmids (one per cell) have been made, each containing a degradative pathway. These plasmids, not naturally occurring, are F'lac and F'gal, wherein the lactose-and galactose-degrading genes were derived from the chromosome of the organism. Such plasmids are described in "F-prime Factor Formation in *E. Coli* K12" by J. Scaife (Genet. Res. Cambr. [1966], 8, pp. 189–196).

If the development of microorganisms containing multiple containing energy-generating plasmids specifying preselected degradative pathways could be made possible, the economic and environmental impact of such an invention would be vast. For example, there would be immediate application for such versatile microbes in the production of proteins from hydrocarbons ("Proteins from Petroleum"—Wang, Chemical Engineering, August 26, 1968, page 99); in cleaning up oil spills ("Oil Spills: An Environmental Threat"—Environmental Sciene and Technology, Volume 4, February 1970, page 97); and in the disposal of used automotive lubricating oils ("Waste Lube Oils Pose Disposal Di-

4,259,444

3

lemma", Environmental Science and Technology, Volume 6, page 25, January 1972).

SUMMARY OF THE INVENTION

A transmissible plasmid has been found that specifies a degradative pathway for salicylate [SAL], an aromatic hydrocarbon. In addition, a plasmid has been identified that specifies a degradative pathway for naphthalene [NPL], a polynuclear aromatic hydrocarbon. The NPL plasmid is also transmissible.

Having established the existence of (and transmissibility of) plasmid-borne capabilities for specifying separate degradative pathways for salicylate and naphthalene, unique single-cell microbes have been developed containing various stable combinations of the [CAM], [OCT], [SAL], and [NPL] plasmids. In addition, stable combinations in a single cell of the aforementioned plasmids together with a non energy-generating plasmid [drug resistance factor RP-1] have been achieved. The versatility of these novel microorganisms has been demonstrated by the substantial extent to which degradation of such complex hydrocarbons as crude oil and Bunker C oil has been achieved thereby.

BRIEF DESCRIPTION OF THE DRAWING

The exact nature of the invention as well as objects and advantages thereof will be readily apparent from consideration of the following specification relating to the annexed drawing in which:

FIG. 1 shows the increase in growth rate in crude oil of Pseudomonas strain bacteria provided with increasing numbers of energy-generating degradative plasmids by the practice of this invention and

FIG. 2 shows the increase in growth rate in Bunker C oil of Pseudomonas strain bacteria provided with increasing numbers of energy-generating degradative plasmids by the practice of this invention.

DESCRIPTION OF THE PREFERRED EMBODIMENT

Microorganisms prepared by the genetic engineering processes described herein are exemplified by cultures now on deposit with the U.S. Department of Agriculture. These cultures are identified as follows:

Pseudomonas aeruginosa (NRRL B-5472) . . . derived from *Pseudomonas aeruginosa* strain 1c (ATCC No. 15692) by genetic transfer thereto, and containment therein, of camphor, octane, salicylate and naphthalene degradative pathways in the form of plasmids.

Pseudomonas putida (NRRL B-5473) . . . derived from *Pseudomonas putida* strain PpGl (ATCC No. 17453) by genetic transfer thereto, and containment therein, of camphor, salicylate and naphthalene degradative pathways and a drug resistance factor RP-1, all in the form of plasmids. The drug resistance factor is responsible for resistance to neomycin/kanamycin, carbenicillin and tetracycline.

A sub-culture of each of these strains can be obtained from the permanent collection of the Northern Marketing and Nutrient Research Division, Agricultural Service, U.S. Department of Agriculture, Peoria, IL, U.S.A.

Morphological observations in various media, growth in various media, general group characterization tests, utilization of sugars and optimum growth conditions for the strains from which the above-identified organisms were derived are set forth in "The Aerobic Pseudomonads: A Taxonomic Study" by Stanier, R.

4

Y. et al [Journal of General Microbiology 43, pp. 159–271 (1966)]. The taxonomic properties of the above-identified organisms remain the same as those of the parent strains.

P. aeruginosa strain 1c (ATCC No. 15692) is the same as strain 131 (ATCC No. 17503) in the Stanier et al study. Later the designation for this strain was changed to *P. aeruginosa* PAO [Holloway, B. W. "Genetics of Pseudomonas", Bacteriological Reviews, 33, 419–443 (1969)]. *P. putida* strain PpGl (ATCC No. 17453) is the same as strain 77 (ATCC No. 17453) in the Stanier et al study.

As will be described in more detail hereinbelow, these organisms thrive on a very wide range of hydrocarbons including crude oil and Bunker C oil. These organisms are non-pathogenic as is the general case with laboratory strains of Pseudomonas.

In brief, the process for preparing microbes containing multiple compatible energy-generating plasmids specifying separate degradative pathways is as follows:

(1) selecting the complex or mixture to be degraded;

(2) identifying the plurality of degradative pathways required in a single cell to degrade the several components of the complex or mixture therewith;

(3) isolating a strain of some given microorganism on one particular selective substrate identical or similar to one of the several components (the selection of the microorganism is generally on the basis of a demonstrated superior growth capability);

(4) determining whether the capability of the given strain to degrade the selective substrate is plasmid-borne;

(5) attempting to transfer this first degradative pathway by conjugation to other strains of the same organism (or to the same strain which has been cured of the pathway) and then verifying the transmissible nature of the plasmid;

(6) purifying the conjugatants (recipients of the plasmids by conjugation) and checking for distinctive characteristics of the recipient to insure that the recipient did, in fact, receive the degradative pathway;

(7) repeating the process so as to introduce a second plasmid to the conjugatants;

(8) rendering the first and second plasmids compatible, if necessary, by fusion of the plasmids and

(9) repeating the process as outlined above until the full complement of degradative pathways desired in a single cell has been accomplished by plasmid transfer (and fusion, when required).

In the first reported instance (Chakrabarty et al article mentioned hereinabove) in which the attempt was made to locate more than one energy-generating degradative pathway in the same cell, it was found that CAM and OCT plasmids cannot exist stably under these conditions. In spite of the implication from these results that multiple energy-generating plasmid content in a single cell could be achieved but not maintained, it was decided to attempt to discover some way in which to overcome this problem of plasmid incompatibility. As noted hereinabove and described more fully hereinbelow with specific reference to energy-generating plasmid transfer in the genus Pseudomonas, the problem of plasmid instability has now been solved by bringing about fusion of the plasmids in the recipient cell.

The development of single cell capability for the degradation and conversion of complex hydrocarbons was selected as the immediate beneficial application with particular emphasis on the genetic control of oil

4,259,444

5

spills by the way of a single strain of Pseudomonas. In order to be able to cope with crude oil and Bunker C oil spills it was decided that the single cells of Pseudomonas derivate produced by this invention should possess degradative pathways for linear aliphatic, cyclic aliphatic, aromatic and polynuclear aromatic hydrocarbons. *Pseudomonas aeruginosa* (NRRL B-5472) strain, which displays these degradative capabilities was thereupon eventually developed.

Massive oil spills that are not promptly contained and cleaned up have a catastrophic effect on aquatic lives. Microbial strains are known that can decompose individual components of crude oil (thus, various yeasts can degrade aliphatic straight-chain hydrocarbons, but not most of the aromatic and polynuclear hydrocarbons). Pseudomonas and other bacteria species are known to degrade the aliphatic, aromatic and polynuclear aromatic hydrocarbon compounds, but, unfortunately any given strain can degrade only a particular component. For this reason, prior to the instant invention, biological control of oil spills had involved the use of a mixture of bacterial strains, each capable of degrading a single component of the oil complex on the theory that the cumulative degradative actions would consume the oil and convert it to cell mass. This cell mass in turn serves as food for aquatic life. However, since bacterial strains differ from one another in (a) their rates of growth on the various hydrocarbon components, (b) nutritional requirements, production of antibiotics or other toxic material, and (c) requisite pH, temperature and mineral salts, the use of a mixed culture leads to the ultimate survival of but a portion of the initial collection of bacterial strains. As a result, when a mixed culture of hydrocarbon-degrading bacteria are deposited on an oil spill the bulk of the oil often remains unattacked for a long period of time (weeks) and is free to spread or sink.

By establishing that SAL and NPL degradative pathways are specified by genes borne by transmissible plasmids in Pseudomonas and by the discovery that plasmids can be rendered stable (e.g. CAM and OCT) by fusion of the plasmids it has been made possible, for the first time, to genetically engineer a strain of Pseudomonas having the single cell capability for multiple separate degradative pathways. Such a strain of microbes equipped to simultaneously degrade several components of crude oil can degrade an oil spill much more quickly (days) than a mixed culture meanwhile bringing about coalescence of the remaining portions into large drops. This action quickly removes the opportunity for spreading of the oil thereby enhancing recovery of the coalesced residue.

Preparation of *P. aeruginosa* [NRRL B-5472]

The compositions of the synthetic mineral media for growth of the cultures were the same for all the Pseudomonas species employed. The mineral medium was prepared from:

6

PA Concentrate . . .
  100 ml of 1 Molar K_2HPO_4
  50 ml of 1 Molar KH_2PO_4
  160 ml of 1 Molar NH_4Cl
100×Salts . . .
  19.5 gm $MgSO_4$
  5.0 gm $MnSO_4.H_2O$
  5.0 gm $FeSO_4.7H_2O$
  0.3 gm $CaCl_2.2H_2O$
  1.0 gm Ascorbic acid
  1 liter H_2O

Each of the above (PA Concentrate and 100×Salts) was sterilized by autoclaving. Thereafter, one liter of the mineral medium was prepared as follows:

PA Concentrate	77.5 ml
100 X Salts	10.0 ml
Agar	15.0 gm
H_2O	to one liter (The pH is adjusted to 6.8–7.0).

All experiments were carried out at 32° C. unless otherwise stated.

It was decided that a very useful hydrocarbon degradation capability would be attained in a single *Pseudomonas aeruginosa* cell, if the degradative pathways for linear aliphatic, cyclic aliphatic, aromatic and polynuclear aromatic hydrocarbons could be transferred thereto. *Pseudomonas aeruginosa* PAO was selected because of its high growth rate even at temperatures as high as 45° C. Four strains of Pseudomonas were selected having the individual capabilities for degrading n-octane (a linear aliphatic hydrocarbon), camphor (a cyclic aliphatic hydrocarbon), salicylate (an aromatic hydrocarbon) and naphthalene (a polynuclear aromatic hydrocarbon).

The specific strains of Pseudomonas able to degrade these hydrocarbons were then treated with curing agent to verify the plasmid-nature of each of these degradative pathways. Of the known curing agents (e.g. sodium dodecyl sulfate, urea, acriflavin, rifampicin, ethidium bromide, high temperature, mitomycin C, acridine orange etc.) most were unable to cure any of the degradative pathways. However, it was found (Table I) that the degradative pathways of the several species could be cured with mitomycin C. Each of the Pseudomonas strains bearing the specified degradative pathways are known in the art:

(a)	CAM $^+$ *P. putida* PpG1	Proc. Nat. Acad. Sci. (U.S.A.), 60, 168 (1968)
(b)	OCT $^+$ *P. oleovorans*	J. Biol. Chem. 242, 4334 (1967)
(c)	SAL $^+$ *P. putida* R-1	Bacteriological Proceedings 1972 p. 60
(d)	NPL $^+$ *P. aeruginosa*	Biochem. J. 91, 251 (1964)

TABLE I

Strain	Degradative Pathway	Mitomycin C Concentration (μg/ml)	Frequency of Curing (Percent)
CAM $^+$ *P. putida* PpG1	cyclic aliphatic hydrocarbon (camphor)	0	<0.01
		10	5
		20	95
OCT $^+$ *P. oleovorans*	aliphatic hydrocarbon (n-octane)	0	<0.1
		10	1.0
		20	3.0
SAL $^+$ *P. putida* R-1	aromatic hydrocarbon	0	<0.1

TABLE I-continued

Strain	Degradative Pathway	Mitomycin C Concentration (μg/ml)	Frequency of Curing (Percent)
	(salicylate)	5	0.7
		10	3.0
		15	4.0
NPL$^+$ *P. aeruginosa*	polynuclear aromatic hydrocarbon (naphthalene)		
		0	<0.1
		5	0.5
		10	1.8

Curing degradative pathways from each strain with mitomycin C was accomplished by preparing several test tubes of L broth [Lennox E.S. (1955), Virology, 1, 190] containing varying concentrations of mitomycin C and inoculating these tubes with suitable dilutions of early stationary phase cells of the given strain to give concentrations 10^4 to 10^5 cells/ml. These tubes were incubated on a shaker at 32° C. for 2–3 days. Aliquots from tubes that showed some growth were then diluted and plated on glucose minimal plates. After growth at 32° C. for 24 hours, individual colonies were split and respotted on glucose-minimal and degradative pathway-minimal plates to give the proportion of CAM$^-$, OCT$^-$, SAL$^-$ and NPL$^-$ in order to determine the frequency of curing. It was, therefore, shown that in each instance the degradative pathway genes are plasmid-borne.

Transductional studies with a number of point mutants in the camphor and salicylate pathways has suggested that the cured segments lost either the entire or the major portion of the plasmid genes. The plasmid nature of the degradative pathways was also confirmed from evidence of their transmissibility by conjugation from one strain to another (Table II). Although the frequency of plasmid transfer varies widely with individual plasmids and although OCT plasmid cannot be transferred from *P. oleovorans* to *P. aeruginosa* PAO at any detectable frequency, most of the plasmids can nevertheless be transferred from one strain to another by conjugation.

The plasmid transfers, instead of being made to other strains could have been made to organisms of the same strain, that had been cured of the given pathway with mitomycin C, acridine orange or other curing agent.

Pseudomonas putida U has been described in the article by Feist et al [J. Bacteriology 100, p. 869–877 (1969)].

The auxotropic mutants (mutants that require a food source containing a particular amino acid or vitamin for growth) shown in Table II as donors were each grown in a complex nutrient medium (e.g. L broth) to a population density of at least about 10^8 cells/ml without shaking in a period of from 6 to 24 hours. The prototropic (cells capable of growing on some given minimal source of carbon) recipients to which degradative pathway transfer was desired were grown separately in the same complex nutrient medium to a population density of at least about 10^8 cells/ml with shaking in a period of from 4 to 26 hours. For each degradative pathway transfer these cultures were mixed in equal volumes, kept for 15 minutes to 2 hours at 32° C. without shaking (to permit conjugation to occur) and then plated on minimal plates containing the particular substrate as the sole source of carbon. This procedure for cell growth of donor and recipient and the mixing thereof is typical of the manner in which conjugation and plasmid transfer is encour-

aged in the laboratory, this procedure being designed to provide a very efficient transfer system. Temperature is not critical, but the preferred temperature range is 30°–37° C. Reduction in the population density of either donor or recipient below about 1,000,000 cells/ml or any change in the optimal growth conditions (stationary growth of donor, agitated growth of recipient, growth in high nutrient content medium, harvest of recipient cells at log phase) will drastically reduce the frequency of plasmid transfer.

The details for preparing and isolating auxotropic mutants is described in the textbook, "The Genetics of Bacteria and Their Viruses" by William Hays [John Wiley & Sons, Inc. (1965)].

TABLE II

Donor	Recipient	Degradative Pathway	Frequency of Transfer
Trp$^-$CAM$^+$	*P. aeruginosa* PAO	CAM	10^{-3}
P. putida PpG1	CAM^{del} *P. putida*	CAM	10^{-2}
Met$^-$OCT$^+$	*P. aeruginosa* PAO	OCT	$<10^{-9}$
P. oleovorans	*P. putida* PpG1	OCT	10^{-9}
	P. putida U	OCT	10^{-7}
His$^-$SAL$^+$	*P. aeruginosa* PAO	SAL	10^{-7}
P. putida R-1	*P. putida* PpG1	SAL	10^{-6}
Trp$^-$NPL$^+$	*P. putida* PpG1	NPL	10^{-7}
P. aeruginosa	NPL^{del} *P. aeruginosa* PAO	NPL	10^{-5}

Abbreviations:
Trp – tryptophane
Met – methionine
His – histidine

Control cultures of donors and recipients were also plated individually on minimal plates containing the requisite substrate in each instance as the sole source of carbon, to determine the reversion frequency of donor and recipient cells.

All plates (including controls) were incubated at 30°–37° C. for several days. In each instance in which colonies appeared in numbers exceeding the colony growth on the reversion plates, it was established that degradative pathway transfer had occurred between the donors and recipients. Such conjugatants were than purified by a series of single colony isolation cultures and checked for growth rates or other distinctive characteristics of the recipient to insure that the recipient actually received the given degradative pathway.

Having determined that the degradative pathways were plasmid-borne and transmissible, the task of transferring the multiplicity of plasmids to a single cell *P. aeruginosa* PAO was undertaken. Prior work (referred to hereinabove) had established that OCT placmids could not be transferred from *P. oleovorans* to *P. aeruginosa* PAO. Therefore, the first task was to discover how (if at all) the OCT and CAM plasmids could be rendered compatible.

4,259,444

9

The CAM plasmid was transferred to a Met⁻ mutant of OCT⁺ *P. oleovorans* strain from a CAM⁺ *P. putida* strain. The conjugatant is, of course, unstable and will segregate either CAM or OCT at an appreciable rate. Therefore, the conjugatant was alternately grown in camphor and then octane as sole sources of carbon to isolate those cells in which both of these degradative pathways were present, even though unstable. The surviving cells were centrifuged, suspended in 0.9% saline solution and irradiated with UV rays (3 General Electric FS-5 lamps providing a total of about 24 watts). Aliquots were drawn from the suspension as follows: one aliquot was removed before UV treatment, one aliquot after UV exposure for 30 seconds and one aliquot after UV exposure for 60 seconds. These aliquots of irradiated cells were grown in the absence of light for 3 hours in L broth and were then used as donors for the transfer of plasmids to the *P. aeruginosa* PAO strain as recipient, selection being made for the OCT plasmid on an octane minimal plate.

As is shown in Table III aliquots of similarly irradiated suspensions for Met⁻OCT⁺CAM^{del} *P. oleovorans* and Met⁻CAM⁺OCT^{del} *P. oleovorans were prepared and used as plasmid donors to*

P. aeruginosa PAO, selection being made for the plasmids shown. The Met⁻CAM⁺OCT^{del} strain was prepared by introducing CAM plasmids into Met⁻OCT⁺ mutant of *P. oleovorans* and selecting for CAM⁺ conjugatants, which have lost the OCT plasmid. The Met⁻OCT⁺CAM^{del} *P. oleovorans* is the Met⁻ mutant of wild type *P. oleovorans*.

The failure to secure determinable transfer of OCT plasmids from Met⁻OCT⁺ *P. oleovorans* to the recipient and the success in securing transfer of CAM plasmids from Met⁻CAM⁺OCt^{del} *P. oleovorans* to the recipient are shown. These results support the theory that the successful transfer of OCT plasmids from the

10

mutant of CAM⁺OCT⁺ *P. aeruginosa* PAO that had been provided with its multiple plasmids by the methods described herein for plasmid transfer and plasmid fusion was used as the donor. After conjugation between the donor and OCT^{del} CAM^{del} *P. putida* PpGl, the resulting culture was plated on minimal plates containing camphor and also on minimal plates containing n-octane. Part of each of 132 colonies growing on the CAM minimal plates were transferred to OCT minimal plates and part of each of 219 colonies growing on the OCT minimal plates were transferred to CAM minimal plates. Each of these transferred portions grew, which tedns to establish that (a) both CAM and OCT plasmids had been transferred to the conjugatant, (b) the transfer had been on a one-for-one basis and, therefore, (c) the CAM and OCT plasmids were fused together.

Similar plasmid transfer was carried out between the Trp⁻CAM⁺OCT⁺ *P. aeruginosa* PAO donor and OCT^{del} CAM^{del} *P. aeruginosa* PAO and similar selection procedures were employed. The results further reinforced the above position as to the fused nature of the transferred CAM and OCT plasmids. When the CAM and OCT plasmids have been subjected to UV radiation as disclosed, if either CAM or OCT plasmid is transferred, the other plasmid will always be associated with it regardless of which plasmid is selected first. If either plasmid of the fused pair is cured from the cell, both plasmids are lost simultaneously. Thus, the conjugatants were treated with mitomycin C and the resultant CAM^{del} segregants were examined. Invariably all CAM^{del} segregants were found to have lost the OCT plasmid as well. Thus, the facts of simultaneous curing of the two plasmids and the co-transfer thereof strongly suggest that incompatible plasmids treated with means for cleaving the DNA of the plasmids results in fusion of the DNA segments to become part of the same replicon.

TABLE IV

Donor	Recipient	Selected Plasmid	Non-selected Plasmid	Total OCT⁺/CAM⁺
Trp⁻CAM⁺OCT⁺	OCT^{del}CAM^{del}	CAM	OCT	132/132
P. aeruginosa PAO	*P. putida* PpG1	OCT	CAM	219/219
	OCT^{del}CAM^{del}	CAM	OCT	107/107
	P. aeruginosa PAO	OCT	CAM	96/96

Met⁻CAM⁻OCT⁺ *P. oleovorans* (that had been irradiated for 30 seconds with UV rays) to *P. aeruginosa* PAO had been made possible by the fusion of the CAM and OCT plasmids in the *P. oleovorans* by the UV exposure and the subsequent transfer of CAM/OCT plasmids in combination (with separate degradative pathways), to the recipient.

Having successfully overcome all obstacles to the formation of a stable CAM⁺OCT⁺SAL⁺NPL⁺ Pseudomonas the several energy-generating degradative plasmids were transferred to a single cell as is shown in Table V by the conjugation techniques described hereinabove. The initial *P. aeruginosa* strain used is referred to herein as *P. aeruginosa* PAO, formerly known as *P.*

TABLE III

Donor	Recipient	Selected Plasmid	Period of UV-Irradiation (Sec)	Transfer of Frequency
Met⁻OCT⁺ *P. oleovorans*	*P. aeruginosa* PAO	OCT	0	$<10^{-9}$
			30	$<10^{-9}$
			60	$<10^{-9}$
Met⁻CAM⁺OCT^{del} *P. oleovorans*	*P. aeruginosa* PAO	CAM	0	10^{-4}
			30	10^{-5}
			60	10^{-7}
Met⁻CAM⁺OCT⁺ *P. oleovorans*	*P. aeruginosa* PAO	OCT	0	$<10^{-9}$
			30	10^{-8}
			60	$<10^{-9}$

Table IV presents verification of this theory of co-transfer of CAM and OCT fused plasmids. A Trp⁻

aeruginosa strain 1c available as ATCC No. 15692 and-

4,259,444

| 11 | 12 |

/ro ATCC No. 17503. This strain of *P. aeruginosa* does not contain any known energy-generating plasmid. The CAM and OCT plasmids exist in the fused state, are individually and simultaneously functional and appear perfectly compatible with the individual compatible SAL and NPL plasmids. Tests for compatibility of obth CAM+OCT+SAL+ *P. aeruginosa* PAO and CAM-+OCT+SAL+NPL+ *P. aeruginosa* PAO revealed that there is no segregation of the plasmids in excess of that found in the donor. Plasmids will be accepted and maintained by *P. acidovorans, P. alcaligenes* and *P. fluorescens.* All of these plasmids should be transferable to and maintainable in these and many other species of Pseudomonas, such as *P. putida, P. oleovorans, P. multivorans,* etc.

Superstrains such as the CAM+OCT+SAL+NPL+ strain of *P. aeruginosa* PAO can grow on a minimal plate of any of camphor, n-octane, salicylate, naphthalene and, because of the phenomenon of relaxed specificity, on compounds similar thereto. Thus, the effectiveness of a given degradative plasmid does not appear to be diminished in its ability to function singly by the presence of other degradative plasmids in the same cell.

strains of *P. aeruginosa* PAO. Curve a shows the cell growth as a function of time of

P aeruginosa without any plasmid-borne energy-generating degradative pathways. Curve b shows greater cell growth as a function of time for SAL+ *P. aeruginosa.* Curve c shows still greater cell growth as a function of time for SAL+NPL+ *P. aeruginosa.* Curve d shows cell growth that is significantly greater still as a function of time for the CAM+OCT+SAL+NPL+ superstrain of *P. aeruginosa.* These results clearly establish that cells artifically provided by the practice of this invention with the genetic capability for degrading different hydrocarbons can grow at a faster rate and better on crude oil as the plasmid-borne degradative pathways are increased in number and variety, because of the facility of these degradative pathways to simultaneously function at full capacity.

Similar results are shown in FIG. 2 displaying the growth capabilities of this same series of organisms utilizing Bunker C oil as the sole source of carbon. Bunker C is (or is prepared from) the residuum remaining after the more commercially useful components have been removed from crude oil. This residuum is

TABLE V

Donor	Recipient	Selected Plasmid	Phenotype of the Conjugatant
Trp⁻CAM+OCT+ *P. aeruginosa* PAO	*P. aeruginosa* PAO	CAM	CAM+OCT+ *P. aeruginosa* PAO
His⁻SAL+ *P. putida* R-1	CAM+OCT+ *P. aeruginosa* PAO	SAL	CAM+OCT+SAL+ *P. aeruginosa* PAO
Trp⁻NPL+ *P. aeruginosa*	CAM+OCT+SAL+ *P. aeruginosa* PAO	NPL	CAM+OCT+SAL+NPL+ *P. aeruginosa* PAO

Indication of the capability of all degradative plasmids to function simultaneously in energy generation is provided by tests in which CAM+OCT+SAL+NPL+ *P. aeruginosa* PAO superstrain was added to separate broth samples each of which contained 1 millimolar (mM) of nutrient (a suboptimal concentration), one set of samples containing camphor, a second set of samples containing n-octane, a third set of samples containing salicylate and a fourth set of samples containing naphthalene, these being the sole sources of carbon in each instance. The superstrain grew very slowly in the separate sole carbon source samples. However, when the superstrain was added to samples containing all four sources of carbon present together in the same (1 mM)

very thick and sticky and without significant use, per se. A small amount of volatile hydrocarbons is often added thereto to lower the viscosity. Curve r reflects the cell growth as a function of time of *P. aeruginosa* cells not having any plasmid-borne energy-degradative pathways. Curve s shows increased cell growth as a function of time for SAL+ *P. aeruginosa.* Curve t shows further increase in cell growth as a function of time for SAL+NPL+ *P. aeruginosa.* Curve u shows still more significant cell growth as a function of time for CAM-+OCT+SAL+NPL+ *P. aeruginosa.*

The SAL+ *P. aeruginosa* and SAL+NPL+ *P. aeruginosa* cultures were prepared as shown in Table VI below:

TABLE VI

Donor	Recipient	Selected Plasmid	Conjugatant
His⁻ SAL+ *P. putida* R-1	*P. aeruginosa* PAO	SAL	SAL+ *P. aeruginosa* PAO
Trp⁻NPL+ *P. aeruginosa*	SAL+ *P. aeruginosa* PAO	NPL	SAL+NPL+ *P. aeruginosa* PAO

concentration of 4 mM, the rate of growth increased considerably establishing that simultaneous utilization of all four sources of carbon had occurred.

Next, the ability of such superstrains to degrade crude oil was demonstrated. Crude oils, of course, vary greatly (depending upon source, period of activity of the well, etc.) in the relative amounts of linear aliphatic, cyclic aliphatic, aromatic and polynuclear hydrocarbons present, although some of each of these classes of hydrocarbons is typically present in some amount in the chemical make up of all crude oils from producing wells.

FIG. 1 shows the difference in growth capabilities in crude oil as the sole source of carbon of four single cell

The experiments providing the data for FIGS. 1 and 2 were conducted in 250 ml Erlenmeyer flasks. To each flask was added 50 ml of mineral medium (described hereinabove) with pH adjusted to 6.8–7.0; 2.5 ml of the sole carbon source (crude oil or Bunker C) and $5 \times 10^6 - 1 \times 10^7$ cells. Growth was conducted at 32° C. with shaking. At daily intervals 5 ml aliquots were taken. The optical densities of these aliquots were determined at 660 nm in a Bausch & Lomb, Inc. colorimeter to determine organism density. Also, viable cell counts were determined by diluting portions of the aliquots and plating on L-agar (L-broth containing agar) plates. The colonies were counted after 24 hours of incubation at 32° C. and these counts were used to construct FIGS. 1

13

4,259,444

14

and 2. Also, the cells were submitted to protein analysis, to be discussed hereinbelow.

The 2.5 ml of crude oil or Bunker C appears to have initially offered an essentially unlimited food supply, but the results shown may well represent less than the full capability of the superstrain, because the relative amounts of the various hydrocarbons (degradable by the CAM+, OCT+, SAL+ and NPL+ plasmids) present in the carbon sources had not been ascertained and after a couple of days the food supply for one or more plasmids may have been limited.

A very significant aspect of the growth of the superstrain in crude and Bunker C oils is the fact that the components, which would spread the quickest on the water's surface from spills of these oils, disappear within 2–3 days and the remaining components of the oil co-

tion of stable plasmids to which the newly introduced plasmid can be fused.

Preparation of *P. putida* [NRRL B-5473]

The mineral medium and the technique for fostering conjugation was the same as described above. A culture of antibiotic-sensitive *P. putida* PpG1 was cured of its CAM plasmids with mitomycin C and was used as the initial recipient. This strain of *P. putida* is sensitive to small (e.g. 25 micrograms/ml) concentrations of neomycin/kanamycin, carbenicillin and tetracycline. As is shown in Table VII below, all the donor strains are auxotropic mutants, because the use of auxotropic mutant donors facilitates counterselection of the conjugatants due to the ease of selecting against such donors.

TABLE VII

Donor	Recipient	Selected Plasmid	Phenotype of the Conjugatant
Trp CAM + *P. putida* PpG1	CAM^{del} *P. putida* PpG1	CAM	CAM+ *P. putida* PpG1
His SAL + *P. putida* R-1	CAM+ *P. putida* PpG1	SAL	CAM+SAL+ *P. putida* PpG1
Trp NPL+ *P. aeruginosa*	CAM+SAL+ *P. putida* PpG1	NPL	CAM+SAL+NPL+ *P. putida* PpG1
Met *P. aeruginosa* Strain 1822 (RP-1)	CAM+SAL+NPL+ *P. putida* PpG1	RP-1	CAM+SAL+NPL+RP-1+ *P. putida* PpG1

alesce to form large droplets that cannot spread out. These droplets can be removed more easily by mechanical recovery techniques as the microbes continue to consume these remaining components.

In practice an inoculum of dry (or lyophilized) powders of these genetically engineered microbes will be dispersed over (e.g. from overhead) an oil spill as soon as possible to control spreading of the oil, which is so destructive of marine flora and fauna and the microbes will degrade as much of the oil as possible to reduce the amount that need be recovered mechanically, when equipment has reached the scene and has been rendered operative. A particularly beneficial manner of depositing the inoculum on the oil spill is to impregnate straw with the inoculum and drop the inoculated straw on the oil spill where both components will be put to use—the inoculum (mass of microbes) to degrade the oil and the straw to act as a carrier for the microbes and also to function as an oil absorbent. Other absorbent materials may be used, if desired, but straw is the most practical. No special care need be taken in the preparation and storage of the dried inoculum or straw (or other absorbent material) coated with inoculum. No additional nutrient or mineral content need be supplied. Also, although culture from the logarithmic growth phase is preferred, culture from either the early stationary or logarithmic growth phases can be used.

It is reasonable to expect that a vast number of plasmid-borne hydrocarbon degradative pathways remain undiscovered. Hopefully, now that a method for controlled genetic additions to the natural degradative capabilities of microbes has been demonstrated by this invention, still more new and useful single cell organisms can be prepared able to degrade even more of the large number of hydrocarbons in crude oil, whether or not the plasmids yet to be found are compatible with each other or with those plasmids present in superstrains NRRL B-5472 and NRRL B-5473.

Both of these superstrains can be used as recipients for more plasmids. The capability for utilizing fusion (by UV irradiation or X-ray exposure) to render additional plasmids compatible is actually increased in a multiplasmid conjugatant, because of the larger selec-

The *P. aeruginosa* RP-1 strain is disclosed in the Sykes et al article [Nature 226, 952 (1970)]. Selection for the RP-1 plasmid was accomplished on a neomycin/kanamycin plate. Further, CAM+SAL+NPL+RP-1+ *P. putida* PpG1 has been determined to be resistant to carbenicillin and tetracycline establishing that the RP-1 plasmid is actually present and that the organisms that survived the selection process were not merely the results of mutant development. Also, the plasmids of this superstrain can be transferred and can be cured. The rate of segregation (spontaneous loss) of plasmids from the superstrain has been found to be the same as in the donors.

Both superstrains can, of course, be used as a source of plasmids in addition to those sources disclosed herein. For example, to transfer CAM, SAL or NPL plasmids from CAM+SAL+NPL+RP-1+ *P. putida* PpG1 to a given Pseudomonas recipient, the procedures for cell growth of donor and recipient and the mixing thereof for optimized conjugation is the same as described hereinabove. These plasmids will have different frequencies of transfer at different times. The order of diminishing frequency of transfer is CAM, NPL, SAL. For the transfer of CAM plasmid, after conjugation, selection is made for CAM. Surviving colonies are subdivided and selection is made for SAL, NPL and CAM plasmids from each colony. Those portions surviving only on camphor as the sole source of carbon will have received the CAM plasmid free of the SAL or NPL plasmids. The same procedure can be followed for the individual transfer of SAL or NPL plasmids.

In addition to the previously discussed capability for improved treatment of oil spills, considerable improvement is now possible in the microbial single-cell synthesis of proteins from carbon-containing substrates. The restriction of having to employ substantially single-component substrates, e.g. alkanes, paraffins, carbohydrates, etc. has now been removed, simultaneously providing the opportunity for increases of 50–100 fold in the amount of cell mass that may be produced by a single cell in a given time period, when the given single cell has been provided with multiple energy-generating

4,259,444

15

plasmids. Also, being able to optimize the protein production of bacteria is of particular interest since bacterial cell mass has a much greater protein content and most bacteria have greater tolerance for heat than yeasts. This latter aspect is of importance since less refrigeration is necessary to remove the heat generated by the oxidative degradation of the substrate.

The general process and apparatus for single cell production of protein is set forth in the Wang article (incorporated by reference) referred to hereinabove. One particular advantage of the multi-plasmid single cell organism of this invention is that after the cell mass has been harvested it can be subjected to a subsequent incubation period in a mineral medium free of any carbon source for a sufficient period of time to insure the metabolism of residual intra-cellular hydrocarbons, e.g. polynuclear aromatics, which are frequently carcinogenic. Presently, treatment of cell mass to remove unattacked hydrocarbons often leads to reduction in the quality of the protein product.

The economics of protein production by single-cell organisms will be further improved by the practice of this invention, because of the reduced cost of substrate (e.g. oil refinery residue, waste lubricating oil, crude oil) utilizable by organisms provided with preselected plasmid content.

Cell mass growth in crude oil using NRRL B-5472 was harvested by centrifugation, washed two times in water and dried by blowing air (55° C.) over the mass overnight. The dried mass was hydrolyzed and analyzed for amino acid content by the technique described "High Recovery of Tryptophane from Acid Hydrolysis of Proteins"-Matsubara et al [Biochem. and Biophys. Res. Comm. 35 No. 2, 175–181 (1969)]. The amino acid analysis showed that the amino acid distribution of superstrain cell mass grown in crude oil is comparable to beef in threonine, valine, cystine, methionine, isoleucine, leucine, phenylalanine and tryptophane content and significantly superior to yeast in methionine content.

Continued capacity for increasing the degrading capability of the superstrains now on deposit has been made possible by the practice of this invention as more plasmid-borne degradative pathways are discovered. To date *P. aeruginosa* strain 1822 has been provided with all four known hydrocarbon degradative pathways (OCT, CAM, SAL, NPL) plus the drug-resistance factor RP-1 found therein. If there is an upper limit to the number of energy-generating plasmids that will be received and maintained in a single cell, this limit is yet to be reached. Attempts to integrate plasmids (CAM, OCT, SAL) with the cell chromosome have been unsuccessful as indicated by failure to mobilize the chromosome. Such results have so far verified the extrachromosomal nature of the energy-generating and drug-resistance plasmids. There is, of course, no reason to expect that the only plasmids are those that specify degradative pathways for hydrocarbons. Conceivably plasmids may be discovered that will provide requisite enzyme series for the degradation of environmental pollutants such as insecticides, pesticides, plastics and other inert compounds.

Energy-generating plasmids in general are known to have broad inducer and substrate specificity [i.e. enzymes will be formed and will act on a variety of structurally similar compounds]. For example, the CAM plasmid is known to have a very relaxed inducer and substrate specificity [Gunsalus et al-Israel J. Med. Sci.,

16

1, 1099–1119 (1965) and Hartline et al-Journal of Bacteriology, 106, 468–478 (1971)]. Similarly, the OCT plasmid has broad inducer and substrate specificity [Peterson et al-J. Biol. Chem. 242, 4334 (1967)]. In the practice of the instant invention it has been demonstrated that plasmids display the same degree of relaxed specificity in the conjugatant as in the donor.

Thus, by the practice of this invention new facility and capability for growth has been embodied in useful single-cell organisms by the manipulation of phenomena that had been previously undiscovered (i.e. the plasmid-borne nature of the degradative pathways for salicylate and naphthalene) and/or had been previously unsuccessfully applied (i.e. rendering stable a plurality of previously incompatible plasmids in the same single cell).

Filed concurrently herewith is U.S. Application Ser. No. 260,488-Chakrabarty, filed June 7, 1972 now U.S. Pat. No. 3,814,474 and assigned to the assignee of the instant invention.

What I claim as new and desire to secure by Letters Patent of the United States is:

1. A bacterium from the genus Pseudomonas containing therein at least two stable energy-generating plasmids, each of said plasmids providing a separate hydrocarbon degradative pathway.

2. The Pseudomonas bacterium of claim 1 wherein the hydrocarbon degradative pathways are selected from the group consisting of linear aliphatic, cyclic aliphatic, aromatic and polynuclear aromatic.

3. The Pseudomonas bacterium of claim 1, said bacterium being of the specie *P. aeruginosa.*

4. The *P. aeruginosa* bacterium of claim 3 wherein the bacterium contains CAM, OCT, SAL and NPL plasmids.

5. The Pseudomonas bacterium of claim 1, said bacterium being of the specie *P. putida.*

6. The *P. putida* bacterium of claim 5 wherein the bacterium contains CAM, SAL, NPL and RP-1 plasmids.

7. An inoculum for the degradation of a preselected substrate comprising a complex or mixture of hydrocarbons, said inoculum consisting essentially of bacteria of the genus Pseudomonas at least some of which contain at least two stable energy-generating plasmids, each of said plasmids providing a separate hydrocarbon degradative pathway.

8. The inoculum of claim 7 wherein the hydrocarbon degradative pathways are selected from the group consisting of linear aliphatic, cyclic aliphatic, aromatic and polynuclear aromatic.

9. The inoculum of claim 8 wherein the bacteria having multiple energy-generating plasmids are of the specie *P. aeruginosa.*

10. The inoculum of claim 8 wherein the bacteria having multiple energy-generating plasmids are of the specie *P. putida.*

11. In the process in which a first energy-generating plasmid specifying a degradative pathway is transferred by conjugation from a donor Pseudomonas bacterium to a recipient Pseudomonas bacterium containing at least one energy-generating plasmid that is incompatible with said first plasmid, said transfer occurring in the quiescent state after the mixing of substantially equal volumes of cultures of said donor and said recipient, each culture presenting the respective organisms in a complex nutrient liquid medium at a population density of at least about 1,000,000 cells/ml, the improvement

4,259,444

17

wherein after conjugation has occurred, the multi-plasmid conjugatant bacteria are subjected to DNA-cleaving radiation in a dosage sufficient to fuse the first plasmid and the plasmid incompatible therewith located in the same cell.

12. The improvement of claim 11 wherein the DNA-cleaving radiation is UV radiation.

13. The improvement of claim 12 wherein the first plasmid provides the degradative pathway for camphor and the recipient Pseudomonas contains the degradative pathway for n-octane.

14. An inoculated medium for the degradation of liquid hydrocarbon substrate material floating on water, said inoculated medium comprising a carrier material able to float on water and bacteria *from the genus Pseudomonas* carried thereby, at least some of said bacteria

18

each containing at least two stable energy-generating plasmids, each of said plasmids providing a separate hydrocarbon degradative pathway and said carrier material being able to absorb said hydrocarbon material.

15. The inoculated medium of claim 14 wherein the carrier material is straw.

16. The inoculated medium of claim 14 wherein the hydrocarbon degradative pathways are selected from the group consisting of linear aliphatic, cyclic aliphatic, aromatic and polynuclear aromatic.

17. The inoculated medium of claim 14 wherein the bacteria are of the specie *P. aeruginosa.*

18. The inoculated medium of claim 14 wherein the bacteria are of the specie *P. putida.*

* * * * *

Index

A

Abortion
  fetal tissue research and, 88
  prenatal genetic testing and, 215
Abstract, evaluation, in peer review, 64
Academic freedom, protection, 144–145
Adolescents, research on, 86–87
Amniocentesis, for genetic testing, 212
Animal experimentation, 101–130
  animal rights in, 103–104
    birth of movement, 101–102
    deontology and, 106–109
    politics of, 119–121
    utilitarianism and, 101–102, 104–106
  case studies, 121–125
  challenges to, 101–103
  cloning, 222–223
  ethical issues in, 104–109
    deontology, 106–109
    utilitarianism, 101–102, 104–106
  historical practices in, 101
  humane care in, 116–118
  institutional committees
    function, 112–115
    protocol review by, 115–116
  legislation on, 110–112
  moral judgments in, 104
  political realities, 119–121
  protocols for
    review, 115–116
    sample, 311–315
  survey form for, 259–260, 270–271
Animal Welfare Act, 110–116
Antivivisectionism, 101–102
Athletes, gene manipulation in, 221
Authorship, 49–60

activities not counting toward, 60
case studies in, 66–70, 277–280
coauthors, 51, 56–59, 188
in collaborative research, 165–166
contribution to, 49, 59–60
copyright for, *see* Copyrights
criteria for, 51
definition, 49–50, 52–53, 56
earned, 56
first (principal) author, 58
honorary, 56
instructions to, 52–55
  authorship definition, 52–53
  conflict of interest, 55
  copyright assignment, 53
  dispute resolution, 55
  manuscript preparation, 52
  manuscript review, 53
  prior publication definition, 53–54
  shared research materials, 54–55
  unpublished information citations, 54
multiple, 51, 56–59
peer review and, 60–65
pressure to publish and, 50–51
principles, 59–60
responsibility in, 49, 59–60
senior (primary) author, 56–58
standards, 292–293
submitting author, 58–59
survey form for, 259, 267–269
Autoradiographs, in data books, 240

B

Bayh-Dole Act, 137
Bentham, Jeremy, utilitarianism theory, 21–22
  animal rights and, 104–105

331